MW01518052

HANDBOOK OF CLIMATE CHANGE AND AGROECOSYSTEMS

Global and Regional Aspects and Implications

Joint Publication with American Society of Agronomy, Crop Science Society of America, and Soil Science Society of America

ICP Series on Climate Change Impacts, Adaptation, and Mitigation

Editors-in-Chief: Daniel Hillel and Cynthia Rosenzweig
(Columbia Univ. and Goddard Institute for Space Studies, USA)

ICP Series on Climate Change Impacts, Adaptation, and Mitigation – Vol. 2

HANDBOOK OF CLIMATE CHANGE AND AGROECOSYSTEMS

Global and Regional Aspects and Implications

editors

Daniel Hillel
Cynthia Rosenzweig

Columbia University and Goddard Institute for Space Studies, USA

Joint Publication with American Society of Agronomy, Crop Science Society of America, and Soil Science Society of America

 |  |

Imperial College Press

ICP

Published by

Imperial College Press
57 Shelton Street
Covent Garden
London WC2H 9HE

Distributed by

World Scientific Publishing Co. Pte. Ltd.
5 Toh Tuck Link, Singapore 596224
USA office: 27 Warren Street, Suite 401-402, Hackensack, NJ 07601
UK office: 57 Shelton Street, Covent Garden, London WC2H 9HE

British Library Cataloguing-in-Publication Data
A catalogue record for this book is available from the British Library.

Cover map: Principal Soil Orders, USDA.

ICP Series on Climate Change Impacts, Adaptation, and Mitigation — Vol. 2
HANDBOOK OF CLIMATE CHANGE AND AGROECOSYSTEMS
Global and Regional Aspects and Implications — Joint Publication with the American Society of
Agronomy, Crop Science Society of America, and Soil Science Society of America

ISBN 978-1-84816-983-8

In-house Editor: Monica Lesmana

Typeset by Stallion Press
Email: enquiries@stallionpress.com

Printed in Singapore by Mainland Press Pte Ltd.

Foreword

This volume is an exciting joint publication of the Imperial College Press (ICP) and the American Society of Agronomy (ASA), the Soil Science Society of America (SSSA), and the Crop Science Society of America (CSSA). It is the first in what is hoped to be an on-going activity, dedicated to elucidating the integrated agricultural impacts of climate change, and to furthering effective responses to this global challenge through agricultural research. The effort is designed to present and integrate the work of leading researchers in the world regarding climate change mitigation and adaptation in agriculture.

The process of climate change, induced by the anthropogenic accumulations of greenhouse gases in the atmosphere, is likely to generate effects that will cascade through the entire biosphere and hydrosphere, impacting all life on earth and bearing upon human endeavors. Of special concern is the potential effect on agriculture and global food security.

Climate change is expected to affect agriculture in complex ways throughout the soil-crop-atmosphere continuum. The effects will modify the production of food, fiber, and fuel, and thus impact the lives and livelihoods of both producers and consumers. Agricultural researchers are playing a crucial role, not just in understanding climate change impacts, but in developing effective responses to them. These responses include mitigation actions aimed at reducing the concentrations of greenhouse gases in the atmosphere (thus avoiding long-term risks) and adaptation strategies designed to accommodate and adjust to climate changes that cannot be avoided.

As the scientific and practical knowledge of the agricultural processes and responses involved in climate change continues to grow, the joint publication series of the ICP, ASA, SSSA, and CSSA will address important aspects of this topic periodically over the coming years. The volumes will encompass cutting-edge research findings from the range of disciplines — agronomy, soil science, and crop science — needed to understand and provide solutions for the challenges posed by climate change. The ultimate goal is to ensure food security sustainably for future generations.

Charles W. Rice

President of the Soil Science Society of America (2011)
University Distinguished Professor and Professor of Soil Microbiology
Kansas State University, Department of Agronomy
Manhattan, Kansas

Jerry L. Hatfield

President of the American Society of Agronomy (2007)
Director, National Laboratory for Agriculture and the Environment
United State Department of Agriculture, Agriculture Research Service
Ames, Iowa

Preface

Climate change is no longer merely projected to occur in the indeterminate future. It has already begun to be manifested in the weather regimes affecting agroecosystems, food production, and rural livelihoods in many regions around the world. It is an actual and growing challenge to the world at large and to the scientific community in particular, which is called upon with increasing urgency to respond effectively.

This is the second volume in the ICP Series on Climate Change Impacts, Adaptation, and Mitigation. The ongoing series is dedicated to elucidating the actual and potential impacts of climate change, and to formulating effective responses to this global challenge. It is designed to inform, spur, and integrate the work of leading researchers in the major regions of the world, and to promote interdisciplinary advances in this crucial field.

Agricultural volumes in this series are now and henceforth to be published jointly by the American Society of Agronomy, the Soil Science Society of America, and the Crop Science Society of America and Imperial College Press. We believe that this fruitful cooperation will further international awareness and synergies within the scientific community and beyond it toward one of the most crucial portending challenges of global importance.

Daniel Hillel and Cynthia Rosenzweig
Editors

About the Editors

Daniel Hillel is a world-renowned environmental scientist and hydrologist who has worked in over 30 countries across North and South America, Europe, Asia, Australasia and Africa. He is the 2012 winner of the World Food Prize, recognizing "individuals who have contributed landmark achievements in increasing the quality, quantity or availability of food in the world." Professor Hillel is receiving the prize for his innovative work in micro-irrigation and his success in bridging cultural gaps to solve a global issue. He has published over 300 scientific papers and manuals on agriculture, the environment, climate change, and water-use efficiency, as well as authored or edited 24 books that have been adopted by many universities as standard teaching and reference textbooks and translated into ten languages. Dr. Hillel is an elected Fellow of the American Association for the Advancement of Science, the American Geophysical Union, the Soil Science Society of America, and the American Society of Agronomy. He was granted the Distinguished Service Award and was the Honoree of a special dedicated Symposium of the latter Society. In addition, he has been the recipient of honorary doctorates by several leading universities in the U.S. and abroad.

Cynthia Rosenzweig is a Senior Research Scientist at NASA Goddard Institute for Space Studies where she leads the Climate Impacts Group. The mission of the Climate Impacts Group is to improve understanding of how climate affects human and natural systems and to develop solutions for adaptation to and mitigation of climate change. A recipient of a Guggenheim Fellowship, Dr. Rosenzweig also holds adjunct positions as Senior Research Scientist at the Columbia University Earth Institute and Professor of Environmental Sciences at Barnard College. She was a Coordinating Lead Author for the Intergovernmental Panel on Climate Change (IPCC) Fourth Assessment Report and is a Co-Leader of the Agricultural Model Intercomparison and Improvement Project (AgMIP).

Acknowledgments

The editors of this book acknowledge with gratitude the following colleagues and organizations:

Dr. Charles W. Rice, former President of the Soil Science Society of America, and Dr. Jerry L. Hatfield, President of the American Society of Agronomy, who helped to plan and implement the Cross-Cutting Symposium that led to this volume at the 2011 Annual Meetings of the ASA, SSSA, and CSSA;

The Agricultural Model Intercomparison and Improvement Project (AgMIP) and the American Society of Agronomy who provided support for international colleagues to participate in the Symposium;

Ms. Andrea Basche, of Iowa State University, who provided much appreciated support during the Symposium;

Ms. Somayya Ali, Mr. Daniel Bader, Ms. Soyee Chiu, Mr. José Mendoza, Ms. Shari Lifson and Ms. Redencion Licardo of Columbia University and the Goddard Institute for Space Studies, who provided excellent assistance in the detailed preparation of the manuscripts and illustrations;

Dr. Zvi Ruder, the Senior Executive Editor on behalf of Imperial College Press, who initiated the series on Climate Change Impacts, Adaptation, and Mitigation; and Mr. Mark Mandelbaum, Director of Publications at the American Society of Agronomy, Soil Science Society of America, and Crop Science Society of America, both of whom worked concertedly to make the joint publication of this volume by ICP and the ASA, SSSA, and CSSA a reality;

And finally, the dedicated staff at Imperial College Press; particularly Ms. Juliet Lee and Ms. Monica Lesmana, for their professional care and expertise in preparing this book for publication.

Contents

Introduction
Climate Change and Agroecosystems:
Global and Regional Perspectives

Daniel Hillel and Cynthia Rosenzweig

Columbia University
and NASA Goddard Institute for Space Studies
2880 Broadway
New York, NY 10025
dh244@columbia.edu

cynthia.rosenzweig@nasa.gov

Much attention has been devoted by the research community to the overarching processes inducing global climate change. It is now time to focus more specifically on how agriculture will be affected both globally and regionally and in turn, how agriculture can contribute to climate change solutions. The response of particular regions and their agroecosystems are certain to vary, and hence require different approaches to both mitigation and adaptation. These solutions span a broad spectrum from genetic improvement to enhanced environmental management to soil carbon sequestration.

This need motivated the convening of a daylong Cross-Cutting Symposium at the 2011 Annual Meetings of the American Society of Agronomy, Soil Science Society of America, and Crop Science Society of America, entitled: *Agriculture's Contributions to Climate Change Solutions: Mitigation and Adaptation at Global and Regional Scales*. The goal of the symposium and this volume is to contribute to fostering a broad range of climate change solutions from agronomy and soil science both globally and regionally. Scientists from North America, Latin America, Europe, Africa, Australia, the Middle East, East Asia, and South Asia participated.

Key issues include the potential for mitigation and adaptation capacities of different regions — including developing or introducing more heat and drought-tolerant crop varieties, management methods for dealing with more extreme climate events, effective means of soil carbon sequestration, soil conservation and fertility enhancement, as well as improved water-use efficiency. Responding to climate change in developing agricultural regions of the world is particularly important because they are projected to experience the greatest impacts of climate change as well as the greatest rates of population growth, thus potentially exacerbating threats to food security.

This volume is organized into three sections, with this *Introduction* setting the stage by highlighting key issues. The first section, *Themes,* presents the contributions to climate change solutions from the agronomy perspective. The second section — *Regions* — includes nine chapters by regional experts focusing on geographically specific impacts, adaptation strategies, and mitigation options. The third section — *Programs and Projects* — describes important activities related to improving climate change solutions in agriculture. Finally, the *Conclusion* section summarizes the main findings and sets research directions.

Section I

Themes

Chapter 1

Agriculture and Environment in a Crowding and Warming World

Daniel Hillel[*] and Cynthia Rosenzweig[†]

The Earth Institute at Columbia University, New York,
and The NASA Goddard Institute for Space Studies, New York
[*]dh244@columbia.edu
[†]cynthia.rosenzweig@nasa.gov

Of the world's land area and soils, only some 25% can be regarded as suitable for agriculture, the remainder having soils that are too dry or too wet, steep, rocky, cold, shallow, acidic, alkaline, or saline to permit the growing of crops. The actual arable land under cultivation[1] (some 12%) is less than half of the potentially cultivable area, with an additional quarter used for grazing livestock (including areas of prairie, savanna, and scrub vegetation) (Hillel, 2008). However, any substantial expansion of cultivation would pose a severe threat to the remaining natural ecosystems and their biodiversity. Hence, to meet the requirements for food security, there is a need to intensify production and to do so sustainably — that is to say, without degrading the resource base of soil, water, and energy. This must be done without changing the climate for the worse, while adapting to its projected changes through the coming decades (Beddington *et al.*, 2011; Hillel, 2009).

At the core of the problem lies the inexorably increasing requirement for agricultural products due to rising populations as well as to the universal desire to attain ever higher living standards. Consequently, our expansive population has been placing ever greater demands on the world's limited and vulnerable soil, water, and biotic resources. At the beginning of the 18th century, three centuries ago, the world's population totaled approximately 600 million (UN, 2004). Since then, with infant mortality reduced and life expectancy prolonged, the population has increased more than tenfold (Hillel, 2008). Although the fertility rate in many countries has

---

[1]Arable land includes land under temporary crops (double-cropped areas are counted once), temporary meadows for mowing or for pasture, land under market or kitchen gardens, and land temporarily fallow. Land abandoned as a result of shifting cultivation is excluded (FAOSTAT, 2012).

been diminishing, the momentum of population growth still continues, due to the increased number of young people of fertile age and the prolonged life expectancy. World population is now projected to stabilize at some 9.3 billion by the middle decade of this century and reach 10.1 billion by 2100, but these are only projections, not certainties (UNPD, 2010).

Another important demographic trend affecting the environment is the changing pattern of human occupation and habitation, specifically the growth of urbanization. Until the start of the Industrial Revolution (about two centuries ago), the majority of workers were subsistence farmers who produced mainly for their own needs. Since that time, as farm work became increasingly mechanized, the portion of the workforce employed on the farm has declined drastically. In some of the industrialized countries (e.g., the United States, Europe, and Japan), it has fallen below 3%.

The vaunted productivity of modern agriculture has its problematic side, however. Farmers, once largely self-sufficient, now rely on external industries for their inputs and tools, including fertilizers, pesticides, electricity, machines, and fuel. The intensive production of marketable farm products has induced a growing industrial dependency, requiring ever-greater quantities of chemical and energy inputs to boost fertility and to control pests and diseases. Accumulating residues of these chemicals tend to contaminate the larger environment, to pollute groundwater aquifers and surface water bodies (streams and lakes), as well as to decimate wildlife, and threaten the health of domestic animals and humans. Large machines, operated conveniently and hence often used excessively, can cause direct damage to the soil — including compaction when wet and pulverization when dry, leading to accelerated erosion — in addition to damage to aquifers by over-pumping, depletion, and pollution.

The enormous increase of labor efficiency resulting from the use of motorized machinery has been purchased at the cost of greatly increased consumption of energy and reliance on external, unstable, and increasingly expensive energy sources. Specifically, agriculture leads to significant emissions of the major greenhouse gases, carbon dioxide (CO_2), methane (CH_4), and nitrous oxide (N_2O) (Hillel and Rosenzweig, 1989, 2011; Rosenzweig and Hillel, 1998; IPCC WGIII, 2007). Carbon dioxide, the most prominent greenhouse gas, is released via agricultural practices related to land-use change, biomass burning, and fertilizer production. Agriculture is a large anthropogenic source of methane, with emissions released from rice paddies, herds of ruminant animals, and biomass burning. Agricultural activities that contribute to nitrous oxide emissions include the use of nitrogenous fertilizers, the burning of biomass, and the conversion of forests to pastures in tropical regions. Furthermore, most energy use for agriculture leads to emissions of carbon dioxide (CO_2), the major greenhouse gas.

The growing cost of fossil fuels and recognition of the negative environmental effects of their excessive consumption, including climate change, must induce a change in the mode of agriculture. Much of the energy used in agriculture (for example, in intensive tillage and irrigation) is commonly wasted. The current trend toward minimum tillage, or even zero tillage, is a positive development. So is the change from excessive flood irrigation to precision sprinkle or drip irrigation. Clearly, much more attention must be devoted to optimizing energy, water, nutrient, and pesticide inputs in agriculture, so as to enhance economic efficiency and avoid environmental degradation and climate change.

Agriculture, being the utilization of land, water, solar energy, and biota for human sustenance, is a fundamental aspect of our civilization upon which depends the

---

The Task of Increasing Food Production

Food production can be increased either by expanding the area under cultivation or by intensifying production on selected tracts without further expansion. For many centuries, the dominant mode was the first option; i.e., the clearing of more land of its native cover, in the process of which natural ecosystems were destroyed on an ever-increasing scale. That process, carried out at the expense of terrestrial biodiversity and with accompanying release of carbon stored in forests and soils, cannot be continued indefinitely. Most of the readily accessible favorable land has already been appropriated, and much of that land has been degraded as a consequence of exploitative and unsustainable modes of management. Henceforth, the main prospect for increasing production is to intensify and sustain production on the most favorable land rather than to appropriate and to degrade more marginal and sub-marginal land.

Intensification of agricultural production involves a package of measures that must be tailored to the specific circumstances in each case. Those measures include the use of high-yielding crops and varieties, and a set of treatments designed to optimize growing conditions. Among such treatments are the provision of adequate water and nutrient supplies; minimizing losses of water and nutrients due to runoff, percolation, or competition by weeds; providing favorable conditions in the root zone; and protecting the crops against diseases and pests. Such management measures must also protect the soil against erosion, waterlogging, nutrient depletion, salinization, acidification, pollution, compaction, or crusting. All this must be accomplished while reducing greenhouse gas emissions and responding to climate change already underway.

quality of human life. If sustainable modes of soil and crop husbandry are further developed and more widely applied, agriculture should become less disruptive and more harmonious within the natural environment and its community of life. At the same time, land and water resources that have been degraded by past abuses should be withdrawn from further exploitation and rehabilitated.

The concentration of population in urban communities and the funneling of products from extensive, sparsely populated hinterlands into densely populated centers have created new problems of transportation, distribution, energy use, waste disposal, as well as regional and global climate impacts. As more and more people are moving into cities, the cities are expanding and usurping some of the best farmland. Worldwide, some 20–30 million hectares are being converted from farmland to urban uses annually at present (Hillel, 2008).

Of the land area, constituting about 30% of the globe's surface, only about 12% is arable, as stated above. However, that proportion varies from one country to another and from one continent to another. In the Ukraine, about 56% of the land is arable. In the US, the fraction is about 18%, whereas in Egypt it is less than 3% (World Bank, 2012). The area per capita devoted to the production of grain (the staple food in many countries) has diminished from 0.2 hectare in 1950 to less than 0.13 hectare in 1990. That decrease — attributable to population growth, land appropriation by expanding urbanization, and soil degradation — is a continuing trend. By the year 2030, the area of grain-producing land per person is expected to average no more than 0.08 hectare (Hillel, 2008). These trends and prospects emphasize the imperative to conserve the remaining agricultural land and to improve and intensify production on a sustainable basis without damage to the natural environment or depletion of its resources.

Although the ongoing growth of population undoubtedly contributes to what is now acknowledged to be an environmental crisis, an even more important factor is the quality of human management of the environment. Some of the most crowded countries in the world (such as the Netherlands and Japan) have managed their environments with greater care, while others have not done so well. Profligate use of energy in industry, transportation, and in domestic life, as well as carelessness in the use of materials and the disposal of wastes, characterize the hasty economic "progress" of some rapidly industrializing countries. The results are pollution of air, soil, and freshwater resources, and even of seas. Not the least of the consequences, now recognized as global, is the threat of climate change.

The dilemma of striving to satisfy increasing demands in the face of limited — and in some cases dwindling — resources has worried many observers ever since the Reverend Thomas Robert Malthus published his *Essay on the Principle of Population* in 1798. In it, he argued that population tends to increase faster than food supply, and that, unless that increase is checked by moral restraint, it must inevitably lead to more wars, famine, and disease. For a time, the Malthusian warnings seemed exaggerated, even wrong. In the last three or four decades of the 20th century, food production grew faster than population in all the continents except Africa. That increase, hailed as the "Green Revolution", resulted from the close cooperation between plant breeders and soil scientists. The plant breeders developed new varieties of grain crops (primarily wheat and rice) with higher yielding potential, which depended, however, on optimal conditions of soil moisture and nutrients. The latter were the contributions of soil and water scientists, whose research resulted in improved soil management practices based on the optimal applications of fertilizers, soil amendments, tillage, and irrigation.

The great improvements of the last few decades, however, may not continue indefinitely. Although new methods of genetic engineering, based on recombinant DNA and other innovations in biotechnology, offer great promise, there are constraints to expanded production resulting from the fact that the most favorable soil and readily accessible water resources (i.e., easily tapped aquifers and favorable sites for damming rivers) have already been appropriated. Further expansion of the area under cultivation and further diversion of water flows can only be achieved at the cost of disrupting the remaining natural ecosystems. Much of the land yet unutilized is marginal in terms of its potential and highly vulnerable to degradation by such processes as erosion, compaction, organic matter and nutrient depletion, pollution, salination, desertification, and the loss of biodiversity.

A major constraint to increasing production is the rising cost of fuel. Mechanized agriculture is highly energy-consumptive. Its increase in output per man-hour of labor has been achieved at the expense of much greater dependence on fuel-driven mechanization. When the cost-benefit ratio is calculated in terms of energy input and output, the energy balance of modern agriculture is, in many cases, negative. With the rising costs of energy, that mode of agriculture may be unsustainable economically. Worse yet, the profligate use of fossil fuels has resulted in a significant and progressive increase in the concentration of radiatively active trace gases in the atmosphere. These atmospheric changes are already linked to a changing climate (IPCC WGI, 2007), and are projected to further raise temperature and very likely the frequency and severity of weather anomalies — storms and droughts — that will diminish rather than enhance food security (IPCC WGII, 2007). Progressive global warming will also result in sea-level rise, which may in turn cause waterlogging and inundation of currently productive coastal lands.

The one continent that has not yet participated fully in the Green Revolution is Africa. The problems of Africa are complex. Much of the continent is arid or semiarid, with fragile soils that are extremely vulnerable to drought and to erosion.

In the parts of the continent that are relatively humid — i.e. the central tropics — the soils are highly weathered, leached of nutrients, and affected by aluminum toxicity. Over large areas, the deep-rooted vegetation that had recycled nutrients from the lower layers of the soil, and that had returned them to the surface zone, was cut and used for construction or burned for heating or cooking. Thus, the soil has been deprived of its protective and self-restorative cover of deeply rooted vegetation, surface mulch, and organic residues.

An effective method to restore soil fertility is the practice known as agro-forestry, by which crops and trees are grown together. Fast-growing nitrogen-fixing (leguminous) trees can be planted in parallel rows, preferably on the contour, with crops grown in between. The tree rows act as nutrient pumps, improving soil fertility and providing forage as well as firewood, while the crops grown between those rows provide food for subsistence and local marketing. In some cases, the tree rows are cut down every few years for timber and fuel, and their strips rotated with alternating cropping strips.

Much more can be done to improve the efficient and sustainable utilization of soil and water resources, both in rain-fed and in irrigated farming. Modern methods of low-volume, high-frequency irrigation can be applied to small-holder farms, using inexpensive, locally-fabricated equipment (Hillel, 1997). Investments are needed, however, in the application of science to the enhancement of food production while protecting the environment and helping to mitigate the portending change of climate. The new efforts to promote a 'Green Revolution' in Africa are highly encouraging.

There is an old adage concerning the difference between the clever and the wise. The clever are those who are able to extricate themselves from situations that the wise would have foreseen and avoided from the outset. Now, however, it seems that our short-sighted cleverness as a species has gotten us into a quandary that mere cleverness can no longer resolve. Wisdom is now needed more than ever before. The wisdom we need can only be developed through interdisciplinary research and active international cooperation in the protection and judicious utilization of our shared environment and its vulnerable resources. The ultimate purpose of environmental activity should be to ensure that each generation bequeaths to its successors a world that has the full range of natural wealth (enhanced, insofar as possible) and the richness of human potentialities that it had received from its predecessors. That range encompasses the land, its soils, its waters, its atmosphere, its energy resources, its raw materials, as well as its variegated and synergistic forms of life.

Present yields in many areas are much below the proven potential yields. Where they can be enhanced substantially, there should be no need to claim new land and to encroach further upon natural habitats and their biodiversity. The possibilities for developing intensive, efficient, and sustainable agriculture can obviate the need for the widespread cultivation and grazing of marginal lands, thus allowing natural habitats to regenerate.

The Population-Food Dilemma in a Warming World

The world's population already exceeds seven billion, and is expected to reach over nine billion by mid-century and over ten billion by 2100 (UNPD, 2010). The task of providing sufficient food for humanity is increasingly constrained by land degradation, water-resource depletion, and shortages of fossil-fuel energy resources.

A problem foreseen by some but ignored by many until recently is that of human-induced climate change. Ever since the Industrial Revolution began two centuries ago, humanity has been progressively changing the climate in which we live together with all biota and in which we raise our crops and livestock. We have been doing this by clearing native vegetation, draining wetlands, cultivating and depleting the organic matter in soils, and burning fossil fuels — thereby releasing increasing quantities of radiatively active gases (carbon dioxide, methane, and nitrous oxide) into the atmosphere.

Agricultural productivity and efficiency need to be sustainable. We can and must achieve this by reducing waste and improving resource and energy-use efficiency all along the chain of food production, processing, storage, distribution, marketing, consumption, and waste recycling, while ensuring food security and maintaining environmental quality. These tasks are made even more daunting since they must be accomplished all the while reducing greenhouse gas emissions and responding to changing climate conditions.

The principal challenge of our time is to achieve harmony between the responsibilities and needs of developing and developed nations, between the needs of our generation and those of future generations, and between the human species as a whole and other species in the community of life on Earth. Ecology teaches us that each member of a community is defined not by its individual traits alone but by the nature of its reciprocal relationships with other sharers of the same domain. The ancient tribal vision of the world is still deeply ingrained in us. Gradually, however, our vision has evolved and our notion of kinship has extended to include first our immediate clan or tribe, and then successively our village, city, country, nation, and — eventually — all of humanity. This expanded perception of kinship and allegiance must now transcend the bounds of the human species and extend to the totality and mutuality of life on Planet Earth. In the powerfully symbolic vision of Genesis 2:15: "The Lord God placed the human Earthlings (Adam and Eve) in the Garden of Delight (Eden) to serve and preserve" its living community. When they betrayed that responsibility by consuming in excess of need, they despoiled the Garden of Delight and in so doing banished themselves, in effect, from its security and comfort. On a larger scale, the biosphere can be our collective Eden, if only we can serve and preserve the sustainability of its functions and of its living community, now and into the future.

References

Beddington, J., *et al.* 2011. Achieving food security in the face of climate change. *CGIAR Research Program on Climate Change, Agriculture and Food Security (CCAFS).* Copenhangen, Denmark.

FAOSTAT, 2012. Available at http://faostat.fao.org/ (Accessed on August 1, 2012).

Hillel, D. 1997. *Small-scale Irrigation for Arid Zones: Principles and Option.* FAO Development Series 2, Rome.

Hillel, D. and C. Rosenzweig. 1989. The greenhouse effect and its implications regarding global agriculture. *Massachusetts Agricultural Experiment Station: Research Bulletin No. 724, 1989,* Amherst.

Hillel, D. and C. Rosenzweig (eds.). 2011. *Handbook of Climate Change and Agroecosystems: Impacts, Adaptation, and Mitigation.* Imperial College Press, London.

Hillel, D. 2009. The mission of soil science in a changing world. *J. Plant Nutr. Soil Sci.,* 10:1–5.

Hillel, D. 2008. *Soil in the Environment: Crucible of Terrestrial Life.* Academic Press/Elsevier, Amsterdam.

IPCC WGI. 2007. *Climate Change 2007: The Physical Science Basis. Contribution of Working Group I to the Fourth Assessment Report of the Intergovernmental Panel on Climate Change,* Solomon, S., D. Qin, M. Manning, Z. Chen, M. Marquis, K.B. Averyt, M. Tignor and H.L. Miller (eds.). Cambridge University Press, Cambridge, United Kingdom and New York, NY, USA, 996 pp.

IPCC WGII. 2007. *Climate Change 2007: Impacts, Adaptation and Vulnerability. Contribution of Working Group II to the Fourth Assessment Report of the Intergovernmental Panel on Climate Change,* M.L. Parry, O.F. Canziani, J.P. Palutikof, P.J. van der Linden and C.E. Hanson, Eds., Cambridge University Press, Cambridge, UK, 976 pp.

IPCC WGIII. 2007. *Climate Change 2007: Mitigation. Contribution of Working Group III to the Fourth Assessment Report of the Intergovernmental Panel on Climate Change,* B. Metz, O.R. Davidson, P.R. Bosch, R. Dave, L.A. Meyer (eds.), Cambridge University Press, Cambridge, United Kingdom and New York, NY, USA.

Rosenzweig, C., and D. Hillel. 1998. *Climate Change and the Global Harvest: Potential Impacts of the Greenhouse Effect on Agriculture.* Oxford University Press.

UN. 2004. *World Population to 2300.* Available at http://www.un.org/esa/population/publications/longrange2/WorldPop2300final.pdf (Accessed on August 1, 2012)

UNPD. 2010. *World Population Prospects, the 2010 Revision.* http://esa.un.org/unpd/wpp/Analytical-Figures/htm/fig_1.htm (Accessed on August 1, 2012).

World Bank. 2012. *Data Table, Arable land* (% of land area). Available at http://data.worldbank.org/indicator/AG.LND.ARBL.ZS/countries?display=default (Accessed on August 1, 2012).

Chapter 2

Contributions to Climate Change Solutions from the Agronomy Perspective

David W. Wolfe

Cornell University Ithaca, NY USA 14853
dww5@cornell.edu

Introduction

The evidence that climate change is already upon us is well documented, including substantial evidence that plants, animals, insects and other living things are already responding. In addition to warming, changes in rainfall patterns and increases in extreme events are being observed or projected for many parts of the world (Field *et al.*, 2012), resulting in higher risks of crop failures, natural disasters, and migration of affected populations. The climate is always changing, but the pace of change projected for this century is far beyond what any previous generation of farmers has had to face. Today's farmers cannot rely on historical climate "norms" or calendar dates for making agronomic decisions such as when to plant, what crop to grow, or how to grow it.

The impacts of climate change on agriculture, food systems, and food security will not be equal across regions or socio-economic groups (Fischer *et al.*, 2005). The inequities can be attributed in part to regional variation in the nature and magnitude of climate change impacts (Fig. 1; Easterling *et al.*, 2007; Lobell *et al.*, 2008), but variation in capacity to adapt and variation in farmer recognition of a climate change signal and the need to adapt also play a role (Adger *et al.*, 2007).

While climate change will create unprecedented challenges, there are likely to be new opportunities as well, such as developing new markets for new crop options that may come with a longer growing season and warmer temperatures at high latitudes (Wolfe *et al.*, 2008). Many farm best-management practices for greenhouse gas mitigation and soil carbon sequestration coincide with a conservation agriculture

D.W. Wolfe

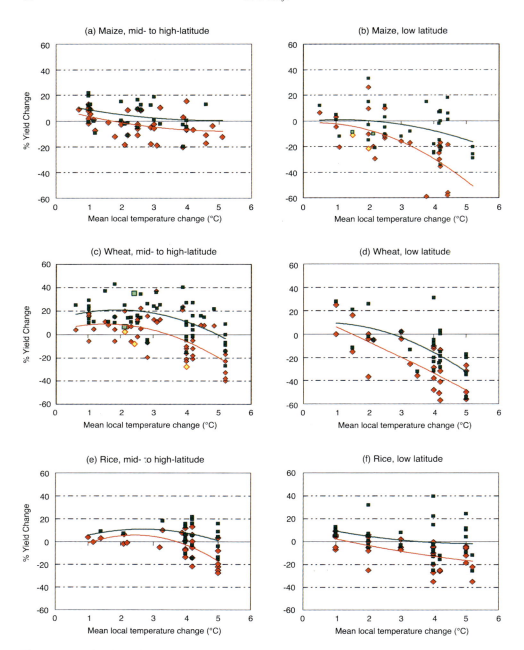

Fig. 1.   Simulated sensitivity of maize, wheat, and rice yields at low vs. mid- to high-latitudes, at varying magnitudes of climate change, and with adaptation (green square symbols) and without (orange diamond symbols). Adaptations included change in planting, cultivar, and irrigation. *Source*: Easterling *et al.* (2007).

approach to farming, which can increase profits, crop productivity, and resilience to climate change if properly implemented (Hobbs and Govaerts, 2010).

Decision Making Under Uncertainty

Climate change assessments often assume farmers will accept climate change information or experience and make so-called "autonomous" adaptations (those using existing knowledge and technology, Easterling *et al.*, 2007) accordingly. However, inequities in adaptive capacity, and the multitude of uncertainties that farmers must weigh in making management and capital investment decisions may prevent timely adaptation. Also, it has become increasingly apparent that individual perceptions and engagement with climate change are inevitably filtered through personal experience and pre-existing cultural worldviews and value systems (Wolf and Moser, 2011).

Before farmers will consider adaptation they must first be convinced the climate is indeed changing. Detecting a climate shift in regions with substantial year-to-year variability is particularly challenging. Maddison (2007) and Meze-Hausken (2004) found that the ability of farmers in Africa to correctly perceive climate shifts was inconsistent. Smit *et al.* (1997) showed that corn variety selection by farmers in Canada was heavily biased by the previous season's weather. An important role for universities and others supporting farmer adaptation is to provide unbiased information that goes beyond anecdotal weather observations, and clarifies the evidence of climate change signals against the background noise of climate variability.

However, while climate scientists are satisfied when well-established statistical methods indicate a high level of certainty regarding recent historical climate change, communicating this certainty to agricultural as well as other audiences has proven complex and frustrating. Even more challenging is communication about future climate scenarios and uncertainties, such as seasonal and regional variation in the magnitude of change and frequency of extreme events. The uncertainties have fueled public debate about whether there is a real threat, and about what type of adaptation or mitigation cost today is warranted to avoid negative economic costs in the future. Some farmers are more concerned about the policy reaction to climate change than they are about the threat of climate change *per se*. The debate has become highly politicized (e.g., Davidson *et al.*, 2003), making it difficult for farmers, the public, and policymakers to sort through the information for decision-making purposes.

Assuming farmers acknowledge a shift in the climate, they must then ascertain their vulnerability to the detected shift, and their capacity to adapt. A survey of wheat farmers in Mexico (Lobell *et al.*, 2005) found that 97% recognized a temperature shift in the climate, but only 38% saw this as affecting their wheat yields. Grothmann and Patt (2005) found that for farmers in Zimbabwe facing drought projections, the divergence between perceived and actual adaptive capacity was a barrier to effective action.

Adaptation Strategies

One general approach to farm management suggested for an uncertain climate is diversification (Reidsma and Ewert, 2008; Fraser et al., 2005; Ellis, 2000). In a diverse system, if one crop or planting date or management approach does not do well due to weather in a given year, all is not lost, and it is even possible that another crop or planting date on the farm may benefit and compensate for losses. The logic seems clear, particularly in poorer countries where an adverse climate event and crop loss directly threaten food supply for the farmer and family. However, empirical evidence of the success of this approach is sparse. A more optimal strategy might be to target a specific agronomic and land-management approach for a given climate change, but for this to be successful many of the uncertainties discussed above must be minimized.

Various adaptation strategies are listed in Table 1. An analysis of 69 published modeling studies (Easterling et al., 2007; Fig. 1) found that effectiveness of adaptations such as those listed in Table 1 varied across regions and magnitude of warming, but on average provided an approximate 10% yield benefit compared to yields without adaptation. This may or may not influence a farmer who must weigh the cost, risk, and effectiveness of any adaptation strategy for particular situation versus the option of doing nothing in response to climate change. This decision must be made within the context of uncertainties specific to climate change risk assessment, as well as uncertainties common to most agronomic decisions, such as future market demands and prices, energy costs, and land-use and energy policies.

Shift/diversify planting dates

Among farmer adaptation options, changing planting date can be an effective, low-cost option to take advantage of an earlier spring and longer growing season (Wolfe et al., 2008), including possibly double-cropping or expanding the use of winter cover crops (Wolfe et al., 2011). Planting date shifts can also be used to avoid crop exposure to adverse climate (e.g., high temperature stress, low rainfall), assuming the timing of adverse events can be predicted. When uncertainty is high, staggered planting dates (i.e., diversification) may be an effective strategy. Effectiveness will depend on region and the magnitude of change in climate and climate variability. In parts of Africa and Asia, farmers routinely modify planting dates in

Table 1.   Agronomic adaptation strategies, and associated benefits, costs, and constraints.

Adaptation	Benefits	Relative costs	Constraints/risks
Shift/diversify planting dates	– Take advantage of longer growing seasons, including possible double-cropping – Avoid dry, wet, hot periods	Low	– Competition to enter market at new harvest times – Hard-to-predict timing of adverse weather events
New/more diversified crop varieties and crops	– Explore new crops and markets for longer growing season; – Use varieties and crops tolerant of new environmental stresses	Low to Moderate	– Competition to enter new markets – New field equipment, infrastructure, transport required for new crops – Stress-tolerant varieties and crops not available or not accepted in marketplace
Improved monitoring of pest and weed populations and range shifts	Better prepared for new pests and weeds	Low to Moderate	– In-field expense low, but requires expensive regional networking to be effective
Chemical, non-chemical control of pests, weeds	Control new pests while minimizing chemical loads to the environment	Low to Moderate	– Chemicals not available, not approved, or expensive – Non-chemical methods not available and take time to develop
Irrigation systems	Maintain yield and quality during dry periods	Moderate to High	– Expensive – Need for adequate water supply
Flood damage protection	Maintain yield and quality during wet periods	Low to High	– Raised bed systems drought-prone in dry periods; – Tile drainage systems expensive – Better-drained or less flood-prone fields not available
Frost and freeze damage protection	Minimize damage to crops due to variable winter and spring temperatures	Moderate to High	Even with forecasting, sprinkler and fan systems not always effective

relation to the onset of the monsoon season (Tadross *et al.*, 2005), although this is likely to become more challenging with climate change (Naylor *et al.*, 2007). In much of the moisture-limited tropics, climate change is expected to reduce the growing season, not lengthen it, and planting date shifts may not remedy the situation. In a maize crop model simulation study, Burke and Lobell (2010) found no shift in optimal planting date with future climate scenarios for southeastern Kenya, where planting dates are based on arrival of the long and short rains.

In high-elevation or high-latitude regions where cool season crops are grown, shifting planting dates may be considered to avoid negative effects from high temperature stress. A potato crop model simulation study (Rosenzweig *et al.*, 1996) found that delaying planting date by two weeks so that tuber development would occur at more favorable cooler temperatures was effective at mitigating yield reductions associated with 1.5 and 2.5°C of warming in some regions (e.g., Caribou, ME) but not others (e.g., Boise, ID and Indianapolis, IN). Regardless of location, delayed planting date could not overcome potato yield reductions with a 5°C warming scenario.

It is important to keep in mind that changing planting date to mitigate yield reductions may not be an adaptive strategy if it means the farmer will be going to market when the supply/demand balance drives prices down. Predicting the optimum planting date for maximum profits will be very challenging in a future with increased uncertainty regarding climate effects not only on local productivity, but also on supply from competing regions.

Change crop varieties and crops

Changing varieties and crops is an off-cited adaptation strategy to take advantage of opportunities and reduce yield losses. It often may go hand-in-hand with changing planting dates, such as switching planting date and crop or crop variety in response to a longer growing season in high-latitude regions. Varieties with improved tolerance to heat or drought will be available for some crop species. New molecular-assisted crop breeding strategies may provide new genetic types more tolerant of environmental stress and pests and pathogens (Reynolds *et al.*, 2010). To date, many such efforts have focused on a few high-caloric major world food crops such as rice and corn, while high-value fruit and vegetable crops important to many regions and to human nutrition have received less attention. Genotypic variation in plant response to elevated atmospheric carbon dioxide (CO_2) exists (Wolfe *et al.*, 1998), but concerted efforts to take advantage of this for yield gains have been few. An evaluation of new wheat varieties released during the twentieth century suggests that responsiveness to CO_2 has not increased but declined with traditional breeding programs (Ziska *et al.*, 2004).

There are a number of situations in which changing varieties or crops might be an expensive or ineffective strategy. A clear case is perennial fruit and nut crops, where changing varieties is extremely expensive and new plantings take several years to reach maximum productivity. Even for annual crops, changing varieties is not always a low-cost option. Seed for new stress-tolerant varieties is sometimes expensive or regionally unavailable, new varieties often require investments in new planting equipment, or require adjustment in a wide range of farming practices. Markets must be found for any venture with a new crop or crop variety. In some cases, it may not be possible to identify an alternative variety that is adapted to the new climate, as well as to local soils and farming practices, *and* meets local market demand regarding timing of harvest and quality features such as size and color.

Improved monitoring and control of pests, pathogens, and weeds

Farmers in many high-latitude regions will experience new challenges with insect management, as longer growing seasons increase the number of insect generations per year, warmer winters lead to larger spring populations of marginally overwintering species, and earlier springs lead to the earlier arrival of migratory insects (Hatfield *et al.*, 2011; Wolfe *et al.*, 2008). Climate change has potential impacts on plant diseases through both the host crop plant and the pathogen. An increase in the frequency of heavy rainfall events projected for many regions will tend to favor some leaf and root pathogens (Coakley *et al.*, 1999, Garrett *et al.*, 2006). The habitable zone of many weed species is largely determined by temperature, and weed scientists have long recognized the potential for northward expansion of weed-species ranges as the climate changes (McDonald *et al.*, 2009). The habitable zone of kudzu (*Pueraria lobata*, var. montana), an aggressive invasive weed that currently infests more than one million hectares in the southeastern United States, is projected to reach into the northeastern part of the country by the end of the 21st century due to climate change (Wolfe *et al.*, 2008). Many C_3 weeds have a stronger growth response to increasing carbon dioxide concentrations than most cash crops (Ziska and George, 2004), and glyphosate (e.g., Roundup) loses its efficacy on weeds grown at the increased carbon dioxide levels likely to occur in the coming decades (Ziska *et al.*, 1999).

An adaptation to increased pest and weed pressure will be increased use of pesticides and herbicides, although for some poor farmers and organic farmers, chemical controls will not be an option. Reduction in the negative economic and environmental impacts of a trend for increased chemical loads will require pre-emptive development of alternative non-chemical weed, insect and disease-control strategies, and/or development of new varieties with resistance or tolerance to agrochemicals.

Those farmers who make the best use of integrated pest management (IPM), such as field monitoring, pest forecasting, recordkeeping and choosing economically and environmentally sound control measures, are most likely to be successful in dealing with a rapidly changing pest, disease, and weed complex. In developed regions, adaptive management is likely to involve increased investment in agricultural consultants and skilled employees by farms, as well as applied research and extension programs by universities and government agencies. Poor farmers and regions lacking funds to support IPM or similar programs will be increasingly at a disadvantage.

Water management under scarcity

Water shortages are projected for many agricultural regions due to reductions in rainfall, and/or reduction in water supplies because of reduced snow pack or gradual loss of glacial mass supplying agricultural regions. Even in many temperate humid regions where growing season rainfall is not projected to decline, crop water demand (i.e., potential evapotranspiration) will increase with warmer summer temperatures and longer growing seasons, increasing the requirement for supplemental irrigation (Hayhoe *et al.*, 2007; Wolfe *et al.*, 2011).

Irrigation systems are a relatively expensive option, and a challenge for farmers will be determining when the yield losses due to summer water deficits have or will become frequent enough to warrant such a capital investment. For some regions, water supplies may be low so farmers will also need to adopt low water-use systems such as drip irrigation, and minimize water use by optimizing irrigation scheduling.

For poor regions or regions where water supplies for agriculture are constrained, maintaining or improving soil water-holding capacity by increasing soil organic matter, conserving soil water by maintaining surface residue, and water "harvesting" are relatively low-cost options that can buffer against short-term water deficits.

Water management under excess

Farms in some coastal zones and flood plains will be subject to increased frequency of severe flooding due to sea level rise. A more widespread problem will be agricultural lands that are projected to continue to be subject to increased frequency of high rainfall events (e.g., more than 5 cm in 48 hours) as a result of climate change. In addition to direct flood damage to crops associated with anaerobic soils, negative economic consequences include delayed spring planting and reduction in the growing season; lack of access to the field during critical periods for farm operations; soil compaction because of tractor use on wet soils; increased crop foliar and root disease; increased soil erosion losses; and increased runoff of chemicals or manures into waterways or crop-growing areas, with negative implications for human health (Hatfield *et al.*, 2011; Wolfe *et al.*, 2011).

Ditch or tile drainage systems are a relatively expensive option, and as in the case of decisions about investment in irrigation, the challenge for farmers will be determining when the frequency of yield losses due to flooding has or will become frequent enough to warrant such a capital investment. In extreme cases, farmers may choose to abandon flood-prone fields, at least for production of high-value crops, and seek higher ground or better-drained soils.

A low-cost option that can buffer against minor or short-term flooding problems is to maintain or improve soil drainage by increasing soil organic matter. Also, maintaining vegetative cover year-round with winter cover crops can minimize soil erosion losses during heavy rainfall events. Changing planting date to avoid wet periods (if they can be predicted) or switching to more flood-tolerant crops or crop varieties will be other relatively low-cost options, when and where they are feasible.

Frost and freeze damage protection

Despite a well-documented trend for warmer winters and earlier springs across the globe, the risk of freeze damage continues, particularly for perennial fruit and nut tree crops, with several damaging events in the past decade. For example, midwinter-freeze damage cost New York Finger Lakes wine-grape growers millions of dollars in losses in the winters of 2003 and 2004 (Levin, 2005). This was likely due to de-hardening of the vines during an unusually warm December, increasing susceptibility to cold damage just prior to a subsequent hard freeze. Another avenue for cold damage, even in a relatively warm winter, is when there is an extended warm period in late winter or early spring causing premature leaf-out or bloom, followed by a damaging frost event, such as that which occurred throughout the Northeast in 2007 (Gu *et al.*, 2008), and again in 2012 when apple, grape, cherry and other fruit crops were hard hit (Halloran, 2012).

Several strategies to avoid damage from spring frost events on strawberries and other perennial crops were reviewed by Poling (2008). These strategies include careful site selection and the use of wind machines, helicopters, heaters and overhead sprinklers. For midwinter freeze problems, approaches might include changes in winter pruning strategies and mulching to insulate the trunks of young plantings. New research will be required to integrate weather forecasts into early-warning systems for extreme events like hard freeze and spring frost events to help perennial fruit-crop growers through a phase of climate change transition that may include increased risk of winter cold-damage.

Adaptations Beyond the Farm Gate

The focus of this chapter is on agronomic adaptations at the farm level. However, climate change impacts on crops will have environmental, human health and political ramifications that cascade beyond the farm gate. For this reason, adaptations that involve societal investment or private-industry responses are also likely to be warranted. Smit and Skinner (2002) described a "typology" of agricultural adaptations that included technological developments, government programs, and farm household financial management, in addition to farm production practices. Below are some specific examples, modified from Wolfe *et al.* (2011):

- *Technological/applied research developments* (e.g., crop breeding for climate stresses, decision-support tools for farmer adaptation, new irrigation technologies)
- *Information delivery/extension systems* (e.g., delivery of real-time local weather data and weather risk forecasts for integration into farm management, better integrated pest management (IPM) monitoring of potential invasives)
- *Locally-available design and planning assistance* for farmers or for water managers in farm regions
- *Disaster-risk management and crop insurance at regional, and national scales*
- *Financial assistance* (e.g., low-cost loans and cost-share programs for adaptation investments)
- *Major capital investments* at a regional or state level (e.g., new dams or reservoirs, new flood-control and drainage systems)
- *Policy and regulatory decisions* (e.g., to facilitate adaptation by farmers, alter regulations, create financial incentives for adaptation or mitigation investment, stimulate local, renewable energy production)
- *Research investment* (e.g., in new crops, new pest and water-management strategies)

Adaptation Capacity

As Table 1 illustrates, agronomic solutions to many of the most likely climate change impacts are reasonably well understood. The challenge is often not that techniques are unavailable, but that the capacity to adapt is lacking. Capacity is often linked with available capital and thus poorer farmers and developing nations are particularly vulnerable (Burke and Lobell, 2010). However, studies in Europe have documented that available capital and access to information do not necessarily translate into successful adaptations (O'brien *et al.*, 2006). Adger *et al.* (2007) summarize the constraints to adaptation according to several categories:

- *Physical and ecological limits* (e.g., the magnitude and/or pace of climate change effects for some vulnerable regions or crops may simply out-pace the ability to adapt).
- *Technological limits* (e.g., suitable varieties or irrigation technologies are unavailable).

- *Financial barriers* (e.g., the individual farmer, or a region or nation, lack the capital for adaptation investments and support for farmers).
- *Informational and cognitive barriers* (e.g., lack of access to information on adaptation, underestimation of the risks of inaction).
- *Social and cultural barriers* (e.g., the social, cultural, political group(s) one belongs to can limit adaptive responses).

Mitigation Strategies

Worldwide, annual agriculture emissions of nitrous oxide (N_2O) are about 2.8 $GtCO_2$-eq yr^{-1} and methane (CH_4) emissions are about 3.3 $GtCO_2$-eq yr^{-1} (Fig. 2), accounting for about 60% and 50% of total global anthropogenic emissions of these two greenhouse gases, respectively. There are very large annual fluxes of CO_2 between the atmosphere and agricultural lands and vegetation, but the net flux is estimated to be approximately balanced, with net CO_2 emissions from soils of only 0.04 Gt CO_2 yr^{-1} (Barker *et al.*, 2007). It is important to note that agricultural emissions as depicted in Fig. 2 do not include CO_2 emissions associated with deforestation for agricultural expansion and other land-use change, which are accounted

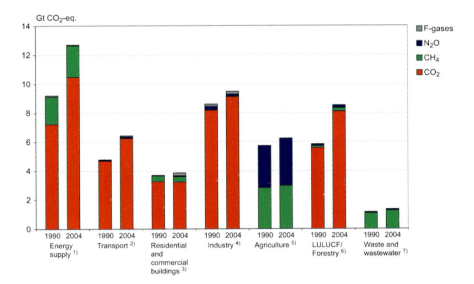

Fig. 2. Greenhouse gas emissions by sector. Deforestation associated with agricultural expansion or slash-and-burn agriculture is accounted for under Forestry. Agricultural CO_2 emissions associated with fossil fuel and electricity use for transportation, buildings, and manufacture of farm inputs are accounted for under those sectors.
Source: Barker *et al.* (2007).

for separately under forestry and contributes over 17% of total global greenhouse gas emissions. Data for the agriculture sector also do not include CO_2 emissions associated with farm fossil fuel use for vehicles, heating and cooling of facilities, or manufacture of fertilizers and other farm inputs, which are accounted for separately under transportation, building, and industry in Fig. 2.

Mitigation management options for agriculture are many, beginning with reducing deforestation and crop residue burning (a major issue in many developing regions using slash-and burn-practices), and improving energy use efficiency and reducing inputs that are energy-intensive to produce, such as synthetic nitrogen (N) fertilizers (a major issue in many intensive operations in developed regions). Other key solutions include reducing tillage, improving fertilizer N use efficiency and manure management, improving water management, winter cover-cropping and, including legumes and perennials in rotation schemes.

Table 2 summarizes key mitigation options associated with the three primary greenhouse gases. Smith *et al.* (2007) and others have provided detailed reviews of this topic. Many of the management options in Table 2 are cost neutral or could potentially increase farm profits, while benefiting the environment and increasing climate change resilience (Scherr and Sthapit, 2009). Many of them involve increasing soil C sequestration, which not only plays an important role in climate change mitigation (Lal, 2004), but also improves soil health, crop productivity, and resilience to climate change (Hobbs and Govaerts, 2010). Corn, wheat and rice production systems are of particular interest because collectively they account for about 50% of all N fertilizer produced worldwide, and there are substantial opportunities for improving N use efficiency (Cassman *et al.*, 2002). Rice production systems are also an important case because of the high methane emissions from flooded rice paddies. There are opportunities for reducing methane emissions by reducing the flooded period by midseason drainage, or various forms of dryland production with irrigation (Ortiz-Monasterio *et al.*, 2010).

The global cropland management mitigation potential by 2030 for all three gases was estimated by Smith *et al.* (2007) at about 760–830 MtCO$_2$-eq yr^{-1}, with an additional 160–190 MtCO$_2$-eq yr^{-1} for rice production (Fig. 3). The range is based on different economic scenarios related to future market values for C. Not included in Fig. 3 are possible mitigation effects from reduction of deforestation rates, and for substituting fossil fuel use on the farm with energy produced from agricultural feedstocks. Smith *et al.* (2007) separately estimated the economic mitigation potential by 2030 for agricultural use of biofuels at 1260 MtCO$_2$-eq yr^{-1} to 2320 MtCO$_2$-eq yr^{-1} at 20 US\$ and 50 US\$ per tCO$_2$-eq, respectively.

Table 2.   Agronomic mitigation strategies for the three primary greenhouse gases.

Greenhouse gas	Primary agricultural sources of emissions	Approaches to mitigation
N_2O	– Excessive or poorly timed N fertilizer applications – Wet soils and manures	– Split fertilizer applications, optimize timing and amount applied by soil tests and new web tools (e.g.:http://adapt-n.cals.cornell.edu/) – Use legumes (biological N fixation) in rotations – Use winter cover crops to scavenge, store N in the root zone – Improve manure management (e.g., keep covered to keep dry) or use manures as energy source in anaerobic digesters – Improve soil drainage
CH_4	– Flooded rice paddies – Enteric fermentation by ruminant livestock – Wet manures	– Drain rice fields intermittently or employ various dryland production options with irrigation where feasible – Incorporating organic materials during dry period of rice production – Develop new feeding strategies and feed amendments to reduce methane emissions from livestock – Use covered or tank storage of manures and store at low temperature
CO_2	– Deforestation and residue burning – Soil organic matter decomposition – Fossil fuel and electricity use for transportation and buildings – Fossil fuel use in manufacture of energy-intensive farm inputs such as synthetic N fertilizers, pesticides, and herbicides	– Reduce or minimize slash-and-burn agricultural practices – Reduce soil tillage (slows organic matter decomposition and reduces fuel use for tractors) – Retain and incorporate crop residues – Increase crop residues by increasing yields and biomass, use high-biomass winter cover crops and rotation crops – Use C-rich sources of fertilizers, composts, or biochar – Improve farm energy use efficiency by building design and insulation, energy-efficient appliances and vehicles – Minimize use of energy-intensive farm inputs (e.g., use organic fertilizers rather than synthetic) – Use alternative energy sources, such as biofuel crops, anaerobic digestion or pyrolysis of farm manures and wastes, solar or wind power.

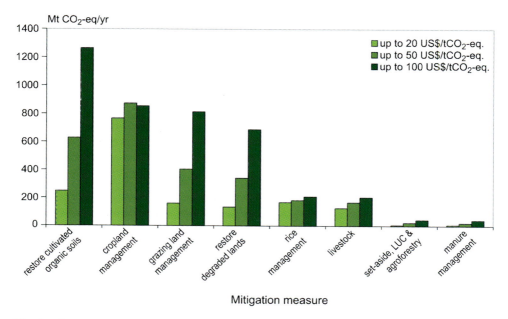

Fig. 3.   Economic potential for agricultural greenhouse gas mitigation by 2030 at a range of C values. Based on the B2 emissions scenario, although the pattern is similar for other scenarios. *Source*: Smith *et al.* (2007).

Conservation Agriculture for Mitigation, Adaptation, and Sustainable Development

Climate change mitigation and sustainable development are intimately linked. Many agricultural mitigation practices also improve resilience to climate change and help meet goals of increasing crop productivity with minimal inputs while protecting natural resources. On the other hand, unmitigated climate change would eventually lead to impacts on crops beyond farmers' ability to adapt, and diversion of funds from agriculture development projects toward disaster relief and emergency food aid. Conservation agriculture (CA) is an example of the possible synergy between agronomic solutions to climate change adaptation and mitigation, and environmental protection, sustainable development, and human well-being.

Conservation agriculture is an ecologically-based approach to farming that attempts to conserve soil, water, and nutrient resources within the farm while main-taining yields and quality. It is a knowledge-intensive management approach, not one fixed set of practices, but typically it involves minimizing tillage, maintain-ing vegetation cover year-round, diversifying crop rotations, re-incorporating crop residues, and use of composts, manures or other organic amendments (Hobbs and Govaerts, 2010; Kassam *et al.*, 2009). All of these practices tend to maintain or build organic matter in the soil, which may have beneficial effects on ecosystem services attributable to soil health such as crop productivity, improved water and nutrient cycling, beneficial soil microbial activity, and improved drainage (Kassam *et al.*, 2009; Wolfe, 2006). Relevant to this chapter, CA practices can increase resilience to climate change (e.g., better soil water-holding capacity, more diverse cropping system), increase soil and biomass carbon (C) sequestration, and reduce requirements for N fertilizers and other inputs, thus contributing to climate change mitigation (Milder *et al.*, 2011; Hobbs and Govaerts, 2010; Scherr and Sthapit, 2009).

Because CA has the potential to reduce the need for external inputs, it is a particularly attractive strategy for poor small-scale land-holders in developing countries with limited access to capital for inputs such as fertilizers (Kassam *et al.*, 2009). In sub-Saharan Africa, fertilizer use averages $13\,kg\,ha^{-1}$ compared to a global average of about $100\,kg\,ha^{-1}$, and irrigation is used on only 3% of farm land (AGRA, 2010). At a broader landscape scale, CA can encompass good agroforestry practices and the avoidance of slash-and-burn clearing of forests. For example, a CA project in Zambia involving over 20,000 small land-holders (Lewis *et al.*, 2011) has recently initiated an ambitious tree-planting project — the intercropping of one million leguminous *Faidherbia albida* trees at a density of 100 trees per hectare. The benefits of *F. albida* as a N-fixing intercrop in Africa are well-documented (Barnes and Fagg, 2003), and over time these trees will also sequester significant C in their biomass.

The benefits of CA and adoption statistics are not that reliable because of differences in how it is defined and lack of good tracking mechanisms in some regions. Derpsch *et al.* (2010) tallied more than 105 million hectares under zero tillage worldwide, primarily in North and South America, Australia and New Zealand, with much less in many developing regions. The data for Africa and other developing regions are likely under-reported (Milder *et al.*, 2011), while the global values are probably an overestimate, since they are based on zero tillage, which is only one component of CA. Although CA techniques will not be suitable for all crops, soils, and climates, there is clearly potential for broader adoption, particularly in some developing regions where the benefits may be high. One constraint to adoption is inadequate farmer knowledge and extension systems to work with farmers during the process of adoption. Also, availability of suitable methods and equipment for planting into no-till fields is lacking for some. At early stages, labor requirements can increase, although in the long term typically decrease (Milder *et al.*, 2011). Fine-tuning CA to specific locations and promoting it to farmers within a package that includes consultation regarding suitable crop varieties, rotation schemes, fertility and water management will be essential.

Research Priorities to Advance Agronomic Solutions to Climate Change

Farmer decision makers need information that can reduce uncertainty about climate change, its impacts on the systems they are managing, and the effectiveness of adaptation and mitigation options. Below are some selected research priorities:

- Adaptation strategies for weed and pest control, such as improve regional monitoring and IPM communication regarding weed and pest range shifts and migratory arrivals; enhance real-time weather-based systems for weed and pest control; develop non-chemical options for new pests; and develop rapid-response action plans to control invasive species.
- Improve water management systems and irrigation scheduling technology.
- Develop better decision tools for determining the optimum timing and magnitude of investments for strategic adaptation to climate change for maintaining/maximizing profits over multiple planning horizons. This will require addressing uncertainties in climate model projections regarding precipitation, frequency of extreme events, and temporal and spatial climate variability. It also will require quantifying costs and benefits of adaptation at the farm level, and for fruit, nut, and vegetable crops and livestock as well as grain crop production systems.
- Integrate cognitive and cultural factors into communication strategies. Advances in the decision sciences on topics such as risk perception, temporal discounting, decision making under uncertainty, participatory processes, decision architecture, equity, and framing should be taken into account in the design of effective adaptation and mitigation programs.
- Improve mitigation efforts in the agriculture sector. Better tools are needed for monitoring, accounting, and management of energy, C, N, and associated greenhouse gases.
- Explore the potential for CA more fully, support networks for farmers to adopt CA practices.

Expected Outcomes with Investments

Sustained major investments are needed in research to develop the new technologies, decision tools, information, and effective communication strategies to transform agriculture into a system that is more resilient and adaptive to climate variability and climate change. With timely and appropriate proactive investment in research the agriculture sector will have the necessary tools for strategic adaptation to meet the challenges and take advantage of any opportunities associated with climate change. Policy-makers will have information to facilitate adaptation and minimize inequities in impacts and costs of adaptation. Farmers will also contribute significantly to greenhouse gas mitigation by having access to new tools and incentives, including new greenhouse gas and soil C accounting tools.

References

Adger, W.N., S. Agrawala, M.M.Q. Mirza, C. Conde, K. O'Brien, J. Pulhin, R. Pulwarty, B. Smit and K. Takahashi. 2007. Assessment of adaptation practices, options, constraints and capacity. pp. 717–743. *In* M.L. Parry, O.F. Canziani, J.P. Palutikof, P.J. van der Linden and C.E. Hanson (eds.), Climate Change 2007: Impacts, Adaptation and Vulnerability. *Contribution of Working Group II to the Fourth Assessment Report of the Intergovernmental Panel on Climate Change.* Cambridge University Press: Cambridge, UK.

AGRA. 2010. *Alliance for a Green Revolution in Africa.* Online: www.agra-alliance.org/section/ work/soils.

Barker, T., I. Bashmakov, L. Bernstein, J.E. Bogner, P.R. Bosch, R. Dave, O.R. Davidson, B.S. Fisher, S. Gupta, K. Halsnæs, G.J. Heij, S. Kahn Ribeiro, S. Kobayashi, M.D. Levine, D.L. Martino, O. Masera, B. Metz, L.A. Meyer, G.-J. Nabuurs, A. Najam, N. Nakicenovic, H.-H. Rogner, J. Roy, J. Sathaye, R. Schock, P. Shukla, R.E.H. Sims, P. Smith, D.A. Tirpak, D. Urge-Vorsatz, D. Zhou. 2007. Technical Summary. *In* B. Metz, O.R. Davidson, P. R. Bosch, R. Dave, L. A. Meyer (eds.), *Climate Change 2007: Mitigation. Contribution of Working Group III to the Fourth Assessment Report of the Intergovernmental Panel on Climate Change.* Cambridge University Press: Cambridge, United Kingdom and New York, NY, USA

Barnes, R.D. and C.W. Fagg. 2003. *Fadherbia albida: Monograph and Annotated Bibliography. Oxford Tropical Forestry Papers No. 41.* Oxford Forestry Institute: Oxford, UK.

Burke, M. and D. Lobell. 2010. Food security and adaptation to climate change: what do we know? *In* Lobell D., M. Burke (eds.) *Climate Change and Food Security.* Springer. New York. Chapter 8.

Cassman, K.G., A. Dobermann and D.T. Walters, 2002. Agroecosystems, nitrogen-use efficiency, and nitrogen management. *Ambio* 31(2):132–140.

Coakley, S.M., H. Scherm and S. Chakraborty. 1999. Climate change and plant disease management. *Annual Review of Phytopathology* 37:399–426.

Davidson, D.J., T. Williamson, J.R. Parkins. 2003. Understanding climate change risk and vulnerability in northern forest-based communities. *Canadian Journal of Forest Research* 33:2252–2261.

Derpsch, R., T. Friedrich, A. Kassam and L. Hongwen. 2010. Current status of adoption of no-till farming in the world and some of its main benefits. *International Journal of Agricultural and Biological Engineering* 3:1–25.

Easterling, W.E., P.K. Aggarwal, P. Batima *et al.* 2007. Food, fibre, and forest products. *In*: Parry M.L., O.F. Canziani, J.P. Palutikof *et al.* (eds.) *Climate Change 2007: Impacts, Adaptation, and Vulnerability. Contribution of Working Group II to the Fourth Assessment Report of the Intergovernmental Panel on Climate Change.* Cambridge University Press: Cambridge, UK.

Ellis, F. 2000. *Rural Livelihoods and Diversity in Developing Countries.* Oxford University Press: Oxford, UK.

Field, C.B., V. Barros, T.F. Stocker, D. Qin, D.J. Dokken, K.L. Ebi, M.D. Mastrandrea, K.J. Mach, G.-K. Plattner, S.K. Allen, M. Tignor, and P.M. Midgley. 2012. *Managing the Risks of Extreme Events and Disasters to Advance Climate Change Adaptation.* A Special Report of Working Groups I and II of the Intergovernmental Panel on Climate Change. World Meteorological Organization: Geneva, Switzerland.

Fischer, G.M., M. Shah, F.N. Tubiello and H. van Velthuizen. 2005. Socio-economic and climate change impacts on agriculture: an integrated assessment, 1990–2080. *Philosophical Transactions of the Royal Society B*. 360:2067–2083.

Fraser, E.D.G., W. Mabee and F. Figge. 2005. A framework for assessing the vulnerability of food systems to future shocks. *Futures* 37:465–479.

Garrett, K.A., S.P. Dendy, E.E. Frank, M.N. Rouse and S.E. Travers. 2006. Climate change effects on plant disease: Genomes to ecosystems. *Annual Review Phytopathology* 44:489–509.

Grotham, T. and A. Patt. 2005. Adaptive capacity and human cognition: The process of individual adaptation to climate change. *Global Environmental Change* 15:199–213.

Gu, L., P.J. Hanson, W.M. Post, D.R. Kaiser, B. Yang, R. Nemani, S.G. Pallardy and T. Meyers. 2008. The 2007 Eastern U.S. spring freeze: Increased cold damage in a warming world? *BioScience* 58(3):253–262.

Halloran, A. May 17, 2012. Growing uncertainties: Climate change is forcing farmers in the Northeast to rethink their seasonal strategies. *Metroland News*. Albany, NY. http://metroland.net.

Hatfield, J.L., K.J. Boote, B.A. Kimball, R.C. Izaurralde, D. Ort, A. Thomson and D.W. Wolfe. 2011. Climate impacts on agriculture: Implications for crop production. *Agronomy Journal* 103:351–370.

Hayhoe, K., C. Wake, T. Huntington, L. Luo, M. Schwartz, J. Sheffield, E. Wood, B. Anderson, J. Bradbury, A. DeGaetano, T. Troy and D.W. Wolfe. 2007. Past and future changes in climate and hydrological indicators in the U.S. Northeast. *Climate Dynamics* 28:381–407.

Hobbs P.R. and B. Govaerts. 2010. How conservation agriculture can contribute to buffering climate change. *In*: Reynolds M. (ed.) *Climate Change and Crop Production*. CABI, Wallingford, UK. Chapter 10.

Kassam, A., T. Friedrich, F. Shaxson and J. Pretty, 2009. The spread of conservation agriculture: justification, sustainability, and uptake. *International Journal of Agricultural Sustainability* 7:292–320.

Lal, R. 2004. Soil carbon sequestration to mitigate climate change. *Geoderma* 123:1–22.

Levin, M.D. 2005. Finger Lakes freezes devastate vineyards. *Wines and Vines*, July 2005 issue.

Lewis, D., S. Bell, J. Fay, K. Bothi, L. Gatere, M. Kabila, M. Mukamba, E. Matokwani, M. Mushimbalume, C.I. Moraru, J. Lehmann, J. Lassoie, D. Wolfe, D. Lee, L. Buck and A.J. Travis, 2011. The COMACO model: Using markets to link biodiversity conservation with sustainable improvements in livelihoods and food production. *Proceedings National Academy of Sciences* 108(34):13957–13962.

Lobell, D.B., M.B. Burke, C. Tebaldi, M.D. Mastrandrea, W.P. Falcon and R.L. Naylor. 2008. Prioritizing climate change adaptation needs for food security in 2030. *Science* 319:607–610.

Lobell, D.B. and J.I. Oriz-Monasterio. Analysis of wheat yield and climate trends in Mexico. *Field Crops Research* 94(2–3):250–256.

Maddison, D. 2007. *The Perception of and Adaptation to Climate Change in Africa*. World Bank: Washington D.C.

McDonald, A., S. Riha, A. Ditommaso and A. DeGaetano. 2009. Climate change and the geography of weed damage: Analysis of US maize systems suggests the potential for significant range transformations. *Agriculture Ecosystems and Environment* 130:131–140.

Meze-Hausken, E. 2004. Contrasting climate variability and meteorological drought with perceived drought and climate change in northern Ethiopia. *Climate Research* 27(1):19–31.

Milder, J.C., T. Majanen and S.J. Scherr. 2011. *Performance and Potential of Conservation Agriculture for Climate Change Adaptation and Mitigation in Sub-Saharan Africa*. WWF-CARE Alliance's Rural Futures Intiative. CARE: Atlanta, GA. (www.careclimatechange.org).

Naylor, R.L., D.S. Battisti, D.J. Vimont, W.P. Falcon and M.B. Burke. 2007. Assessing risks of climate variability and climate change for Indonesian rice agriculture. *Proceedings National Academy of Sciences* 104(19):7752–7757.

O'Brien, K., S. Eriksen, L. Sygna and L.O. Naess. 2006. Questioning complacency: Climate change impacts, vulnerability, and adaptation in Norway. *Ambio* 35:50–56.

Ortiz-Monasterio, I., R. Wassmann, B. Govaerts, Y. Hosen, N. Katayanagi and N. Verhulst. 2010. Greenhouse gas mitigation in the main cereal systems: Rice, wheat, and maize. *In* Reynolds M. (ed.) *Climate Change and Crop Production.* CABI, Wallingford, UK. Chapter 10.

Poling, E.B. 2008. Spring cold injury to winegrapes and protection strategies and methods. *HortScience* 43(6):1652–1662.

Reidsma, P. and F. Ewert. 2008. Regional farm diversity can reduce vulnerability of food production to climate change. *Ecology and Society* 13(1): article 38, online: www.ecologyandsociety.org/.

Reynolds, M.P., D. Hays and S. Chapman. 2010. Breeding for adaptation to heat and drought stress. Chapter 5. In: Reynolds, M. (ed.) *Climate Change and Crop Production.* CABI, Wallingford, UK.

Rosenzweig, C., J. Phillips, R. Goldberg, J. Carroll J and T. Hodges. 1996. Potential impacts of climate change on citrus and potato production in the U.S. *Agricultural Systems* 52:455–479.

Scherr, S.J. and Sthapit S. 2009. *Mitigating Climate Change Through Food and Land Use.* Worldwatch Report179. Worldwatch Inst., Wash. D.C.

Smit, B., R. Blain and P. Keddle. 1997. Corn hybrid selection and climate variability: Gambling with nature? *Canadian Geography* 41(4):429–438.

Smit, B. and M.W. Skinner. 2002. Adaptation options in agriculture to climate change: A typology. *Mitigation and Adaptation Strategies for Global Change* 7:85–114.

Smith, P.D., Z. Martino, Z. Cai *et al.* 2007. Agriculture. *In*: Metz B., O.R. Davidson, P.R. Bosch *et al.* (eds.) *Climate Change 2007: Mitigation. Contribution of Working Group III to the Fourth Assessment Report of the Intergovernmental Panel on Climate Change.* Cambridge University Press. Cambridge, UK. Chapter 8.

Tadross, M.A., Hewitson B.C. and Usman M.T. 2005. The interannual variability of the onset of the maize growing season over South Africa and Zimbabwe. *Journal of Climate*, 18(16):3356–3372.

Wolf, J. and S.C. Moser. 2011. Individual understandings, perceptions, and engagement with climate change: Insights from in-depth studies across the world. *WIREs Climate Change* 2(4):547–569.

Wolfe, D.W. 2006. Approaches to monitoring soil systems. *In*: Uphoff N. *et al.* (eds.) *Biological Approaches to Sustainable Soil Systems.* CRC Press. Boca Raton, FL, USA. Chapter 47.

Wolfe, D.W., J. Comstock, A. Lakso, L. Chase, W. Fry, C. Petzoldt, R. Leichenko and P. Vancura. 2011. Agriculture. *In*: Rosenzweig C., W. Solecki, A. DeGaetano *et al.* (eds.) *Responding to Climate Change in New York State.* pp. 217–254. New York Academy of Sciences. Blackwell Pub., Boston, MA. Chapter 7.

Wolfe, D.W., R. Gifford, D. Hilbert and Y. Luo. 1998. Integration of acclimation to elevated CO_2 at the whole-plant level. *Global Change Biology* 4:879–893.

Wolfe, D.W., L. Ziska, C. Petzoldt, A. Seaman, L. Chase and K. Hayhoe. 2008. Projected change in climate thresholds in the Northeastern U.S.: Implications for crops, pests, livestock, and farmers. *Mitigation and Adaptation Strategies for Global Change* 13:555–575.

Ziska, L.H. and K. George. 2004. Rising carbon dioxide and invasive, noxious plants: Potential threats and consequences. *World Resource Review* 16(4):427–447.

Ziska, L.H., C.F. Morris and E.W. Goins. 2004. Quantitative and qualitative evaluation of selected wheat varieties released since 1903 to increasing atmospheric carbon dioxide: can yield sensitivity to carbon dioxide be a factor in wheat performance? *Global Change Biology* 10:1810–1819.

Ziska, L.H., J.R. Teasdale and J.A. Bunce. 1999. Future atmospheric carbon dioxide may increase tolerance to glyphosate. *Weed Science* 47:608–615.

Section II

Regions

Chapter 3

North American Perspectives on Potential Climate Change and Agricultural Responses

Jerry L. Hatfield

USDA-ARS
National Laboratory for Agriculture and the Environment
Ames, Iowa
jerry.hatfield@ars.usda.gov

Introduction

Climate impacts agriculture in all aspects of crop growth and livestock production. The magnitude of these impacts is of growing interest in view of projected climate change across North America. The potential impacts of climate change on crops were summarized by Hatfield *et al.* (2011a) and on rangeland and pastures by Izaurralde *et al.* (2011). These summaries demonstrate the importance of understanding the linkage between climate change and agricultural systems. Understanding these impacts will also provide a framework for the assessment of potential adaptation and mitigation strategies. In this chapter, climate change across North America is discussed along with a framework for the assessment of climate impacts on agriculture and linkages to adaptation and mitigation practices.

Climate change scenarios for the next 30–50 years vary throughout the world. Across North America, there are projected to be regional and seasonal differences in precipitation and temperature. These scenarios have been developed from multiple climate models assembled by the Intergovernmental Panel on Climate Change (IPCC, 2007). Temperature and precipitation are two critical parameters affecting plant growth. Their changes may induce plant stress either through direct effects on plant growth, increased water use rates caused by increased evaporative demand, or reduced or altered precipitation patterns. Meehl *et al.* (2007) stated on a global basis that:

> "It is very likely that heat waves will be more intense, more frequent, and longer lasting in a future warmer climate. Cold episodes are projected to decrease significantly in a future warmer climate. Almost everywhere, daily minimum temperatures are projected to increase faster than daily maximum temperatures, leading to a decrease in diurnal temperature range. Decreases in frost days are projected to occur almost everywhere in the middle and high latitudes, with a comparable increase in growing season length."

In their assessment on precipitation, Meehl *et al.*, 2007 found that:

> "For a future warmer climate, the current generation of models indicates that precipitation generally increases in the areas of regional tropical precipitation maxima (such as the monsoon regimes) and over the tropical Pacific in particular, with general decreases in the subtropics, and increases at high latitudes as a consequence of a general intensification of the global hydrological cycle. Globally averaged mean water vapour, evaporation and precipitation are projected to increase."

These indications confirm an earlier observation by Katz and Brown (1992) that the increased variability in climate attributes may be more important than the increase in mean values. These summaries represent the current thinking on climate scenarios pertaining to the expected global change in temperature and precipitation. There are efforts to update the climate scenarios so as to provide new scenarios to assess potential future impacts for the IPCC Fifth Assessment Report. Projections of climate change are often met with doubt because of the natural variation that occurs in seasonal weather and between successive years; however, there are definite signals in the observations that climate is indeed changing (Karl *et al.*, 2009). There is an unmistakable increase in carbon dioxide (CO_2) concentrations in the atmosphere along with nitrous oxide (N_2O) and methane (CH_4) (IPCC, 2007). These levels have reached concentrations exceeding observed values in the history of cultivated agriculture.

For North America, regional differences are expected in both temperature and precipitation (Christensen *et al.*, 2007). Temperatures are expected to increase, with the largest projected increases expected in northern latitudes above 50°N and more moderate increases across the remainder of North America. Projected increases for mean temperature over the mid-continent are 4°C by 2100 with an uncertainty range of ±2°C. In the northern latitudes there would be greater warming during the winter months with less seasonal variation in the mid-continent regions (Christensen *et al.*, 2007). An important change across the mid-latitudes is the projection of a greater increase in the nighttime minimum temperatures than in the daytime maximum temperatures. These changes may already be evident in the climate record. Indeed, across the Midwest, there has been an increase in the summer minimum temperatures

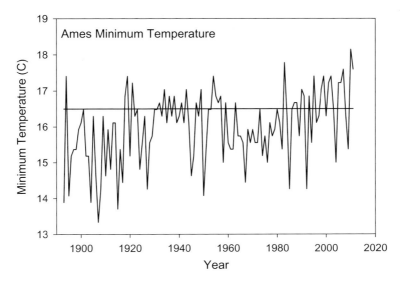

Fig. 1.   Minimum summer (July-August-September) temperatures from 1890 through 2011 for Ames, Iowa. Long-term average minimum is shown as the horizontal line.

as illustrated by this example from Ames, Iowa where the summer minima (June-July-August) have shown a steady increase since 1990 and have exceeded the long-term average (Fig. 1). Minimum summer temperatures in 2010 were the highest on record for the Midwest and although there was a decrease in 2011, the values were significantly above the long-term average. Over these months there was no discernible trend in the maximum temperatures for Ames.

While not yet discernible in the Midwest, a significant potential impact for plant growth is the projected increase in the number of days with high temperatures, defined as those days that are significantly above the average temperature. Battisti and Naylor (2009) utilized global climate models to demonstrate a probability $>90\%$ of the growing season temperatures in the sub-tropics and tropics exceeding the most extreme temperatures observed in the 1900–2006 period by the end of the 21st century. This is reinforced by the recent findings from Rahmstorf and Coumou (2011), which showed that extreme heat events are a consequence of climate change and that the heat extremes in Russia during 2010 were likely due to climate change. Recent heat waves in the southern United States during 2011 demonstrate the potential for extremes. Growing season weather in 2012 has demonstrated that combinations of extreme heat and drought can occur across the United States. Record temperatures have been set coupled with record low precipitation amounts across the United States. These conditions demonstrate the rapid change which can occur in growing season conditions and how much of the production area can be affected. These conditions being experienced in 2012 are not unlike the projections for future climate from climate scenarios. These increases in temperature will exacerbate plant stress because of the effect on evaporative demand. The increase in water demand will have direct implications for plant growth and development.

Meehl *et al.* (2004) also projected a decrease in the number of frost days in the year; however, the largest projected decrease was observed to occur in the spring compared to the fall. There would be regional differences in these patterns because of the impact of local factors, e.g., regional atmospheric circulation patterns and pressure systems (Meehl *et al.*, 2004).

In addition to the projected changes in temperature there is the projected increase in CO_2 concentrations from the 2012 value of nearly $390 \, \mu g \, m^{-3}$ to approximately $550 \, \mu g \, m^{-3}$ by 2050. These may well have a positive impact on plant growth, as indicated in the reviews by Hatfield *et al.* (2011a) and Izaurralde *et al.* (2011).

The climate change that has occurred to date may have already begun to impact crop production. Lobell *et al.* (2011) observed that the global temperature record trends from 1980 to 2008 have exceeded one standard deviation of the prior year-to-year variation. From their analysis, they concluded that maize (*Zea mays* L.) yields have already declined 3.8% and wheat (*Triticum aestivum* L.) declined 5.5% compared to the yields without climate trends. An important aspect of this research was the conclusion that climate trends had a large enough effect to offset the yield gains from advancing technology and CO_2 increases. Analyses such as these and the results reported by Hatfield (2010) reveal that climate change has indeed started to impact crop production. The expectations are that a continuing change of both means and extremes of temperature will cause progressive impacts on agriculture. The challenges being faced by agriculture relative to climate change are pertinent and therefore require the development of a framework for the integration of adaptation and mitigation strategies.

Precipitation, as it varies seasonally, is a critical determinant in rainfed agricultural systems. One of the projected changes in climate for the United States is the increase in spring precipitation across the northern section and a reduced summer precipitation total for all of the country (Karl *et al.*, 2009). High-precipitation events (e.g., >5 cm in 48 hours) are predicted to increase and are of great concern in many parts of the U.S. because of the inability of the soil to maintain infiltration rates high enough to absorb high-intensity rainfall events (Hayhoe *et al.*, 2007). This trend is projected to apply to many regions (Lettenmaier *et al.*, 2008). The increased variability in rainfall during the summer months will lead to increased variability in available soil water. The signals in the climate record for precipitation are not as easily discernible compared to temperature. There has been a slight but significant increase in the precipitation amounts in some regions, as shown, for example by

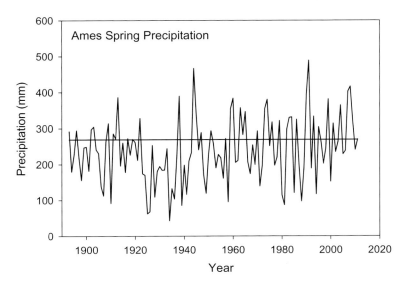

Fig. 2. Total spring precipitation (April-May-June) temperatures from 1890 through 2011 for Ames, Iowa. Long-term total precipitation is shown as the horizontal line.

the record from Ames, Iowa. (Fig. 2). There has been an increase in the number of days with rainfall totals exceeding 30 mm day^{-1}, demonstrating that more intense storms are occurring. Management of soil water must become an integral part of the overall management system in order to increase the resilience of cropping systems to climate change.

Agricultural productivity is sensitive to all the manifestations of climate change. The positive effects of climate change (particularly increased atmospheric CO_2) may be partially or totally offset by the negative impacts due to the higher temperatures shortening grain-fill duration and increasing evapotranspiration rates (Adams *et al.*, 1990). The projected increases in temperatures for the U.S. would result in an increased rate of soil water evaporation and crop transpiration (ET). An increase in ET leads to an increase in soil water deficits and hence production losses. These losses could translate into economic losses unless reduced by other factors, e.g., increase in precipitation, increase in crop water use efficiency, or alteration of leaf area or planting density to reduce ET, or the use of supplemental irrigation.

Hayhoe *et al.* (2007) projected a significant increase in summer soil water deficits by mid-century for the relatively humid Northeastern United States region, with little change in total annual precipitation. Reduction in snow pack and earlier snow melt in the western U.S. will exacerbate the potential threat of drought for farmers because of the reduction in the reservoir of water available for irrigation (Lettenmaier *et al.*, 2008). Wang (2005) suggested that increases in atmospheric concentrations of greenhouse gases will cause a worldwide increase in the occurrence of agricultural droughts. These climate model results were consistent in predicting drier soil over the Southwest U.S. during all seasons. Across the Midwest, Mishra and Cherkauer (2010) found that droughts actually decreased in the last half of the 20th century, with the last significant widespread droughts having occurred in the 1930s. They found that maize and soybean (*Glycine max* (L.) Merr.) yields were negatively correlated with meteorological drought and maximum daily temperature during the grain-filling period.

Framework for Climate Assessment

The impacts of climate on agriculture vary by crop and cropping system and the pattern of weather within the growing season. There is a growing interest in assessing the role of climate change and agricultural system performance relative to environmental endpoints or product quality in addition to crop yield. A proposed framework to integrate climate changes across cropping systems into outputs is shown in Fig. 3. On the input side of this framework there is long-term climate, seasonal weather, soils, nutrient management, agronomic management, and irrigation as examples of the types of inputs to the cropping system. Across North America, there is a wide range of different crops and each has its own production system. This framework can be expanded to include livestock production systems; however, in the scope of this chapter we focus on crop production systems. The output side of this framework adds

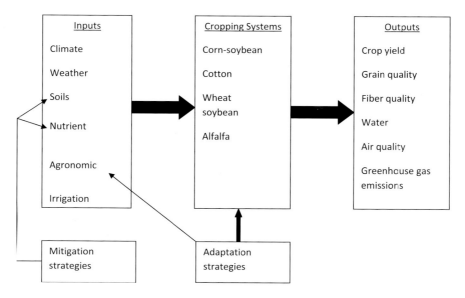

Fig. 3.   Schematic diagram of the inputs, systems, and outputs for the assessment of climate change on cropping system performance and the role of adaptation and mitigation strategies.

the dimensionality of all types of products from a cropping system. This includes grain, forage, and fiber; the quality of these products (protein, oil, starch, etc.); and environmental goods and services (e.g., water quality, air quality and greenhouse gas emissions). The latter components are beginning to be considered as part of the overall system in the same context as crop productivity. Two additional components in this framework are climate change adaptation strategies that directly act on the cropping systems component (e.g., planting date, cultivar) and mitigation strategies that are directly related to soil management and nutrient management. Throughout this chapter, this framework is developed to show how impacts, adaptation, and mitigation are integrated into a unifying concept.

Impacts

Impacts of climate on crop production can be considered in terms of the seasonal weather variation within a cropping system, and reflected in the annual variation in crop production (expressed as yield) as shown for wheat yields in Mexico (Fig. 4). However, weather variation among years may also be expressed in terms of the planted or harvested area for a particular crop and region (Fig. 5). In Mexico, there has been a continual increase in wheat production over the past 50 years, with a noticeable decline in the rate of yield increase in recent decades. Annual yield

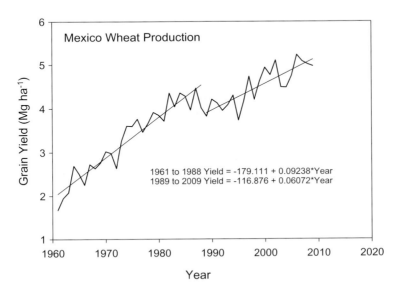

Fig. 4.   Annual variation in wheat yields for Mexico (FAOSTAT, accessed 27-October-2011 at http://faostat.fao.org).

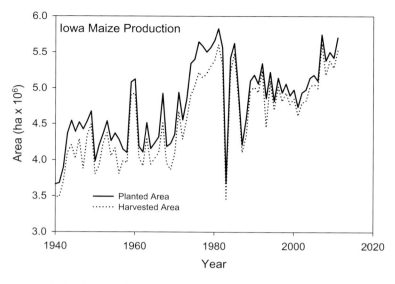

Fig. 5.   Annual variation in area of maize planted and harvested for Iowa from 1940 through 2011. (Source of data: USDA-NASS, http://quickstats.nass.usda.gov, accessed 27-October-2011).

variation among years is not as large as in other countries because of the large area of irrigated wheat production within Mexico.

An interesting comparison is the change in the planted versus harvested maize area for Iowa from 1940 to the present. In the early period, from 1940–1980, there had been a 6% loss of planted area during the growing season, which decreased to a 3% loss of planted area from 1980 to the present. The reduction of the lost area by half over the years could be attributed to improved technology in terms of seed protection, subsurface drainage, or improved hybrids. Although the planted-to-harvested maize area ratio has changed on the multi-decadal timeframe, the variation in crop production remains affected by the weather conditions among the years. Climate variation induces variability in maize yields under both irrigated and rain-fed production systems suggesting that the addition of water does not eliminate yearly variation but reduces its magnitude (Fig. 6). These observations suggest that temperature may induce a portion of the annual variation in maize yields. The slope of the yield trend line is nearly three times larger with irrigated than with non-irrigated environments (Fig. 6).

In a general sense, climate variation over time induces variation in crop yield because of the weather variation within the growing season. This is also evident in the wheat yields from Canada, which represent non-irrigated conditions in the central area of Canadian Plains (Fig. 7). Annual yield variation for Canadian wheat cannot be solely attributed to variations in precipitation because some of the effect is caused by frost damage either early or late in the growing season. What is striking

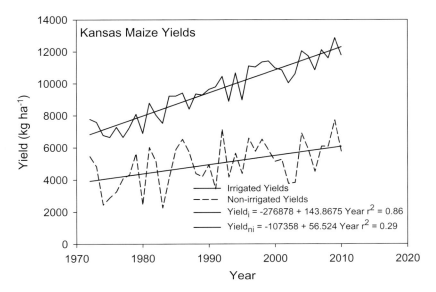

Fig. 6. Annual variation and the yield trends in the maize grain yields under both irrigated and non-irrigated conditions for Kansas from 1972 through 2010. (Source of data: USDA-NASS, http://quickstats.nass.usda.gov, accessed 27-October-2011).

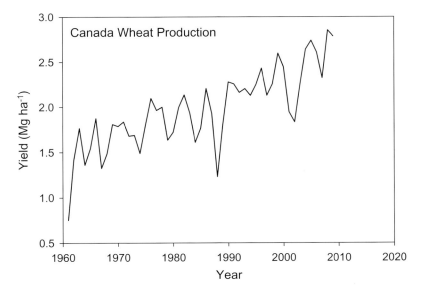

Fig. 7. Annual variation in wheat yields for Canada (FAOSTAT, accessed 27-October-2011 at http://faostat.fao.org).

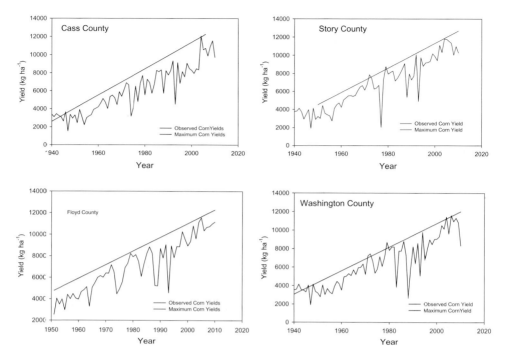

Fig. 8.   Annual variation in maize yields across four counties in Iowa along with the upper boundary of maize yield observed during the sequence. (Source of data: USDA-NASS, http://quickstats.
nass.usda.gov, accessed 27-October-2011).

in these data is the magnitude of the variation in yields among years induced by the
seasonal weather variation. As shown in the framework in Fig. 3, it is apparent that
variation in the inputs causes variation in the yield outputs.

Examination of yield trends and the variation among years provides a valuable
overview of the linkages between climate and crop production. A critical analysis of
the impacts on crop production from climate change will provide insights into the
potential effectiveness of adaptation strategies. To address this challenge requires a
quantitative approach to evaluating the impacts. The approaches taken to date have
typically been a combination of statistical methods and crop simulation models.
An example of the interannual variation in maize grain yield is shown in Fig. 8 for
four counties in Iowa. There is a large variation in yield over the years and there
are differences in the years with the low yields, demonstrating that local weather
during the growing season is a primary determinant of yield variation. If we take
the upper straight line as representing the upper bound of maize yield, then the
assumption would be that weather was not the limiting factor in crop production,
a concept developed by Hatfield (2010). Applied to these county-level data, the
reduction in grain yield from the upper boundary for all counties averaged 15%

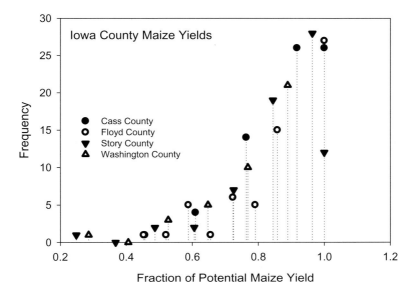

Fig. 9.   Frequency distribution of maize yield deviations from the upper boundary of yield for Cass, Floyd, Story and Washington counties in Iowa.

with losses in excess of 75% for Story County in 1977 (Fig. 9). Examination of the state or national-scale yields provides an indication of the large-scale climatological impacts; however, these events must occur at a large scale to perturb the overall yield trend line. Examining the deviation from an upper yield boundary provides a method to classify years in which weather was not a limiting variable and to identify the causal factors for climate impacts on crop production. Downscaling yield data to the county level can offer insights in terms of utilizing finer resolution data to determine the potential efficacy of both adaptation and mitigation strategies.

Examination of the frequency of yield loss from the upper boundary yields shows that the yield loss is typically less than 20% of the potential yield for a given year, with less frequent occurrences of yield loss exceeding 50% of the potential yield. This is similar to earlier observations by Hatfield (2010). If these frequency distributions are combined with frequency distributions of temperature and precipitation deviations from the normal, then the cause of the extreme yield losses can be determined. Using these four counties as examples, the extremes in the yield losses can be related primarily to deviations from normal precipitation with reduced summer precipitation in the majority of the times, with the exception of 1993 and 2010 in which there was excess precipitation during the growing season. The deviations in 2010 were also related to the largest deviation of summer minimum temperatures from the normal, based on 1980–2010 average.

These results are very similar to findings reported by Runge (1968). He observed that maize yields were affected by a combination of daily maximum temperature and precipitation during a 25-day period prior to and extending to 15 days after anthesis. When seasonal rainfall was low (0–44 mm per eight days), yield was reduced by 1.2–3.2% per 1°C rise in temperature. Alternately, if the temperature was warm (T_{max} of 35°C), yield was reduced by 9% per 25.4 mm rainfall decline. The temperature effect was confirmed through an analysis of maize yields across locations reported by Muchow *et al.* (1990), who found the highest observed and simulated grain yields at locations with relatively cool temperature (growing season mean of 18.0–19.8°C at Grand Junction, CO) compared to warmer sites (e.g., Champaign, IL 21.5–24.0°C), or warm tropical sites (26.3–28.9°C).

Increasing attention has been given to temperature impacts on crop yields in recent years. The recent study by Schlenker and Roberts (2009) discusses the potential nonlinear effects of warming temperatures on crop yields in the U.S. Lobell *et al.* (2011) have suggested that observed temperature trends have already impacted global maize and wheat production levels. This has induced a greater sense of urgency to understand the impacts of past climate on crop production and to develop a more robust observational framework for the assessment of agricultural impacts across North America. A combination of experimental observations and enhanced crop simulation models is required to define the complex interactions among rising CO_2 levels, extreme temperature events, and more variable precipitation patterns. If we examine the interrelationships among inputs and outputs in Fig. 3, the role of different soil characteristics becomes evident in terms of affecting the degree of variation in the outputs. Interactions of soil with weather during the growing season will cause spatial variations in the yields and these patterns will change from year-to-year, presenting an additional challenge when we consider the potential of different management practices to increase crop resilience.

In summary, temperature effects on crop production have already been observed and are projected to increase. (The review provided by Hatfield *et al.* (2011a) gives an indication of the general impacts of increasing temperatures on crop production.) The combined effects of increased temperatures on the phenological development of the crop, coupled with the increased ET demand, along with the potential for extreme events during critical growth stages (e.g., pollination) suggest that there should be a more concentrated effort to evaluate these effects because the seasonal patterns of temperatures and the potential for extreme events may be more predictable than precipitation patterns. Development of adaptation strategies may help to offset the deleterious impacts of temperature change on the growth and yield of various important crops.

Climate Change Effects on Product Quality

The focus of climate change impacts research for agriculture has been on overall plant productivity; however, there is also a need to consider the quality of grain, forage and fiber. The increasing world population will require not only an increase in production but also an increase in product quality. Aspects of quality are associated with the interactions between CO_2 and nitrogen (N) and their uptake by plants. A meta-analysis performed by Luo *et al.* (2006) concluded that the response of plants to N under increasing CO_2 varies among ecosystems and is not well-defined. This effect may explain the observations by Conroy and Hocking (1993) who found a steady decline in grain protein from 1967 to 1990 in wheat in Australia. They proposed that this change could not be specifically linked to rising CO_2; however, CO_2 increases may have contributed indirectly to this change in grain protein. Part of this change may be related to nutrient status in plants, as proposed by Luo *et al.* (2006). Kimball *et al.* (2001) found that increasing CO_2 concentrations caused a reduction in grain quality when coupled with low N in the crops because of the interaction between N status and grain quality. Erbs *et al.* (2010) found that increasing CO_2 to 550μ mol mol^{-1} with two rates of N [adequate and half of the recommended N in wheat and barley (*Hordeum vulgare* L.)], affected crude protein, starch, total and soluble β-amylase, and single kernel hardness. Increasing CO_2 reduced crude protein by 4–13% in wheat and by 11–13% in barley, but increased starch by 4% when half-rate N was applied. The combination of elevated CO_2 with low N fertilization would diminish the nutritional and processing quality of flour. Tester *et al.* (1995) had earlier shown that a combination of elevated CO_2 from 350–700 μg m^{-3} with an increase in temperature of 4–7°C above ambient caused a decrease in grain starch content in winter wheat. In their study, however, there was no effect of the increased CO_2 on any of the grain quality parameters.

A framework to explain the relationships between the effects of CO_2 and N status in plants was proposed by Morgan (2002). He proposed that increasing CO_2 concentrations would enhance plant growth and increase the litter and exudates from decaying plants and roots. These nutrients would be unavailable until broken down by soil microbial systems in the soil; in soils with limited microbial activity, this would decrease available nutrients to the growing plant. In rangeland species there has been accumulating evidence that protein content in forage declines with rising CO_2.

Increasing temperatures have also been observed to have an effect on grain quality. Hurkman *et al.* (2009) observed that a high-temperature regime after anthesis in wheat affected the accumulation of proteins during the grain-filling period. They observed that the specific protein response depended upon when the high temperature stress was imposed during the grain-filling period. Zhang *et al.* (2010) found exposure of spring wheat to high temperatures shortened the time to maturity and increased the grain arabinoxylan concentrations. Arabinoxylans are a major constituent of wheat grain; the water-extractable arabinoxylans are important in the baking quality of wheat because of their effect on the viscosity of dough. Coles *et al.* (1997) observed that concentrations of arabinoxylan in wheat increased under mild drought conditions but declined under severe drought. In the Zhang *et al.* (2010) study, they found that arabinoxylan fractions were correlated with water use efficiency and concluded that these fractions could be increased in the grain by selecting plants for increased water use efficiency. Moldestad *et al.* (2011) studied temperature variation during grain filling in winter wheat and observed gluten quality varied with mean temperature. Their analysis revealed that a higher mean temperature was positively related to gluten quality. A reduced gluten quality was found when the mean temperature was lower from heading to midway through the grain filling period was relatively low.

Weightman *et al.* (2008) observed that drought affected grain hardness in soft and hard winter wheat; however, vitreosity and protein content were poorly related to drought severity even though protein content increased with drought. Grain quality is evidently sensitive to weather patterns. Kettlewell *et al.* (1999) found that wheat quality was related to the North Atlantic Oscillation (NAO), which has been related to the strength and direction of the westerly winds and storm tracks across the North Atlantic. This influence was due to the rainfall during the grain-filling period in August, which affected the starch content of the grain is measured by the Hagberg falling number and specific weight of the grain. In their analysis, the weather during the vegetative period, however, had little influence on the Hagberg falling number and specific weight. Collectively these observations show the relationships between grain quality and the environmental conditions during the grain-filling period. These effects must be considered as the climate continues to change because of their potentially negative impacts on grain quality.

The possible effects of changing temperatures on yield quality are not confined to grain and forage. Pettigrew (2008) evaluated the effect of increased temperature on the physiological performance of two cotton (*Gossypium hirsutum* L.) genotypes. Temperatures were increased only 1°C above ambient throughout the growth cycle. Both genotypes responded similarly to increased temperatures. In two out of the three years, lint yield decreased 10%. This yield reduction was a result of both smaller boll mass and reduced seed per boll. However, fiber produced in the warmer environment was stronger. An increase in temperature compromises ovule fertilization, resulting in less seed per boll and reduced boll mass, the end result being reduced lint yield.

These studies highlight the need to improve our understanding of the interactions between increasing CO_2 and temperature, and more variable soil water supplies. A better understanding of these interactions will be necessary to provide insights into genetic x management interactions. In addition to the emphasis given to the environmental variables, it will be necessary to consider the effect of N-management and status in plants in order to increase production efficiency and to maintain yields and protein concentration. In the context of the assessment framework (Fig. 3), it is evident that variation in the climate inputs will induce variation in the outputs in terms of both quantity (yields) and quality (protein, starch, oil). Understanding the linkages between inputs and outputs for different cropping systems then allows the development and integration of adaptation and mitigation strategies.

Adaptation Strategies

Adaptation to climate is necessary to ensure agricultural productivity and increase resilience to climate variability (Easterling, 2011). It will be critical to address adaptation in agronomic management of cropping systems as shown in Fig. 3. There are several elements to be considered in the process of developing and evaluating potential adaptation strategies. Falloon and Betts (2010) stated that there is a need for an integrated approach to agricultural and water management in the context of adaptation to climate change. Their argument for an integrated approach is the crucial role that water management plays in determining the effectiveness of agricultural adaptation strategies.

Adaptation is driven by many different factors primarily at local to regional scales. Examination of the 2011 map of maize production across the United States reveals an expansion of the production area west of the Missouri river compared to 10 years ago (Fig. 10). This shift in production area is in response to an increase in more favorable climatic conditions. Across North Dakota and South Dakota in the United States, crop distribution includes more soybean and maize, evident in the amount of land designated to these crops in recent years (Figs. 11 and 12). The area alloted to both of these crops has increased since the mid-1990s and continues to rise. This is due primarily to a shift in precipitation patterns increasing the amount of summer rainfall. This precipitation increase has been coupled with the expansion of no-till cropping practices, to more effectively increase soil water availability to the crop; however, there has not been a significant change in temperature. In North Dakota, the shift to maize and soybean has not been as large, although there has been an increase in the last ten years (Fig. 12). Part of the limitation to the expansion is

J.L. Hatfield

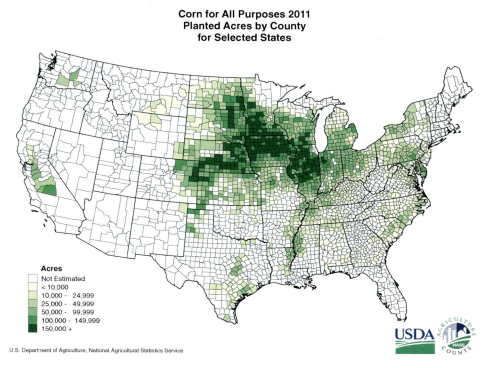

Fig. 10.   Distribution of maize (corn) production across the United States in 2010. (Source of data: USDA-NASS, URL accessed 27-October-2011).

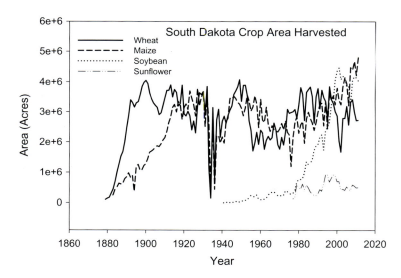

Fig. 11.   Area of land planted to maize, wheat, soybean, and sunflower in South Dakota, United States. (Source of data: USDA-NASS, http://quickstats.nass.usda.gov, accessed 27-October-2011).

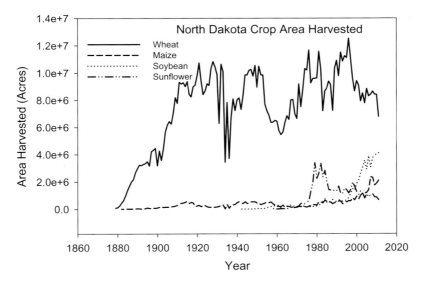

Fig. 12. Area of land of maize, wheat, soybean, and sunflower planted in North Dakota, United States. (Source of data: USDA-NASS, http://quickstats.nass.usda.gov, accessed 27-October-2011).

related to temperature. This is because the length of the growing season has not yet changed enough to support the profitable cultivation of these warm season crops. These shifts in production areas indicate the dynamic nature of agriculture and the continual shifts that occur as a result of small changes in the climate.

Another example of a potential positive benefit on crop distribution may occur in the northwestern United States in the winter wheat regions of Oregon and Washington. The projected shifts in precipitation patterns identified by Lettenmaier *et al.* (2008) to more rainfall and less snowfall during the winter would potentially benefit winter-wheat production because of the increased availability of soil water during the winter wheat growing season. This shift in precipitation could have positive benefits in this production area. Changes in irrigation area would represent an adaptive change to climate change, to address limitation of water availability. At the present time, there is no clear signal in the amount of irrigated area shifting in response to climate change; however, it could be expected that land under irrigation would rapidly change as water supplies from precipitation shifts become more evident.

Adaptation in cropping systems may employ many different strategies ranging from change in planting date, planting density, hybrid or cultivar maturity, hybrid or cultivar selection, or crop rotations. These are general proposals as strategies; however, an important aspect that needs to be considered is to determine through the framework shown in Fig. 3 if there is a reduction in the variability in the production levels with the adoption of these strategies. Challinor *et al.* (2007) demonstrated this for groundnut (*Arachis hypogaea* L.) using a crop simulation model in which the genetic coefficients were modified and then used to evaluate the potential response to climate scenarios. They evaluated the potential significance of genotypic manipulation to offset climate change impacts. This type of research will become increasingly important in devising management practices to offset climate change stresses.

Mitigation Strategies

Mitigation strategies for agriculture focus on practices to enhance the capacity of the soil to increase carbon (C) storage, thereby reducing CO_2 concentrations in the atmosphere; as well as on N-management practices capable of reducing nitrous oxide (N_2O) emissions. Implementation of mitigation strategies will affect the status of the inputs in the framework (Fig. 3) and thus simultaneously affect the variation in the outputs due to climate change. This demonstrates the necessity for mitigation and adaptation to be addressed synergistically.

Modification of soil C through soil management offers one option as a mitigation strategy. These practices are related to tillage and residue manipulation. Modification of soil-management practices will affect soil water balance, soil temperature, soil biological activity and gas exchange between the soil and the atmosphere. The linkage between soil management and soil organic carbon (SOC) is related to tillage systems; amount of crop cover; and efficiency of input use (N and water) by the cropping system (Follett, 2001). Several reviews have been published over the past decade outlining the role of soil management on mitigation of greenhouse gases (GHG) (e.g., Martens *et al.*, 2005, West and Marland, 2002, Franzluebbers, 2005). These reviews indicate that to quantify the effects of soil management practices on soil C mitigation strategies, there needs to be an improved understanding of agricultural impacts on soil C cycling. West and Marland (2002) found that the amount of reduction in C emissions from agriculture would vary by cropping system and climate regime. Franzluebbers (2005) concluded that reducing tillage intensity would increase SOC content, however, the effects of reduced tillage on other GHG emissions were not as well-defined. The benefit for soil C management may be larger in terms of affecting adaptation through improved soil water-holding capacity and reductions in soil water evaporation.

Reduced tillage intensity increases the water content in the soil profile thanks to the supression of soil water evaporation by the presence of the crop residue on the surface and the reduction in episodic bursts of soil water evaporation due to tillage-induced exposure of moist soil to the atmosphere. Maintenance of crop residues increases infiltration into the soil because of the reduction in the rainfall intensity striking the surface, and decreases soil water evaporation. These concepts were outlined by Steiner (1994). Ritchie (1971) developed one the first quantitative approaches linking soil water evaporation with surface soil water content and amount of surface residue and plant cover. Sauer *et al.* (1996) observed that surface residue reduced soil water evaporation by 34–50%, and that creating a 15 cm wide simulated tillage strip increased soil water evaporation 7% compared to full residue cover. Manipulation of the surface residues affects both the soil water evaporation rate and the infiltration rate into the soil.

The increasing potential for variation in precipitation amounts[1] and frequency of storms with climate change suggest that efforts will be needed to manage soil water in the most effective way possible so as to increase soil water availability to the crop. With an increase in water availability, there will be a potential increase in crop yield. Since most of the variation among years can be related to variation in precipitation, management of crop residue to increase water availability to the crop will increase water use efficiency and thus crop yield.

Mitigation strategies associated with N management are related to reductions of N_2O emissions. Such strategies involve changes in N fertilizer (rates, sources, and timing) as well as the use of controlled release fertilizers and of nitrification inhibitors to reduce N_2O emissions (Snyder *et al.*, 2009). The use of improved N management to increase N-use efficiency will increase climatic resilience because of the greater stability of yields and quality under more variable climate; however, there are no current studies investigating these linkages. Linkages between soil water availability and N_2O emissions are not clearly defined. Chatskikh *et al.* (2008) observed in oilseed rape (*Brassica napus* L.) and wheat rotations that soil water was the primary factor affecting N_2O emissions. In a recent study, Chen and Huang (2009) found no effect of tillage systems on N_2O emissions from fields in southeast China; however, the N_2O emission rates were dependent upon the soil water regime during the study period.

In summary, soil management practices are capable of reducing CO_2 emissions and increasing soil C storage and N management practices can reduce N_2O emissions in a changing climate (Hatfield *et al.* 2011b). To achieve the goals of mitigation, strategies will require changes in management practices, e.g., reduced tillage, residue management, and improved N management. Mitigation of soil CO_2 emissions will increase soil water availability in the short-term through reduced soil water evaporation and increased infiltration, and in the long-term through improved soil water-holding capacity facilitated by the increase in SOC content. Enhanced soil water availability under conditions of increased variability of precipitation is advantageous in terms of increasing resilience to climate change. Reduction of N_2O

---

[1] Observations and projections both show increased frequency of high precipitation events in some parts of the U.S. (Karl *et al.*, 2009).

emissions through improved N management offers the potential for a two-fold ben-efit. One benefit will be to the atmosphere through reductions in GHGs; the second benefit will be to crop production through reduced variability in production levels and improved product quality.

Challenges

The structure proposed in Fig. 3 as a framework for coupling weather, climate, and production inputs through cropping systems to achieve a variety of outputs demonstrates a way to address these linkages. The linkages between climate change and agricultural production are complex because of the interaction of temporal scales, from within-season weather to long-term climate change. Weather and climate affect all crops, and there are very few years in which weather anoma-lies do not have an effect on production amounts of any commodity grown in North America. Within the records on crop production, it is relatively easy to identify the factors causing major losses in production; however, it is more difficult to ascer-tain factors that contribute to the smaller losses between 5–15% of the potential yield. One of the challenges is to focus research on evaluating the physiologi-cal responses occurring under field conditions and to identify adaptive manage-ment strategies that could reduce this yield variation and positively impact product quality.

The largest challenge will be to evaluate the range of adaption and mitigation strategies applicable to crop, fruit, fiber and forage production with an emphasis on the interactions among the factors of water and nutrient management across a range of cropping systems and genetic material. The scientific community must derive solutions that can be implemented by producers to increase food security in North America and elsewhere.

In recognition of the importance of this issue, a working group of the American Society of Agronomy, Crop Science Society of America, and Soil Science Society of America developed a position statement on climate change and agriculture to foster a greater exchange of information regarding the impacts of climate change on agriculture (Climate Change Position Statement Working Group, 2011). These efforts addressed many of the issues outlined in this chapter and offer a challenge to the research community in terms of knowledge gaps that need immediate attention.

References

Adams, R.M., C. Rosenzweig, R.M. Peart, J.T. Richie, B.A. McCarl, J.D. Glyer, R.B. Curry, J.W. Jones, K.J. Boote and L.H. Allen, Jr. 1990. Global climate change and US agriculture. *Nature* 345:219–224.

Battisti, D.S. and R.L. Naylor. 2009. Historical warnings of future food insecurity with unprecedented seasonal heat. *Science*. 323:240–244.

Challinor, A.J., T.R. Wheeler, P.Q. Craufurd, C.A.T. Ferro and D.B. Stephenson. 2007. Adaptation of crops to climate change through genotypic responses to mean and extreme temperatures. *Agric. Ecosys. Environ.* 119:190–204.

Chatskikh, D., J.E. Olesen, E.M. Hansen, L. Elsgaard and B.M. Petersen. 2008. Effects of reduced tillage on net greenhouse gas fluxes from loamy sand soil under winter crops in Denmark. *Agric. Ecosys. Environ.* 128:117–126.

Chen, S. and Y. Huang. 2009. Soil respiration and N_2O emission in croplands under different plough-ing practices: A case study in south-east China. *Aust. J. Soil Res.* 47:198–205.

Christensen, J.H., B. Hewitson, A. Busuioc, A. Chen, X. Gao, I. Held, R. Jones, R.K. Kolli, W.-T. Kwon, R. Laprise, V. Magaña Rueda, L. Mearns, C.G. Menéndez, J. Räisänen, A. Rinke, A. Sarr and P. Whetton. 2007. Regional Climate Projections. pp. 847–940. *In* Solomon, S., D. Qin, M. Manning, Z. Chen, M. Marquis, K.B. Averyt, M. Tignor and H.L. Miller (eds.), *Climate Change 2007: The Physical Science Basis. Contribution of Working Group I to the Fourth Assessment Report of the Intergovernmental Panel on Climate Change.* Cambridge University Press: Cambridge, United Kingdom and New York, NY, USA.

Climate Change Position Statement Working Group. 2011. *Position Statement on Climate Change Working Group Rep.* ASA, CSSA, and SSSA, Madison, WI, May 11, 2011.

Coles, G.D., S.M. Hartunian-Sowa, P.D. Jamieson, A.J. Hay, W.A. Atwell and R.G. Fulcher. 1997. Environmentally-induced variation in starch and non-starch poly-saccharide content in wheat. *J. Cereal Sci.* 26:47–54.

Conroy, J. and P. Hocking. 1993. Nitrogen nutrition of C-3 plants at elevated atmospheric CO_2 con-centrations. *Physiologia Plantarum.* 89:570–576.

Easterling, W.E. 2011. Guidelines for adapting agriculture to climate change. pp. 269–286. *In* D. Hillel and C. Rosenzwieg (eds.), *Handbook of Climate Change and Agroecosystems: Impact, Adap-tation and Mitigation.* Imperial College Press: London, UK.

Erbs, M., R. Manderscheid, G. Jansen, S. Seddig, A. Pacholski and H.-J. Weigel. 2010. Effects of free-air CO_2 enrichment and nitrogen supply on grain quality parameters of wheat and barley grown in a crop rotation. *Agric. Ecosys. Environ.* 136:59–68.

Falloon, P. and R. Betts. 2010. Climate impacts on European agriculture and water management in the context of adaptation and mitigation- The importance of an integrated approach. *Science of the Total Environ.* 408:5667–5687.

Follett, R.F. 2001. Soil management concepts and carbon sequestration in cropland soils. *Soil Till. Res.* 61:77–92.

Franzluebbers, A.J. 2005. Soil organic carbon sequestration and agricultural greenhouse gas emissions in the southeastern USA. *Soil Till. Res.* 83:120–147.

Hatfield, J.L. 2010. Climate impacts on agriculture in the United States: The value of past observa-tions. pp. 239–253. *In* D. Hillel and C. Rosenzweig (eds.), *Handbook of Climate Change and Agroecosystems: Impact, Adaptation and Mitigation.* Imperial College Press: London, UK.

Hatfield, J.L., K.J. Boote, B.A. Kimball, L.H. Ziska, R.C. Izaurralde, D. Ort, A.M. Thomson and D.W. Wolfe. 2011a. Climate Impacts on Agriculture: Implications for Crop Production. *Agron. J.* 103:351–370.

Hatfield, J.L., T.B. Parkin, T.J. Sauer and J.H. Prueger. 2011b. Mitigation Opportunities from Land Management Practices in a Warming World: Increasing Potential Sinks. *In* Liebig, M.A., R.F. Follett and A.J. Franzluebbers (eds.), *Managing Agricultural Greenhouse Gases.* Else-vier Inc. New York. (In Press).

Hayhoe, K., C. Wake, T. Huntington, L. Luo, M. Schwartz, J. Sheffield, E. Wood, B. Anderson, J. Bradbury, A. Degaetano, T. Troy and D. Wolfe. 2007. Past and future changes in climate and hydrological indicators in the U.S. Northeast. *Clim. Dynam.* 28(4): 381–407.

Hurkman, W.J., W.H. Vensel, C.K. Tanaka, L. Whitehand and S.B. Altenbach. 2009. Effect of high temperature on albumin and globulin accumulation in the endosperm proteome of the developing wheat grain. *J. Cereal Sci.* 49:12–23.

Intergovernmental Panel Climate Change (IPCC), 2007. *Climate change 2007: Impacts, Adaptation and Vulnerability: Contribution of Working Group II to the Fourth Assessment Report of the Intergovernmental Panel on Climate Change.* Cambridge University Press, Cambridge, U.K.; New York.

Izaurralde, R.C., A.M. Thomson, J.A. Morgan, P.A. Fay, H.W. Polley and J.L. Hatfield. 2011. Climate Impacts on Agriculture: Implications for Forage and Rangeland Production. *Agron. J.* 103:371–380.

Karl, TR., J.M. Melillo and T.C. Peterson. (eds.), 2009. *Global Climate Change Impacts in the United States.* Cambridge University Press.

Katz, R.W. and B.G. Brown. 1992. Extreme events in a changing climate: Variability is more important than averages, *Climatic Change* 21:289–302.

Kettlewell, P.S., R.B. Sothern and W.L. Koukkari. 1999. U.K. wheat quality and economic value are dependent on the North Atlantic oscillation. *J. Cereal Sci.* 29:205–209.

Kimball, B.A., C.F. Morris, P.J. Pinter, Jr., G.W. Wall, D.J. Hunsaker, F.J. Adamsen, R.L. LaMorte, S.W. Leavitt, T.L. Thompson, A.D. Matthias and T.J. Brooks. 2001. Elevated CO_2, drought and soil nitrogen effects on wheat grain quality. *New Phytologist.* 150:295–303.

Lettenmaier, D.P., D. Major, N.L. Poff and S. Running. 2008. Water resources. pp. 121-150. *In* M. Walsh *The Effects of Climate Change on Agriculture, Land Resources, Water Resources and Biodiversity in the United States.* A report by the U.S. Climate Change Science Program and the Subcommittee on Global Change Research: Washington, DC.

Lobell, D.B., W. Schlenker and J. Costa-Roberts. 2011. Climate trends and global crop production since 1980. *Science.* 333:616–620.

Luo, Y., D. Hui and D. Zhang. 2006. Elevated CO_2 stimulate net accumulations of carbon and nitrogen in land ecosystems: A meta-analysis. *Ecology* 87:53–63.

Martens, D.A., W. Emmerich, J.E.T. McLain and T.N. Johnsen. 2005. Atmospheric carbon mitigation potential of agricultural management in the southwestern USA. *Soil Till. Res.* 83:95–119.

Meehl, G.A., C. Tebaldi and D. Nychka. 2004. Changes in frost days in simulations of twenty first century climate. *Climate Dynamics* 23:495–511.

Meehl, G.A., T.F. Stocker, W.D. Collins, P. Friedlingstein, A.T. Gaye, J.M. Gregory, A. Kitoh, R. Knutti, J.M. Murphy, A. Noda, S.C.B. Raper, I.G. Watterson, A.J. Weaver and Z.-C. Zhao. 2007. Global Climate Projections. *In* Solomon, S., D. Qin, M. Manning, Z. Chen. M. Marquis, K.B. Averyt, M. Tignor and H.L. Miller (eds.), *Climate Change 2007: The Physical Science Basis. Contribution of Working Group I to the Fourth Assessment Report of the Intergovernmental Panel on Climate Change.* Cambridge University Press: Cambridge, United Kingdom and New York, NY, USA.

Mishra, V. and K.A. Cherkauer. 2010. Retrospective droughts in the crop growing season: Implications to corn and soybean yield in the Midwestern United States. *Agric. Forest Meteorol.* 150:1030–1045.

Moldestad, A., E.M. Fergestad, B. Hoel and A.O. Skjelvag. 2011. Effect of temperature variation during grain filling on wheat gluten resistance. *J. Cereal Sc.* 53:1–8.

Morgan, J.A. 2002. Looking beneath the surface. *Science* 298:1903–1904.

Muchow, R.C., T.R. Sinclair and J.M. Bennett. 1990: Temperature and solar-radiation effects on potential maize yield across locations. *Agron. J.* 82: 338–343.

Pettigrew, W.T. 2008. The effect of higher temperature on cotton lint yield production and fiber quality. *Crop Sci.* 48:278–285.

Rahmstorf, S. and D. Coumou. 2011. Increase of extreme events in a warming world. PNAS. doi/10.1073/pnas.1101766108.

Ritchie, J.T. 1971. Dryland evaporative flux in a subhumid climate. 1. Micrometeorological influences. *Agron. J.* 70:723–728.

Runge, E.C.A. 1968. Effect of rainfall and temperature interactions during the growing season on corn yield. *Agron. J.* 60: 503–507.

Sauer, T.J., J.L. Hatfield and J.H. Prueger. 1996. Corn residue age and placement effects on evaporation and soil thermal regime. *Soil Sci. Soc. Am. J.* 60:1558–1564.

Schlenker, W. and M.J. Roberts. 2009. Nonlinear temperature effects indicate severe damages to U.S. crop yields under climate change. *PNAS.* 106:15594–15598.

Snyder, C.S., T.W. Bruulsema, T.L. Jensen and P.E. Fixen. 2009. Review of greenhouse gas emissions from crop production systems and fertilizer management effects. *Agric. Ecosys. Environ.* 133:247–266.

Steiner, J.L. 1994. Crop residue effects on water conservation. pp. 41–76. *In* Unger, P.W. (ed.), *Managing Agricultural Residues*. Lewis Publishers: Boca Raton, FL.

Tester, R.F., W.R. Morrison, R.H. Ellis, J.R. Piggott, G.R. Batts, T.R. Wheeler, J.I.L. Morison, P. Hadley and D.A. Ledward. 1995. Effects of elevated growth temperatures and carbon dioxide levels on some physicochemical properties of wheat starch. *J. Cereal Sci.* 22:63–71.

Wang, G. 2005. Agricultural drought in a future climate: Results from 15 global climate models participating in the IPCC 4th assessment. *Climate Dynamics* 25:739–753.

Weightman, R.M., S. Millar, J. Alava, M.J. Foulkes, L. Fish and J.W. Snape. 2008. Effects of drought and the presence of the 1BL/1RS translocation on grain vitreosity, hardness and protein content in winter wheat. *J. Cereal Sci.* 47:457–468.

West, T.O. and G. Marland. 2002. A synthesis of carbon sequestration, carbon emissions, and net carbon flux in agriculture: Comparing tillage practices in the United States. *Agric. Ecosys. Environ.* 91:217–232.

Zhang, B., W. Liu, S.X, Chang and A.O. Anyia. 2010. Water-deficit and high temperature water use efficiency and arabinoxylan concentration in spring wheat. *J. Cereal Sci.* 52:263–269.

Chapter 4

Latin American Perspectives on Adaptation of Agricultural Systems to Climate Variability and Change

Walter E. Baethgen* and Lisa Goddard[†]

International Research Institute for Climate and Society
The Earth Institute
Columbia University
*baethgen@iri.columbia.edu
[†]goddard@iri.columbia.edu

Introduction

The Latin America and Caribbean (LAC) region encompasses areas with pronounced year-to-year climate variability. Such variability and the expected longer-term changes in climate are putting significant pressures on agricultural production and water resources management.

The LAC region is a land of contrasts and disparities. Consider the following environmental contrasts (UNDP, 2011; Magrin *et al.*, 2007):

- The LAC region is endowed with one of the richest natural resource bases of the world, but environmental degradation in the region has been steadily worsening over the past 40 years.
- It includes some of the main food baskets for the world and has the world's largest reserves of arable land. But it also contains 16% of the world's total degraded land and it is home to large food-insecure populations.
- It is one of the most important forested regions of the world, with nearly 25% of the Earth's forest cover. But the rate of deforestation in the region is one of highest on the planet (in spite of important advances in this respect such as those seen in Brazil, where deforestation rate has decreased by 70% since 2005).
- The region is rich in renewable water resources, and includes vast river basins such as the Amazon and La Plata. However, water resources are not distributed evenly throughout the region, and a significant proportion of the population does not have adequate access to water. Moreover, populations dependent on glacier

melting as their main source for water are facing huge threats due to climate variability and change.

Socioeconomic contrasts and disparities are also a typical feature of the Latin American and Caribbean region. The UN-Human Development Index (HDI; UNDP 2011) for the LAC region is estimated at 0.73, considerably above the values for other regions of the developing world (e.g., 0.55 for South Asia, 0.46 for Sub-Saharan Africa). This regional HDI value may suggest that the region is in a relatively good situation within the development world. However, an outstanding characteristic of the LAC region is its wide socioeconomic disparity. Thus, the region includes countries with HDI values as high as 0.79–0.80 (Chile, Argentina, Uruguay) and countries with values of 0.57–0.58 (Guatemala, Nicaragua) and 0.45 (Haiti). Furthermore, compound indexes such as the UN-HDI often mask inequalities that exist within each of the region's societies.

Considering the agricultural production systems, the LAC region is also heterogeneous. It includes small-scale farmers that mainly produce food for their own community sustenance (e.g., NE Brazil, Central America, and the Andes), and regions where commercial farmers and large national and international corporations farm millions of hectares and constitute key food producers for the international market (e.g., in the Pampas, Uruguay, southern Brazil, eastern Paraguay).

Given these contrasts and disparities, it is difficult to discuss in a single chapter the expected impacts of climate change and to consider adaptive measures for the entire LAC region. On the other hand, the reports of the Intergovernmental Panel on Climate Change (IPCC) have produced chapters that specifically describe some of the expected impacts in different socioeconomic sectors (Magrin *et al.*, 2007; Mata *et al.*, 2001). This chapter therefore discusses some concepts that are relevant and applicable to the regional agricultural production in general. Specifically, it covers some issues regarding climate information that, in the authors' opinion are relevant to the Agricultural Model Intercomparison and Improvement Project (AgMIP) efforts, as well as some associated methodological challenges.

As stated in a paper published 15 years ago (Baethgen, 1997):

"the vulnerability of the agricultural sector in any region of the world to future possible climate change scenarios is determined to a great extent by the vulnerability of the sector to current climatic, economic and policy scenarios. Agricultural systems that are currently subject to extreme climatic interannual variability (droughts, floods, storms, etc.) are likely to become even more vulnerable under the (. . .) expected scenarios of climate change (i.e. increased temperatures, increased rainfall variability). Similarly, agricultural systems that are currently subject to drastic changes in economic and policy scenarios (large changes in input/output price ratios, changes in subsidy-related policies, modifications in rural credit policies, etc.) are also prone to become more vulnerable under any situation which could result in higher yield variability and/or require increased investment, such as expected conditions under climate change. Consequently, studies of agricultural systems'

vulnerability to climate change must be conducted in the context of current vulnerability to these different scenarios."

This chapter does not deal with the current or expected economic or policy context of the region, although it is based on the same premise: agricultural production systems that are now vulnerable to current climate variability and its extremes will likely become even more vulnerable to changes in climate that may deliver more frequent and more damaging climatic events. The chapter discusses the nature of climate information, products, and tools that intend to provide assistance to decision-makers, including policy-makers, for improving adaptation in agricultural production.

Adaptation and Decision-Making: Uncertainties and Conflict of Scales

Individuals and institutions of the private and public agricultural sectors are expected to elaborate policies and make decisions to respond to immediate socioeconomic challenges and pressures. Public policy-makers are often especially interested in actions whose impacts can be documented during their own terms of activity (typically 1–5 years, sometimes up to 10 years). Consequently, issues targeting impacts that may only be seen in the long term (e.g., 50 or more years) often attract less attention and become a lower priority.

Conversely, the scientific community working in climate change has been producing scenarios for a relatively far future (e.g., 50–100 years ahead). For example, the assessment reports that the IPCC has been producing since 1990 provide the best available projections of the world's climatology based on the anticipated changes in the composition of the atmosphere and its impact on the Earth's energy balance. The original intent of the consensus reports of the IPCC was not primarily to study the expected impacts of future climate on socioeconomic activities. The IPCC Assessment Reports were intended as an effort of the international scientific community to create consensus regarding the role of human activities in climate change and to encourage mitigation policies and actions (IPCC, 1990). Distant time horizons and broad (i.e. global) spatial coverages allow for a clearer signal of anthropogenic climate change to be detected against the large local manifestations of short-term climate variability.

Climate Change Information for Adaptation — Shortcomings of Common Practice

The direct use of climate model projections for studying the expected impacts of future climate on socioeconomic activities was not the original intention of the IPCC reports. Rather, the intention of the IPCC projections, as they are presented in the Assessment Reports, is to inform long term decisions and policies. Firstly, they project changes in climatology (i.e., the means and statistics of 20–30 year periods), which provide only a limited amount of information to assist in actual decision-making and planning. Secondly, the spatial scale of the projections is too coarse to effectively inform decisions and policies at national or sub-national levels, the scale at which most decision-makers work. Furthermore, the projections contain uncertainty levels that are usually too large for them to be considered in the elaboration of policy (Baethgen, 2010). It is only logical that efforts towards circumventing these shortcomings should have evolved. However, the efforts that promise increased accuracy or specificity in space and time also have shortcomings, which also must be communicated to those who would use this additional information.

Accuracy and uncertainty of the global climate models

Even in the interest of obtaining information on change of the mean climate, a number of limitations exist in the direct use of global climate model projections. The most fundamental of these is that the models are not perfect. The climate models are based on physical equations of the ocean, atmosphere, cryosphere, biosphere and their interactions, but many important processes that happen on small scales are represented in aggregate over 100km+ grid cells, and some of those processes are not well measured or understood. Clouds, for example, remain an outstanding uncertainty in climate models. Small errors in a single process can be amplified through feedbacks with other processes, until certain aspects of the climate system become unrealistic. Although many advances have been made in climate models over the last 20 years (Reichler and Kim, 2008; Achuta Rao and Sperber, 2006), no single model is best in every respect (Gleckler *et al.*, 2008) and as a result, multi-model ensembles that combine the results of the different models show better performance than any individual model (Reichler and Kim, 2008). Thus, projections either given on the average across models, or over their range, are typical formats for climate change projections.

The suite of climate model projections (more than 20 in the IPCC 4th Assessment Report) provides a fairly good representation of the global mean changes in temperature over the 20th century, and thus there is confidence in its range of projections of global mean temperatures for the 21st century. However, at regional scales — especially for societally important variables like precipitation — model deficiencies become a problem. For example, the regional pattern of trends in near-surface air temperature and precipitation over the last half of the 20th Century does not compare well between the multi-model mean and observations (Fig. 1). This appears to be related to differences between the models and observations in the trends in tropical sea surface temperatures (SSTs) (Shin and Sardeshmukh, 2011), which is related in

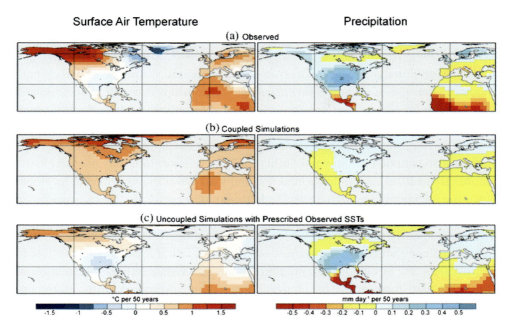

Fig. 1.   Trends of annual-mean surface air temperature (left) and precipitation (right) over 1951–1999 derived from (a) observations, (b) multi-model ensemble-mean coupled climate model simulations, and (c) multi-model ensemble-mean uncoupled atmospheric model simulations with prescribed observed time varying SSTs. Annual averages are over July to June.
Source: Shin and Sardeshmukh (2010).

part to model deficiencies in representing the dynamics of the upper ocean (Davey *et al.*, 2002). Climate model deficiencies and performance are often not appreciated or even investigated before the models are used, as they should be since this knowledge can guide more appropriate use of the models.

It is generally acknowledged, however, that the multi-model mean across the entire suite of models is better than the use of any single model (Reichler and Kim, 2008). To date, there is no agreed-upon approach to weight or select models, although in some cases it may be possible to exclude certain poorly performing models. Ultimately, the IPCC suite represents an 'ensemble of opportunity', rather than a collection of models that deliberately represents a scientifically designed range of physical configurations. Therefore, the apparent uncertainty quantified by model disagreement should not be interpreted as actual uncertainty.

Spatial resolution

The shortcoming that is most readily acknowledged by those who would use the IPCC projections is the coarse resolution of the climate models relative to the scale of sectoral models, such as those used in agriculture or hydrology (e.g. Barnett *et al.*, 2008; Soussana *et al.*, 2010). Researchers turn to downscaling (Wilby and Wigley, 1997), which can be accomplished by statistical means or by nesting a higher-resolution dynamical model over a smaller region within the global climate model. In the case of statistical downscaling, relationships between large-scale and local-scale climate are determined based on past observations, and these are assumed to hold into the future. The limitations of statistical downscaling include short observational histories with which to work, especially for developing countries. The limitations also include the likely possibility that past relationships may not hold in the future due to local effects from land use and land cover change, or due to global warming effects such as higher specific humidity (Greene *et al.*, 2011b), which can affect the relationships between large-scale circulation and rainfall amounts, intensity and frequency or duration. Dynamical downscaling, on the other hand, requires substantial computing and storage resources. As a result, it is often only possible to downscale a few of the global model projections, and often only to a resolution of 20–50 km (e.g. Jones *et al.*, 2011), which may still be too coarse for the aforementioned sectorial models. With either approach to downscaling, the starting point is the global model projection, so again, errors in the large-scale response of the global model will at best be made more spatially detailed. More desirable approaches may first consider selection of a sub-set of models that may be better suited to representing the relevant processes that impact the region of interest. It may also be appropriate to consider approaches to first correct large-scale systematic biases of the climate models, particularly in the case of statistical downscaling.

Temporal downscaling

Time scale, or the time horizon of the projections, is another shortcoming of the IPCC scenarios. While agricultural decision-makers are typically interested in the expected climate for the next 10–30 years (Vera *et al.*, 2010), IPCC projections consider changes in the mean climate up to the 21st Century. This shortcoming is one that many users of the information believe they can easily circumvent. Since 'climate change' is viewed as trends, the solution usually involves either extrapolation of recent trends, which would also seem to avoid model deficiencies, or interpolation of centennial-scale climate projections to represent those of the next few decades. The problem with either of these approaches is that they neglect low-frequency variability that is intrinsic to the climate system, (i.e., not explicitly due to anthropogenic climate change). Particularly at the regional scale, decadal-scale climate variability can enhance or moderate experience of trends due to human-made climate change. Extrapolating observed trends, especially those from short observational histories, may under- or over-estimate anthropogenic climate change trends. Also, the use of

IPCC climate model projections, even if they perform well over a specific region, are not configured to capture the timing of natural decadal variability, although such processes are contained in the models. The result is that the natural decadal-variability superposed on the climate change trends will either be ill-timed in the case of a single projection, or the range of possibility may be missed entirely in the case of multi-model averages.

Both statistical and dynamical downscaling have their advantages relative to each other for providing more detailed climate information in space and time (Wilby and Wigley 1997). Dynamical downscaling is based on physical equations through which the climate model solution evolves. Local physical responses to the large-scale climate changes that may not have existed or been recorded in the observational record can be realized. However one of the aspects most desired from dynamical downscaling is information on precipitation extremes, and these daily rainfall statistics are not captured even at very fine-resolution downscaling, on the order of a few kilometers. Even the daily observations averaged over spatial scales of just a few kilometers begin to lose the distributional properties of the extremes observed at the station scale.

The advantage of statistical downscaling is that the magnitude of variation of the observed climate, which is often obscured by the coarse resolution of climate models, can be reinstated by the statistics. Also, the methods can be used at any resolution for which data are available, including the station scale. Many of the subtleties and difficulties in statistical downscaling arise from treatment of the human-made trends relative to the natural climate variability. Most statistical approaches treat these separately — whether implicitly or explicitly, though the separation between these is not easily accomplished, and in some cases perhaps not even possible.

Statistical and dynamical downscaling techniques may have different advantages, but they share a common shortcoming: they depend on the large-scale anthropogenic climate changes as simulated by global climate models and whether these are captured correctly. No regionally- or locally-focused downscaling approach can reverse a wrongly projected trend.

Climate Information for Adaptation — Promising Approaches

For the next assessment report of the IPCC (AR5), attention is being given to prospects for information development that are more appropriate for adaptation, such as decadal-scale predictions (Taylor *et al.*, 2011) and perhaps even seasonal predictions (Kirtman and Pirani, 2009). In particular, as with the climate change projections coordinated for the IPCC-AR4, a set of experiments has been coordinated across the international climate modeling centers to produce decadal predictions from the same set of models (Taylor *et al.*, 2011). The difference in AR5 is that for the decadal predictions the modeled climate system, particularly the oceans, is initialized close to the observed conditions. It is thought that the seeds of decadal predictability lie in the slow changes of the ocean circulation and upper ocean heat content (Hurrell *et al.*, 2010). Some pioneering work towards decadal predictions has demonstrated improved representation of global, and even regional, temperature changes on decadal timescales compared to the uninitialized projections (Smith *et al.*, 2007). However, the ability of these predictions to capture the evolution of natural decadal variability, or merely improve the projection of climate change trends, still appears limited (Goddard *et al.*, 2010). It must be emphasized that these are simulation experiments, and that the uncertainties and unknowns associated with these predictions may be even greater than those associated with the climate change projections (Meehl *et al.*, 2009). Additionally, because these are predictions that are initialized with observational information, as is done for weather and seasonal predictions but very different from climate change projections, there are additional considerations to the treatment of these data before one can arrive at the decadal forecast. Thus, it may take some time before these predictions can be used to provide reliable information to the adaptation community, but the experiments represent an important advancement by the climate-research community to develop more actionable information to inform decisions and improve adaptation.

While the science and practice evolve toward production and use of dynamical decadal prediction, much information can be gained from analysis of past decadal-scale variability. Statistical approaches that replicate the characteristics of year-to-year variability, and thus simulate persistently wet or dry periods, can be employed to indicate this range of possibilities and their likelihoods (e.g., Prairie *et al.*, 2008; Greene *et al.*, 2012). Such methodology has been employed by the U.S. Bureau of Reclamation and by the International Research Institute for Climate and Society (IRI) for water-resource management in South Africa to examine the vulnerability to low-frequency climate variability and change. These approaches carry some of the same caveats mentioned for statistical downscaling, one of the most restrictive being that associated with short historical records.

Beyond information of the slowly evolving climate state due to natural decadal variability and man-made climate change, it is also important to consider the risks and opportunities that arise from year-to-year climate variability. In terms of climate change, one should investigate how the characteristics of year-to-year variability may change in the future as well as how that variability intersects with the slow evolution of overall climate change (Coelho and Goddard, 2009). It is these additive effects that can produce the greatest impacts, such as the 2003 European Heat Wave that killed tens of thousands of people (Stott *et al.*, 2004). Such crisis events suggest that climate adaptation involves not just plans for the next ten years but also decision systems and strategies that can incorporate seasonal-to-interannual forecasts of the coming three to six months.

Information on Climate Variability and Change for Assisting Adaptation

Institutions and individuals acting in the agricultural sector of LAC are interested in future climate scenarios because they believe that those scenarios will assist them in assessing the expected impacts on agricultural production. It is further believed that the scenarios can promote improved adaptive practices and technologies that will assist the agricultural sector to cope with the expected climate conditions and will improve farmers' income and food security.

One analysis that can help to identify the timescale(s) of information that are most needed to assist decisions and policies is provided by an estimation of the relative magnitude of climate variability at different temporal scales (seasonal, decadal, longer-term) observed in LAC throughout the 20th century. Relying on model projection scenarios of future climate requires the consideration of these relative magnitudes in order to take into account how climate variations at different time scales may combine, particularly in the case where they lead to adverse impacts. For example, sub-regions with a large observed component of decadal variability will need to factor the uncertainty in evolution of climate due to this variability into their adaptation plans.

Greene *et al.* (2011a) developed a web-based tool entitled "Time Scales Map Room" (http://iridl.ldeo.columbia.edu/maproom/.Global/.Time_Scales/), which describes the characteristics of historical temperature and precipitation variability. The tool provides estimates of the relative contribution of interannual, decadal and longer-term climate variability to the observed historical climate (globally or regionally). The observed data partitioned into these scales come from the TS2.1 data product (Climatic Research Unit, University of East Anglia), a global monthly, quality-controlled, gridded data set that covers all land areas between 1901 and 2002. The data set has no missing values but includes grids that were filled in with climatological values in regions with insufficient weather station data. However, TS2.1 also provides a time history of the numbers of stations contributing data in each grid box, which is used by the Map Room to mask grids with insufficient station reports.

The IRI's Time Scales Map Room has been used here to partition the total variability of the observed monthly precipitation and temperature in LAC over the 20th century into the three components mentioned above: (a) long-term trend ("climate change"), (b) decadal variability, and (c) interannual variability (Fig. 2). The results indicate that the long-term trend or "climate change" component explains up to

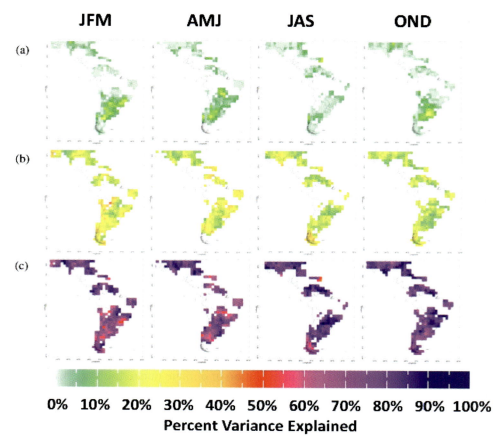

Fig. 2. Percent of precipitation variance observed in 1920–2000 explained by climate variability at three temporal scales: (a) long-term trend, (b) decadal, and (c) interannual.

20% of the total precipitation variance over the 20th century, although in most of the LAC sub-regions, present long term trends account for 5–10% of the variance. The decadal component explains 10–30% of the total variance in most sub-regions and seasons. As expected, the remaining 50% or more of the total precipitation variance was explained by the interannual component.

In order to compare the magnitude of the precipitation variation (in mm/month) observed at the three temporal scales, two small regional boxes were selected: one in which the long-term trend accounts for a relatively large percentage of the variance, and the other in which the decadal variability is relatively large. The results indicated that even in these selected sites, the precipitation variability at the interannual scale was always larger than the variability observed at the other scales (Fig. 3).

The same analyses were performed with temperatures, and in this case, there were sub-regions (especially between 20°S and 20°N) where the long-term trend

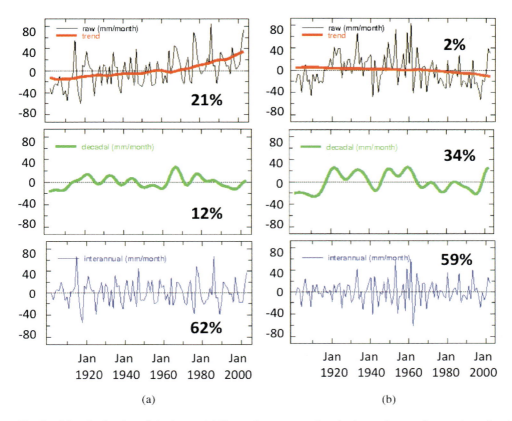

Fig. 3.   Magnitude of precipitation variability at three temporal scales in two boxes of approximately 10° × 10°: (a) Southern Chile (centered at 61°W–36°S) with a high long-term trend component, and (b) Central-Western Argentina (centered in 74°W–44°S) with high decadal component. The numbers in the figures represent the proportion of total variance explained by the corresponding temporal scale (long-term trend, decadal and interannual).

and decadal variability explained the majority of the total variance (50–60%) over the 20th century (Fig. 4). In order to assess the magnitude of the variations in temperature observed in those sub-regions, again two small regional boxes were selected: one box in eastern Brazil with a high long-term trend, and another box in western Peru with high decadal variability (Fig. 5). However, it should be noted that in most sub-regions (especially in latitudes higher than 20°) the interannual variability explains 60–90% of the total variance in temperature.

These findings emphasize the importance of the interannual climate variability as compared to either of the other two timescales. They also stress the shortcomings of considering only observed or projected long-term trends in sub-regions with relatively large decadal variability.

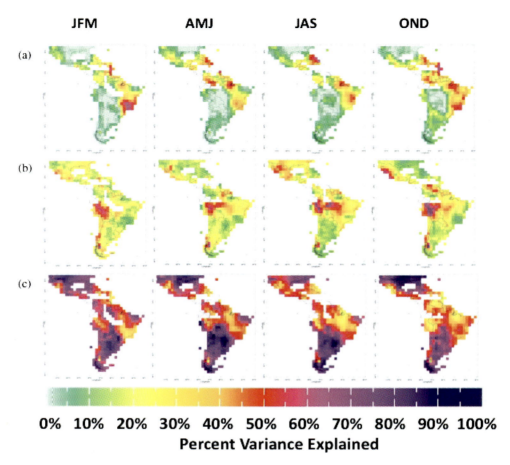

Fig. 4.   Percentage of temperature variance observed in 1920–2000 explained by climate variability
at three temporal scales: (a) long-term trend, (b) decadal, and (c) interannual.

Discussion

With these results in mind, and targeting the goal to improve decisions and policies
for the agricultural production systems of LAC, two issues become crucial:

1. Agricultural production systems in the LAC region should build resilience to the
   impacts associated with current climate variability. This is an obvious starting
   point for informed adaptation to future changes in climate means and variability.
   It is primarily the year-to-year climate variability, which may be worsened in the
   presence of trends, that will lead to the greatest socio-economic impacts over the
   coming near-term decades. The majority of the existing agricultural production
   systems in LAC, where droughts, floods, storms, untimely frosts have had — and

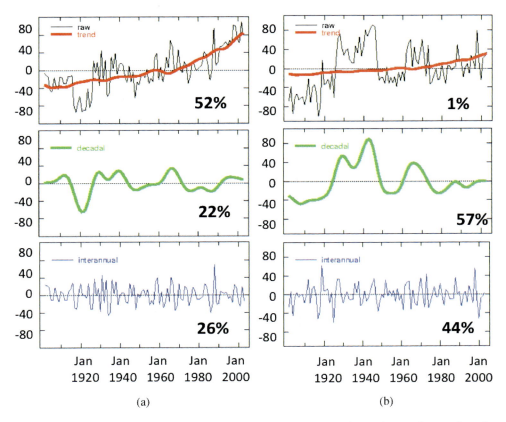

Fig. 5. Magnitude of temperature variability at three temporal scales in two boxes of approximately $10° \times 10°$: (a) Eastern Brazil (centered at $48°W - 3°S$) with a high long-term trend component, and (b) Peru (centered at $71°W - 12°S$) with high decadal component. The numbers in the figures represent the proportion of total variance explained by the corresponding temporal scale (long-term trend, decadal and interannual).

    will have — profound impacts on agricultural production throughout the region, remain vulnerable.

2. Planning for possible climate evolution over the next 10–30 years ("near-term climate change") must consider decadal variability. The magnitude of this variability, and associated impacts, can be larger than the longer-term changes in climate caused by the alteration of the chemical composition of the atmosphere at this timescale.

In summary, the appropriate use of climate information must take into account the limitations as well as opportunities of both climate model and observational data. Important to the provision of information is quantification of uncertainty, which is extremely difficult, given imperfect models and short observational records. The uncertainty measured by differences in projections across climate models or scenarios of greenhouse gas increases does not necessarily capture the uncertainty in future man-made climate changes since climate models are imperfect. Additionally, natural variations in climate over those timescales will be part of the experience of climate changes. However, while models may simulate decadal variability, they are not synchronized in time, since these models result from naturally occurring interactions in the climate system rather than forced by increasing greenhouse gases or other anthropogenic driving forces, such as land-use change.

Observational analyses can provide a context within which the climate models can be interpreted. The observational analyses (1) can indicate the relative magnitude of past variability, as an initial consideration for possible year-to-year or more persistent impacts; (2) can show whether observed 20th century climate trends conform to model-based expectations of the future; and (3) can help diagnose whether a recent 20–30 year "trend" has been dominated by low-frequency variability that is very possibly part of natural fluctuations rather than associated with man-made change. The observational analyses will be limited by the length of the record, the spatial representation of the data, and the quality of the observations. With our understanding of limitations of climate information, it should yet be possible to make informed use of the opportunities provided by climate information from both model data and observations to better plan for the future of agriculture in the LAC region.

References

Achuta Rao K and K.R. Sperber. 2006. ENSO simulation in coupled ocean–atmosphere models: are the current models better? *Clim Dyn.* 27:1–15.

Baethgen, W.E. 1997. Vulnerability of the agricultural sector of Latin America to climate change. *Climate Res.* 9:1–7.

Baethgen, W.E. 2010. Climate Risk Management for Adaptation to Climate Variability and Change. *Crop Sci.* 50(2):70–76.

Barnett, T.P., D.W. Pierce, H.G. Hidalgo, C. Bonfils, B.D. Santer, T. Das, G. Bala, A.W. Wood, T. Nozawa, A.A. Mirin, D.R. Cayan and M.D. Dettinger. 2008. Human-Induced Changes in the Hydrology of the Western United States, *Science*, 319(5866), 1080–1083, DOI: 10.1126/science.1152538.

Coelho, C.A.S. and L. Goddard. 2009. El Nino-induced tropical droughts in climate change projections, *J. Climate* 22: 6456–6476.

Davey M., M. Huddleston, K. Sperber, P. Braconnot, F. Bryan, D. Chen, R. Colman, C. Cooper, U. Cubasch, P. Delecluse, D. DeWitt, L. Fairhead, G. Flato, C. Gordon, T. Hogan, M. Ji, M. Kimoto, A. Kitoh, T. Knutson, M. Latif, H. Le Treut, T. Li, S. Manabe, C. Mechoso, S. Power, E. Roeckner, L. Terray, A. Vintzileos, R. Voss, B. Wang, W. Washington, I. Yoshikawa, J. Yu, S. Yukimoto, S. Zebiak and G. Meehl. 2002. STOIC: a study of coupled model climatology and variability in tropical ocean regions. *Clim Dyn.* 18:403–420.

Gleckler, P.J., K.E. Taylor and C. Doutriaux. 2008. Performance metrics for climate models. *J. Geophys. Res.* **113**, D06104, doi:10.1029/2007JD008972.

Goddard, L., Y. Aichellouche, W. Baethgen, M. Dettinger, R. Graham, P. Hayman, M. Kadi, R. Martinex and H. Meinke, 2010. Providing seasonal-to-interannual climate information for risk management and decision-making, *Procedia Environ. Sci.* 1:81–101.

Greene, A.M, L. Goddard and R. Cousin. 2011a. Interactive "Maproom" Provides Perspective on 20th-Century Climate Variability and Change, *EOS*, Transactions American Geophysical Union, 92(45):397, doi:10.1029/2011EO450001.

Greene, A.M., A.W. Robertson, P. Smyth and S. Triglia. 2011b. Downscaling projections of Indian monsoon rainfall using a nonhomegneous hidden Markov model, *Quarterly Journal of the Royal Meteorological Society* (137), 347–359. doi: 10.1002/qj.788m.

Greene, A.M., M. Hellmuth and T. Lumsden. 2012. Stochastic decadal climate simulations for the Berg and Breede Water Management Areas, Western Cape province, South Africa. *Water Resourc. Res.*, in revision.

Hurrell J.W., T. Delworth, G. Danabasoglu, H. Drange, S. Griffies, N. Holbrook, B. Kirtman, N. Keenlyside, M. Latif, J. Marotzke, G.A. Meehl, T. Palmer, H. Pohlmann, T. Rosati, R. Seager, D. Smith, R. Sutton, A. Timmermann, K.E. Trenberth and J. Tribbia. 2010. Decadal climate prediction: Opportunities and challenges. *In* Hall, J., Harrison, D.E., Stammer, D. (eds) *Proceedings of OceanObs'09: sustained ocean observations and information for society, vol 2*. ESA Publication WPP-306: Venice, 21–25 Sep 2009. Available at http://www.oceanobs09.net/blog/?p=97.

IPCC. 1990. Climate Change: The IPCC Scientific Assessment. *In* Houghton, J.T., G.J. Jenkins, and J.J. Ephraums (eds). *Intergovernmental Panel on Climate Change*. Cambridge University Press. Cambridge, 365 pp.

Jones, C., F. Giorgi, G. Asrar. 2011. The Coordinated Regional Downscaling Experiment: CORDEX, an international downscaling link to CMIP5. *CLIVAR Exchanges* 16: 34–39.

Kirtman, B. and A. Pirani. 2009. The state of the art of seasonal prediction: outcomes and recommendations from the First World Climate Research Program Workshop on seasonal prediction. *Bull Am Meteorol Soc.* 90:455–458.

Magrin, G., C. Gay García, D. Cruz Choque, J.C. Giménez, A.R. Moreno, G.J. Nagy, C. Nobre and A. Villamizar. 2007. Latin America. Climate Change 2007: Impacts, Adaptation and Vulnerability. pp. 581–615. M.L. Parry, O.F. Canziani, J.P. Palutikof, P.J. van der Linden and C.E. Hanson, (Eds.) *In Contribution of Working Group II to the Fourth Assessment Report of the Intergovernmental Panel on Climate Change*, Cambridge University Press, Cambridge, UK.

Mata, L.J., M. Campos, E. Basso, R. Compagnucci, P. Fearnside, G. Magrin, J. Marengo, A.R. Moreno, A. Suárez, S. Solman, A. Villamizar, L. Villers (Mexico). 2001. Latin America region: Impacts, Adaptation and Vulnerability. pp. 693–734. J.J. McCarthy, O.F. Canziani, N.A. Leary, D.J. Dokken and K.S. White, (Eds.) *In Contribution of Working Group II to the Third Assessment Report of the Intergovernmental Panel on Climate Change*. Cambridge University Press, Cambridge, UK.

Meehl, G.A., L. Goddard, J. Murphy, R.J. Stouffer, G. Boer, G. Danabasoglu, K. Dixon, M.A. Giorgetta, A.M. Greene, E. Hawkins, G. Hegerl, D. Karoly, N. Keenlyside, M. Kimoto, B. Kirtman, A. Navarra, R.S. Pulwarty, D. Smith, D. Stammer and T. Stockdale. 2009. Decadal prediction: can it be skillful? *Bull Am Meteorol Soc.* 90:1467–1485.

Prairie, J., K. Nowak, B. Rajagopalan, U. Lall and T. Fulp. 2008. A stochastic nonparametric approach for streamflow generation combining observational and paleoreconstructed data. *Water Resour. Res.* 44, doi:10.1029/2007WR006684.

Reichler, T. and J. Kim. 2008. How well do coupled models simulate today's climate? *Bull. Amer. Meteorol. Soc.* 89:303–311.

Shin, S.-I. and P. D. Sardeshmukh, 2011. Critical influence of the pattern of Tropical Ocean warming on remote climate trends. *Clim. Dyn..* 36:1577–1591.

Smith D., S. Cusack, A. Colman, C. Folland, G. Harris and J. Murphy. 2007. Improved surface temperature prediction for the coming decade from a global circulation model. *Science* 317:796–799.

Soussana, J.-F., A.-I. Graux and F.N. Tubiello. 2010. Improving the use of modelling for projections of climate change impacts on crops and pastures, *J. Exp. Bot.* 61(8):2217–2228. doi: 10.1093/jxb/erq100.

Stott, P.A., D.A. Stone and M.R. Allen. 2004. Human contribution to the European heatwave of 2003. *Nature* 432:610–614.

Taylor K.E., R.J. Stouffer and G.A. Meehl. 2011. Bulletin of the American Meteorological Society. http://dx.doi.org/10.1175/BAMSD-11-00094.1.

United Nations Development Programme (UNDP). 2011. *Human Development Report 2011. Sustainability and Equity: A Better Future for All.* UNDP: New York.

Vera, C., M. Barange, O.P. Dube, L. Goddard, D. Griggs, N. Kobysheva, E. Odada, S. Parey, J. Polovina, G. Poveda, B. Seguin, K. Trenberth. 2010. Needs assessment for climate information on decadal timescales and longer, *Procedia, Environ. Sci.* 1:275–286.

Wilby, R.L. and T.M.L. Wigley. 1997. Downscaling general circulation model output: A review of methods and limitations. Progress in *Phys. Geog.* 21(4): 530–548, doi: 10.1177/030913339702100403.

Chapter 5

European Perspectives: An Agronomic Science Plan for Food Security in a Changing Climate

John R. Porter*, Jean-Francois Soussana[†], Elias Fereres[‡], Stephen Long[§], Fritz Mohren[¶],
Pirjo Peltonen-Sainio[‖], and Joachim von Braun**

*Faculty of Science, University of Copenhagen, Denmark
[†]INRA, France
[‡]Institute of Sustainable Agriculture, University of Cordoba, Spain
[§]Department of Plant Biology and Crop Sciences, University of Illinois, USA
[¶]Centre for Ecosystem Studies, Wageningen University, The Netherlands
[‖]MTT Agrifood Research, Finland
**ZEF, Department of Economic and Technological Change,
University of Bonn, Germany
*jrp@life.ku.dk

Introduction

Today's agriculture is at a crossroads. Climate change is already calculated to be having a negative impact on food production in some areas of the world (Lobell, 2011), while there are expectations for the sector to meet a rise in demand by 70 to 100% (FAO, 2011) within the next 40 years. Failure to reach this target would reduce food security, while success may commit the world to further warming by accelerating greenhouse gas emissions. Can we avoid a spiral of poverty, hunger, environmental degradation and conflicts by developing climate-smart agriculture and stabilising world food systems? The challenge is to increase the global food supply to accommodate a world growing to 10 billion or more people while avoiding dangerous environmental change (Rockström *et al.*, 2009). Land management for food production is a fundamental human activity, supporting the livelihood and the fundamental basis and stability of human societies. Although there should be sufficient food for all, a billion people are currently undernourished, since 40% of grain is used to feed livestock and an additional 6.5% to fuel cars and trucks. Furthermore, a large amount of food is either lost after harvest or wasted (FAO, 2011).

Until recently, it was expected that despite climate change and increasing world population, there would be several decades with food surplus — and low prices — ahead (Easterling *et al.*, 2007). Contrary to this expectation, the first decade of the new century has witnessed large spikes in international grain prices with some grains more than doubling in cost (von Braun, 2009). This unforeseen onset of a global food crisis reveals our poor understanding of the complex interactions between food systems and climate, and demonstrates the fragile nature of global food production. Increasing demands for land are exacerbating pressures on biodiversity and natural resources. An estimated one third of the world's cropland is losing topsoil due to erosion. Unprecedented water shortages are also increasingly apparent in many parts of the developed world, including southern Europe. Species-rich regions of the world are undergoing loss of biodiversity caused by intensive agriculture (TEEB, 2010).

The Role of Europe

A number of recent studies (Paillard, 2011; FAO, 2011; SCAR) have indicated the needs for increasing research efforts in the area of agriculture, food security and climate change, mostly focussing either on a global scale, or specifically on developing countries. Recently, international research programs (e.g. CCAFS — www.ccafs-climate.org) have been initiated to address these needs for the developing world. However, the links between agriculture, food security and climate change also concern developed regions, such as Europe, which play a large role in the international trade of agricultural and food products and are contributing prominently to research in the fields of agricultural, plant, soil and animal sciences.

The European Union (EU-27) accounts for 20% of global cereal production and 19% of global meat production (www.faostat.fao.org). Current trends show an intensification of agriculture in northern and Western Europe and a decline and abandonment in some parts of the Mediterranean and south-eastern regions of Europe (Stoate *et al.*, 2009). The yield gap in areas in Eastern Europe is greater than in many of Western European prime lands, where farm yields are approaching the biological potential of present cultivars. The food system (including pre-chain inputs, agriculture, food processing and retailing) is the largest industrial sector in Europe and has the potential to generate economic growth and employment. Moreover, there are large prospects for economic growth based on the bioeconomy. Agricultural policies, which contribute to the dual objectives of rural development and sustainability, will largely determine to what extent and how the sustainable intensification of production will be achieved.

The variability of crop yields has already increased in Europe as a consequence of extreme climatic events, such as the summer heat wave of 2003 that led to uninsured economic losses for the agriculture sector in the European Union, estimated at 36 billion Euros (Pachauri *et al.*, 2007). For two decades, there has been a decline in the growth trend of cereal yields, especially wheat, in many European countries (Olesen *et al.*, 2011). While the EU is likely to be able to offset internal reductions in yield under climate change by increasing imports, this would further increase pressure on already food-insecure, low-income, and low-yielding regions of the world (Porter *et al.*, 2012, pending).

The EU has been a global leader in policy and action to decrease greenhouse gas emissions. Not only does the EU's own agricultural production result in significant emissions, but as a net importer of primary agricultural products, EU causes significant emissions elsewhere (Davis and Caldeira, 2010). We contend that judicial use of the EU's land resources, planning, and agricultural sciences could adapt production to climate change, lower emissions, and eliminate net imports, thereby increasing global food security. Across the biophysical and social sciences, the EU is well placed to address these issues and its society recognizes the significance of global climate change, thus providing an example and leadership to other developed regions.

Recognizing that research will be a key to meeting these global challenges in the coming decades, twenty European countries have developed a Joint Programming Initiative on Agriculture, Food Security and Climate Change (FACCE JPI: www.faccejpi.com). The joint programming of EU research is a new process aimed at combining a strategic framework, a bottom-up approach, and high-level commitment from member states to work together on important issues such as food security and climate.

The interactions among agriculture, food security and climate change have been envisioned highlighting the three binary interactions and the single ternary interaction, which are at the heart of the FACCE JPI. The complex system formed by each of these components and by their interactions is under multiple pressures from external drivers, such as the rising food and fibre demand, globalisation and global environmental changes and is moreover constrained by planetary boundaries such as land and water limits (Fig. 1). Such an integrated research agenda will deliver key outputs by contributing: (i) to raising the biological efficiency of European agriculture; (ii) to responding to a globally increased food demand without increasing the demands of food products on other (e.g., developing) world regions; (iii) to operating agriculture within greenhouse gas, energy, biodiversity and contaminant limits and; (iv) to building resilience in agricultural and food systems (Fig. 2). To deliver this agenda, five evidence-based interdisciplinary core research themes were

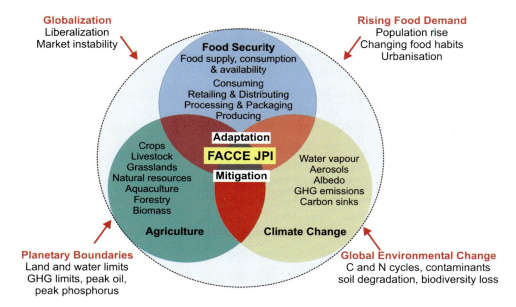

Fig. 1. A vision of research areas in the FACCE JPI showing drivers (in red) and highlighting interactions between agriculture, food security and climate change.

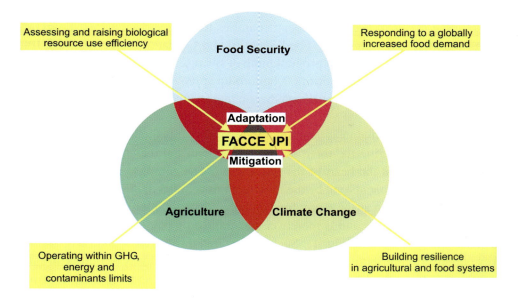

Fig. 2. A vision of key outputs (underlined in yellow) of FACCE JPI for Europe.

defined, which together are expected to provide high returns with the potential to enhance the contribution of Europe to global public goods.

Sustainable food security under climate change

Scenarios of sustainable food security under climate change need to be developed based on an integrated food-systems perspective. This will require an integrated risk analysis of European agriculture (and food systems) under climate change: (i) testing responses to volatility both from natural and market phenomena and to higher commodity and fossil fuel price levels; (ii) understanding global change impacts and resilience of food systems (through the value chain and to the consumer); (iii) studying Europe's role in international markets, price volatility, and global food security impacts; (iv) developing contrasted scenarios involving perceptions and policy dialogs, including scenarios on reduction in agricultural wastes and losses and changes in diets; (v) combining observations, experiments, and modelling through the development of appropriate European research infrastructures. An innovative research approach that integrates a wide range of concepts and methods is needed to provide scientific support for policy formulation and resource management.

Environmentally sustainable growth and intensification of agriculture under current and future climate and resource availability

Current high input levels used in European agriculture demand improvements in resource use efficiency by reducing fertilizer, water, and other inputs and by increasing feed use efficiency and waste reuse in livestock systems. Water scarcity in agriculture will increase, demanding new approaches to manage the limited water supplies at different scales from farms to regions. In some areas, restoring degraded soils, minimizing erosion, and build-up of soil organic matter will be essential, all while improving the understanding and control of soil functioning and biotic interactions at field-to-landscape scales. Crop and animal production systems of increased productivity with reduced environmental impact per unit product should be developed. This will require accurate benchmarking of the main drivers of current agroecosystems (economics, genotype × environment × management interactions), and the design, experimentation, and modelling of alternative systems. Knowledge-based IT innovations, increased use of biodiversity, and plant and animal breeding are likely to be major pillars of the new crop and livestock systems.

Assessing and reducing tradeoffs: food production, biodiversity and ecosystem services

The agroecosystem service approach offers a multi-criteria framework to describe the state of agroecosystem and natural capital, and to assess the effects of different management and policy options in terms of gains and losses. Agroecosystem services that include regulation of GHG emissions and food production are benefits that humans derive from ecological processes and ecosystem functions. They include regulating services, such as pollination, which was estimated worldwide at 9.5% of world agricultural output in 2005 (Gallai *et al.*, 2009). Non-market valued ecological wealth, such as avoidance and absorption of GHG emissions, underpins our market-based economic wealth (DEFRA, 2011). Research is required to develop quantitative methods for assessing and valuing ecosystem services in agricultural systems and to develop approaches to optimise trade-offs among agriculture, land use, and ecosystem services at the farm and landscape scale taking account of the positive and negative contributions that agriculture can make to ecosystem services. There is also a need to develop evidence-based knowledge of the provision of public goods and services by European agriculture, so that ecosystem services are enhanced, valued, and delivered equitably.

Adaptation to climate change

Planned adaptation in agriculture will require a large coordinated research effort to develop seeds and breeds adapted to the unchartered climatic conditions projected for the end of this century and to design resilient and eco-efficient crop and livestock systems, while ensuring the dynamic conservation of soil, water and genetic resources. Agro-ecological engineering through the increased use of genetic and species diversity at field and landscape scales and eco-technologies to recycle farm wastes, to monitor greenhouse gases, verify soil carbon stocks and to adapt water management will play a key role. Investments in monitoring crop and animal diseases and invasive species will be required to preserve plant, animal and human health. However, the results of adaptation will depend on the likely technical effectiveness of adaptations and their adoption rate. Given the generation times involved in breeding crop and livestock species, to obtain adaptable commercial lines for 2030 will require at least 15 years of development. Therefore, it falls to the public sector to undertake the development of suitable genotypes now. This requires realistic field facilities for selection and testing. The technologies for such open-air facilities which can control the key aspects of climate change — elevated carbon dioxide, elevated temperatures, and soil moisture–exist, but are not of the scale that will be needed for crop adaptation.

Greenhouse gas mitigation

Understanding, predicting and mitigating climate change requires accurate knowledge of the emissions and natural sinks of greenhouse gases. A European greenhouse-gas tracking system is required that has the capability to separate the influence of anthropogenic emissions from natural fluxes and thereby to improve reporting to the United Nations Framework Convention on Climate Change. The development of such a system (www.icos-infrastructure.eu) will facilitate the recognition of mitigation efforts by agriculture. Many agricultural practices can potentially mitigate GHG emissions, the most prominent of which are improved cropland and grazing land management (including restoration of degraded lands and cultivated organic soils). Lower, but still significant, mitigation potential is provided by water and rice management, set-asides, land use changes, agro-forestry, livestock management, and manure management (Smith *et al.*, 2008). In addition, GHG emissions could be reduced by substitution of fossil fuels. The links between production of biofuels from feedstock (in many cases subsidised), consequent land use changes, and the rise in food prices demonstrate the importance of foreseeing the range of consequences (Searchinger *et al.*, 2009).

Some agricultural GHG mitigation options are cost-competitive with a number of non-agricultural options in achieving long-term climate objectives. Such options should not reduce agricultural productivity, but rather improve the eco-efficiency of agricultural systems by reducing GHG emissions per unit of crop and animal products. The mitigation potential should be assessed based on cost-benefit approaches taking into account environmental constraints, land and labour requirements, demands for food and non-food products, and biodiversity issues, thereby accounting for regional differences in site conditions as well as for climate change impacts over different time-spans.

The Role of Agronomy

On October 21 2009, the Royal Society of London (2009), presented a report on global food security defining the role of science in the sustainable intensification of global agriculture (see References list). In their recommendations, they stressed the need to integrate the disciplines of agronomy and crop physiology — i.e. the sciences of crop production in the field. As representatives of global agronomy societies and members of institutions that teach and research these subjects, we must indeed reexamine what we mean by scientific agronomy and its role in ensuring sustainable food production in the future.

Agronomy is the applied science of crop and plant production for food, fibre and energy. It is intrinsically multi-disciplinary — it encompasses plant genetics, crop physiology, climate and meteorology, and soil science, and it expresses the interactions as genotype × environment × management × technology (GEMT), of which it is possible to coalesce M and T to give GEM. Agronomists need to have knowledge of biology, chemistry, ecology, soil and earth sciences and genetics. In addition to understanding inter-relationships among biotic and abiotic ecosystem components, agronomy focuses on ways to predict the responses of food producing systems by using models and other tools, such as statistical analysis, that had their birth within agronomy. Finally, agronomy tries to improve the systems that humans use to produce food, feed, fuel, and fibre. A distinctive feature of agronomy is that it is both a science and an accredited profession.

Agronomy is always faced with the GEM composite that stresses the interactions between its individual components. Plant science per se has had difficulty dealing with interactions, especially those where changes in scale are needed, as is the case in agronomy — from leaf to individual plant to population. If agronomy were termed *agronomics* and considered a field in biology as are other — *omics* (e.g. genomics, proteomics) then it is likely that the subject would preoccupy itself largely with the development and application of methods and techniques, molecular breeding, details of plant processes and the description of crop structure but not function. The fields in which crops are grown are the only meaningful sites for the integration of otherwise reductionist plant sciences.

An acute challenge for agronomy is to provide the basic science and practice to raise the productivity of cropping in the face of climate change and more variable conditions for crop growth. How might agronomy approach this issue? Taking a starting point in GEM, the answerable question can be phrased in terms of how the respective contributions of G, E and M can be deconstructed and partitioned over a given distribution of yields (Fig. 3). In an analysis of Australian wheat yields over the past 100 years (Fischer, 2009) for which total yield change was an average increase of 1.3% per year, climate and environment (E) contributed 0.1%, but can be ignored as this is cancelled out by the expansion of cultivated area onto mainly drier land. Increase in CO_2 level (E) contributed about 0.2% of the increase, G plus (G × M), especially as G × fertilizer response, was worth about 0.5%, which leaves M on its own to contribute 0.6%. Expressed in percentages relative to the total yield increase — G on its own contribute about 40%, E about 15%, with M having contributed about 55% of the increase in yield. An agronomist would also note that the G contribution is mostly at the high-end of the distribution (Fig. 3) and thus raises the already highest yields in the distribution, whereas M operates mainly at the lower end and contributes to raising the lowest yields. It is clearly desirable to have a situation in which a yield distribution has both a high mean value and a narrow distribution, both temporally and spatially.

Using this analysis to consider the impacts of a more variable climate, one would expect a lower mean and a widened distribution of yields, giving an expanded importance to management in adapting crops to such changing climates. Assuming that agronomic adjustment can help buffer yields against mean climate change, then

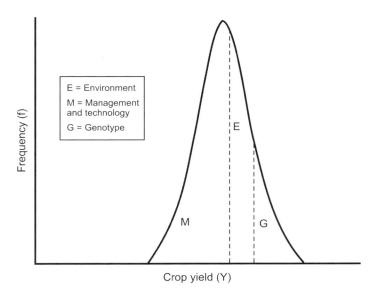

Fig. 3.   A crop yield (Y) frequency (f) distribution showing posited contributions of genotype (G), environment (E) and management and technology (M) to yield variation. The figure shows that the contribution of management to yield variation accounts for about 55%, environment about 15% and genotype about 30% of the yield variation (based on data in Fischer, 2009).

adaptation to increased weather variation will require reliable seasonal forecasts. If it is getting hotter in an irregular pattern, earlier planting of crops and thereby earlier flowering should help. In the longer term, adopting earlier flowering varieties may be a feasible option, especially in areas that had been prone to spring frosts. More attention to soil water conservation would also help if the weather is getting periodically drier, although the near constant gain in yield per unit increase in evapo-transpiration means that greater water use still generates higher yields. Only if warming and water supply permits a farmer to switch from a C_3 crop to a C_4 one that has higher water use efficiency, or from a spring-down cereal to a winter-sown one, would there be a yield gain. Crop breeding and varietal choice are perhaps simpler to apply as they permit one to target the specific bottlenecks in the yield process, which are likely to be further constrained by climate change. Breeding wheat in rain-fed environments outside Europe is already confronting water shortage as a constraint. High temperature effects may be somewhat different, especially if the likelihood of extreme maximum temperature occurrences increases and yields responses are nonlinear (Semenov, 2009). We need to learn how sensitivity varies with stage of development (Porter and Gawith, 1999) and identify the current maximum temperature thresholds and their variability. The general policy message from the above is that future agricultural R&D needs to pay at least as much, if not more, attention to crop management and agronomy as to breeding to cope with future extreme conditions.

Developing countries can be very different from the above, with often large yield gaps arising because of lack of application of existing improved management and often lack of use of (or access to) improved varieties. In such situations, the adoption of improved but already known management can play a bigger role well into the future, but G progress should still have an important positive effect, both via potential yield and via yield gap-closing breeding. However, the balance between G and M in developing countries is likely to side more with M than G, at least in the short-term. The prospect that agronomy as the science of G × E × M will make a larger contribution to food production than biotechnology in Europe in the future needs to be recognised in future research funding.

Dealing with the above balances and trade-offs within the GEM combination and how they differ according to regional and geographical context has to become the core of the science of agronomy. Processes that can be studied at the cropping level are just as intellectually challenging as gene sequencing and micro-studies of metabolism. However, agronomy must be improved. It should anticipate and welcome the contributions that can be made from new developments in other disciplines, such as gene technology, remote sensing, systems theory and software developments, as they are important for predictive simulation modelling. Agronomy has to move towards more integrated systems approaches and focus more on the multiple functions of agro-systems rather than on individual aspects, while retaining the primary focus of understanding, describing and predicting the consequences of sustainable primary production. Agronomy would not gain from being re-titled 'agronomics' as this may not capture the need for the cross-disciplinary insights needed to advance future food production. Such cross-disciplinarity needs to be built on firm disciplinary pillars.

Agronomic insights have had direct impacts on the development of agricultural policies in many parts of the world. There are two of many examples. The first concerns silvo-arable agroforestry (SAF) in Europe, where a EU-FP5 research programme (SAFE: http://www.ensam.inra.fr/safe) that studied the effect of combining trees and crops on production per unit area and their environmental impacts has led to a new European directive (CR 1698/2005 — Art. 44) and policy change. By providing experimental and model-based evidence of the mutual benefits of SAF systems for food and wood production, as well as quantification of economic costs and environmental impacts and services, this research showed the conflicting nature of current sectoral policies and provided quantitative information to build up a new policy to stimulate this type of cropping systems in various regions of Europe. A second example from China concerns how agronomic insight on crop yield responses to nitrogen application has led to revision of the policy on subsidies to nitrogen fertiliser use (www.csiro.au). This followed the finding by agronomists that nitrogen application could be markedly reduced without sacrifice of yield but with benefits to China's agricultural economy and the environment. Agronomic knowledge as formulted in crop and cropping system models also plays an essential role in climate change impact assessments and pest risk analysis.

The latest FAO estimates show that world agricultural production needs to increase by 70% over the next 40 years to keep up with growing demand, and the needed increase is nearer 100% in developing countries. This requires an estimated increase in agricultural investments of 50% (www.fao.org). The situation is exacerbated by the fact the share of public spending on agriculture has fallen to an average of around 7% in developing countries, and the share of official international development assistance directed to agriculture has fallen to as little as 4%. The Declaration of the World Food Summit in November 2009 therefore called for a reinforcement of all efforts to greatly reduce the current burden of more than 1 billion people now suffering from hunger and poverty. The Declaration also calls for augmenting research for food and agriculture, including research to mitigate and adapt to climate change. This is a clear call to strengthen agronomic research, requiring large financial and intellectual investments in this vital area. Unfortunately, education at the university level in agronomy has declined drastically since the 1980s. As an example, the country from which the Royal Society report emanated could invest profitably in chairs in agronomy and research into agronomy and crop physiology as much as it has invested in biotechnology. Agronomy needs to be reinvented as it has a crucial role to play in helping to solve the inter-linked problems that include increasing food demand, energy provision, and climate change, which affect the future of agriculture.

References

Davis, S.J. and K. Caldeira. 2010. Consumption-based accounting of CO_2 emissions. *Proceedings of the National Academy of Sciences of the United States of America* 107:5687–5692.

DEFRA. 2011. UK National Ecosystem Assessment (2011) *The UK National Ecosystem Assessment: Synthesis of Key Findings*. UNEP-WCMC, Cambridge.

Easterling, W.E., P.K. Aggarwal, P. Batima, K.M. Brander, L. Erda, S.M. Howden, A. Kirilenko, J. Morton, J.-F. Soussana, J. Schmidhuber and F.N. Tubiello, 2007: Food, fibre and forest products. pp. 273–313. In: M.L. Parry, O.F. Canziani, J.P. Palutikof, P.J. van der Linden and C.E. Hanson (Eds.), *Climate Change 2007: Impacts, Adaptation and Vulnerability. Contribution of Working Group II to the Fourth Assessment Report of the Intergovernmental Panel on Climate Change.* Cambridge University Press, Cambridge, UK.

Fischer, R.A. 2009. Farming systems of Australia: exploiting the synergy between genetic improvement and agronomy. pp. 23–54. *In* V. Sadras and D. Calderini (Eds.), *Crop physiology: applications for genetic improvement and agronomy.* Academic Press: San Diego, CA.

Food and Agriculture Organization of the United Nations. 2011. *The State of Food insecurity in the world 2010.* FAO, Rome, Italy.

Food and Agriculture Organization of the United Nations. 2011. *The State of Food and Agriculture 2010–2011.* Available at http://www.fao.org/publications/sofa/en/.

Gallai, N., J. Salles, J. Settele and B.E. Vaissiere. 2009. Economic valuation of the vulnerability of world agriculture confronted with pollinator decline. *Ecological Economics* 68:810–821.

Lobell, D.B., W. Schlenker and J. Costa-Roberts. 2011. Climate Trends and Global Crop Production since 1980. *Science* 333:616–620.

Olesen, J.E., M. Trnka, K.C. Kersebaum, A.O. Skjelvag, B. Seguin, P. Peltonen-Sainio, F. Rossi, J. Kozyra and F. Micale. 2011. Impacts and adaptation of European crop production systems to climate change. *European Journal of Agronomy* 34:96–112.

Pachauri, R.K. *et al.* 2007. *Climate Change 2007: Synthesis Report. Contribution of Working Groups I, II and III to the Fourth Assessment Report of the Intergovernmental Panel on Climate Change.* IPCC: Geneva, Switzerland.

Paillard, S., S. Treyer and B. Dorin. 2011. Agrimonde: scenarios and challenges for feeding the world in 2050. *Editions Quae, Versailles, France.* Porter, JR. and Gawith, M. 1999. Temperatures and the growth and development of wheat: a review. *European Journal of Agronomy.* 10:23–36.

Porter, JR., R. Dyball, D. Dumaresq, L. Deutsch and H. Matsuda. 2012. Feeding capitals: urban food security and sovereignty in Canberra, Copenhagen and Tokyo.

Rockstrom, J., W. Steffen, K. Noone, A. Persson, F.S. Chapin III, E.F. Lambin, T.M. Len-ton, M. Scheffer, C. Folke, H.J. Schellnhuber, B. Nykvist, C.A de Wit, T. Hughes, S. van der Leeuw, H. Rodhe, S. Sorlin, P.K. Snyder, R. Costanza, U. Svedin, M. Falkenmark, L. Karlberg, R.W. Corell, V.J. Fabry, J. Hansen, B. Walker, D. Liverman, K. Richardson, P. Crutzen and J.A. Foley. 2009. A safe operating space for humanity. *Nature* 461:472–475.

Royal Society of London. 2009. *Reaping the Benefits.* Available at www.royalsociety.org

Searchinger, T.D., S.P. Hamburg, J. Melillo, W. Chameides, P. Havlik, D.M. Kammen, G.E. Likens, R.N. Lubowski, M. Obersteiner, M. Oppenheimer, G.P. Robertson, W.H. Schlesinger and G.D. Tilman. 2009. Fixing a Critical Climate Accounting Error. *Science* 326:527–528.

Semenov, M.A. 2009. Impacts of climate change on wheat in England and Wales. *Journal of the Royal Society Interface* 6:343–350.

Smith, P., D. Martino, Z. Cai, D. Gwary, H. Janzen, P. Kumar, B. McCarl, S. Ogle, F. O'Mara, C. Rice, B. Scholes, O. Sirotenko, M. Howden, T. McAllister, G. Pan, V. Romanenkov, U. Schneider, S. Towprayoon, M. Wattenbach and J. Smith. 2008. Greenhouse gas mitigation in agriculture. *Philosophical Transactions of the Royal Society B-Biological Sciences* 363:789–813.

Stoate, C., A. Baldi, P. Beja, N.D. Boatman, I. Herzon, A. van Doorn, G.R. de Snoo, L. Rakosy and C. Ramwell. 2009. Ecological impacts of early 21st century agricultural change in Europe — A review. *Journal of Environmental Management* 91:22–46.

TEEB. 2010. The Economics of Ecosystems and Biodiversity: Mainstreaming the Economics of Nature: A synthesis of the approach, conclusions and recommendations. The Economics of Ecosystems and Biodiversity TEEB. UN Environment Programme, Japan.

von Braun, J. 2009. Addressing the food crsis: Governance, market functioning, and investment in public goods. In *Food Security* 1:9–15.

Chapter 6

African Perspectives on Climate Change and Agriculture: Impacts, Adaptation and Mitigation Potential

[1]Jesse Naab, [2]Andre Bationo, [3]Benson M. Wafula, [4]Pierre S. Traore,
[5]Robert Zougmore, [6]Mamadou Ouattara, [7]Ramadjita Tabo, and [8]Paul L. G. Vlek

[1]*West African Science Service Center on Climate Change and Adapted Land Use
(WASCAL) Ouagadougou, Burkina Faso
jessenaab@gmail.com*
[2]*Alliance for a Green Revolution in Africa (AGRA), PMB KIA 114, Accra, Ghana
ABationo@agra-alliance.org*
[3]*Kenya Agricultural Research Institute (KARI), P.O. Box 27, Embu, Kenya
bensonwafula@gmail.com*
[4]*International Crops Research Institute for the Semi-Arid Tropics
(ICRISAT) BP 320, Bamako, Mali
p.s.traore@cgiar.org*
[5]*International Crops Research Institute for the Semi-Arid Tropics
(ICRISAT) BP 320 Bamako, Mali
R.Zougmore@cgiar.org*
[6]*West African Science Service Center on Climate Change and Adapted Land Use
(WASCAL) KIA 114, Accra, Ghana
mouattara@agra-alliance.org*
[7]*Forum for Agricultural Research in Africa (FARA), PMB CT 173, Cantonments
rtabo@fara-africa.org*
[8]*West African Science Service Center on Climate Change and Adapted Land Use
(WASCAL) KIA 114, Accra, Ghana
p.vlek@uni-bonn.de*

Introduction

Agriculture plays a dominant role in supporting rural livelihoods and economic growth over most of Africa. Agriculture is the backbone of most African economies employing between 60–90% of the total labour force and accounting for as much as 35–40% of the total export earnings (UNEP, 2006). With regard to development status, Africa contains the poorest and least developed nations of the world, (though some countries in Africa have been making good socio-economic progress) with 41.1% of the population living on less than $1 a day, as compared to 29.5% in Southern Asia and 9.9% in Eastern Asia (UN, 2007). Africa is also the one continent in the world where per capita food production has been declining or stagnant at a level that is less than adequate (Scholes and Biggs, 2004). While Asia achieved a green revolution within a short time, Africa is yet to meet the Comprehensive African Agricultural Development Programme (CAADP) 6% growth and Millennium Development Goal 1 (MDG1) target. This trend is linked to low and declining soil fertility (Sanchez, 2002), and inadequate fertilizer inputs. Over the past two to three decades land degradation in sub-Saharan Africa has relentlessly continued, with 10% of the most productive regions visibly affected as observed from space (Vlek *et al.*, 2010). Africa loses the equivalent of $4 billion annually due to nutrient mining. Another $42 billion is lost in income and six million hectares of productive land is threatened each year due to land degradation. About 55% of land in Africa is unsustainable for crop production.

The agricultural sector faces the challenge of increasing production to provide food security for the projected human population of two billion by mid-century while protecting the environment and the functioning of its ecosystems. At the same time climate change threatens production's stability and productivity as Africa is thought to be the region most vulnerable to the impacts of climate variability and change. Smallholder subsistence farmers in African rainfed farming systems are considered the most vulnerable to climate change impacts (Jarvis *et al.*, 2011).

Agriculture in Africa must undergo a significant transformation in order to meet the related challenges of achieving food security and responding to climate change. It is possible to lift millions of people out of poverty and hunger by increasing the productivity and profitability of small scale farmers in Africa. Increasing productivity to achieve food security is projected to entail a significant increase in emissions from the agricultural sector in developing countries (IPCC, 2007b). Achieving the needed levels of growth, but on a lower emissions trajectory will require a concerted effort to maximize synergies and minimize tradeoffs between productivity and mitigation (FAO, 2010). Ensuring that institutions and incentives are in place to achieve climate-smart transitions, as well as adequate financial resources, is thus essential to meeting these challenges.

The main objective of this chapter is to discuss the potential impact of climate change on African agriculture, feasible and sustainable adaptation and mitigation options available to African countries. Finally, this chapter examines some institutional, policy and financial responses required to achieve this transformation.

Africa's Contribution to Greenhouse Gases (GHGs) Emissions

Africa contributes very little to global climate change, with low carbon dioxide emissions from fossil fuel use and industrial production in both absolute and per capita terms (UNECA, 2002). Average total anthropogenic carbon emissions in Africa were 500 Tg C yr^{-1} for the period 2000–2005; 260 Tg C yr^{-1} from combustion of fossil fuels and 240 Tg C yr^{-1} from land-use change (Canadell *et al.*, 2009). Over this period, the African share of global emissions from land-use change and fossil fuels were 17% and 4% respectively. At the national level, countries with high fossil fuel emissions (but low emissions from land use change) are South Africa, Egypt, Algeria and Libya while countries with high emissions from land use change are DR Congo, Zambia, Cameroon, Tanzania and Nigeria (Canadell *et al.*, 2009).

A recent study (Brown *et al.*, 2012) in four East African and five West African countries shows that the major sources of GHG emissions that exist arise from the livestock sector and land-use changes (Table 1). The total amount of GHG emissions from the nine African countries was almost 129 million t CO_2e yr^{-1} in the mid-2000s (Table 1). The largest amount of GHG emissions was from the livestock sector, mostly methane from enteric fermentation as expected (84% of the total), followed by emissions from soil only due to the conversion of native ecosystems to cropland (11% of the total). Emissions from use of fertilizer were lower than all other sources and represent just 0.7% of the total emissions. Despite the large area of grazing

Table 1.   Total annual GHG emissions (x1000 t CO_2e) from land-use change, livestock, nitrogen fertilizer consumption and fires in grazing lands from nine East and West African countries.

Region	Country	Land-use Change	Livestock	Nitrogen fertilizer	Grazing area burnt	Total
East Africa	Ethiopia	7,339	41,966	356	1,254	50,915
	Kenya	1,812	11,988	339	232	14,372
	Tanzania	1,833	13,935	44	1,736	17,548
	Uganda	1,112	6,204	23	524	7,863
	Sub-total	12,097	74,093	762	3,745	90,697
West Africa	Burkina Faso	273	8,779	19	306	9,377
	Ghana	1,664	1,865	58	491	4,079
	Mali	440	9,270	65	241	10,016
	Niger	31	10,405	15	9	10,461
	Senegal	369	3,364	88	249	4,070
	Sub-total	2,778	33,683	245	1,297	38,003
	Total	14,874	107,776	1,009	5,043	128,699

Source: Brown *et al.*, 2012.

lands burned each year (about nine million ha), the emissions of CH_4 and N_2O, as CO_2e, represent about 4% of total emissions.

With the fastest population growth in the world and rising per capita GDP, Africa is likely to increase its share of global emissions over the coming decades although emissions from Africa will remain low compared to other continents (Darwin et al., 1995; Canadell et al., 2009).

Projected Climate Change in Africa

There are many model-based projections of climate change across Africa. Generally, climate change due to GHG emissions is expected to increase temperature and alter precipitation patterns. The range of the projected changes is however considerable and arises because of the different input assumptions (namely GHG emission levels) and range of climate models. The results reported in IPCC (2001a,b) suggest temperature changes over the coming decades for Africa of between 0.2 and 0.5°C per decade, with the greatest warming in interior regions. A warming of approximately 0.7°C was recorded over most of the African continent during the 20th century (Elasha et al., 2006). In the past three decades, a 25% decrease in rainfall occurred over the Sahel and since mid-1970s a 2.4% per decade decrease in precipitation has happened in tropical rainforest regions in Africa (Elasha et al., 2006). The sign of changes in mean precipitation in many parts of Africa varies across climate models. Of three macro-regions of sub-Saharan Africa (West, East and Southern) reviewed in IPCC (2001b) only one shows consistent temperature and precipitation projections across climate models (the West region shows consistent changes for December–January; the Southern region for June–August). More recent studies also show conflicting evidence: for example, Held et al. (2005) show a drier Sahel in the late 21st century, whilst Kamga et al. (2005) show a wetter Sahel. These results reflect the uncertainty described above. The magnitude of projected rainfall changes for 2050 is small in most African areas, but can be up to 20% of 1961–1990 baseline values (IPCC, 2001b). The uncertainty in data, methodologies or operations of mechanisms is, however, no reason for Africa to be complacent while the rest of the world searches for solutions.

Projected Impacts of Climate Change on Agriculture in Africa

Despite producing the least GHGs, Africa is predicted to be the continent that will suffer the most. Most studies show a negative impact of climate change on crop productivity in Africa. Projections suggest that among main world crops, African millet and sorghum production will witness, by far, the fastest growth rate for the

2000–2050 period (Nelson *et al.*, 2009). A review of projected effects on pearl millet and sorghum yields indicates 10–15% losses that would be statistically significant by 2050 (Knox *et al.*, 2011). Roudier *et al.* (2011) used a meta-database of future crop yields from 16 recent studies, to assess the potential impact of climate change on yields in West Africa and reported yield changes ranging from −50% to +90% with a median yield loss of −11%. The predicted impact was larger in the Sudano-Sahelian countries (−18%) than in the southern Guinean countries (−13%) likely due to drier and warmer conditions in the former countries. Jones and Thornton (2003) analyzed the possible impacts of climate change on maize production in Africa and Latin America and reported yield decreases to the year 2055, in three-quarters of the countries, as a result of temperature increases and rainfall becoming less conducive to maize production.

The livestock-based systems will be especially affected by changes in length of growing period (LGP), as these systems are predominantly in marginal areas. Losses of more than 20% in LGP are expected in countries such as Eritrea, Ethiopia, Kenya and Sudan.

The scientific evidence leaves little room for doubt that our climate is changing and it will have a significant impact on agriculture. This makes adapting agriculture to climate variability and change an essential component of agricultural research and development programs.

Adaptation and Mitigation Potential

Climate change brings new challenges but it also highlights the need to address more comprehensively the same old problems that agriculture in the continent is struggling to cope with. The sustainable intensification of production can ensure food security and contribute to mitigating climate change by reducing deforestation and the encroachment into natural ecosystems (Burney *et al.*, 2010; Bellassen *et al.*, 2010). The overall efficiency, resilience, adaptive capacity and mitigation potential of the production systems can be enhanced through improving its various components, some of the key ones are highlighted below.

Addressing current challenges to agriculture

Addressing current challenges to agriculture in Africa through natural resource management provides a good starting point to deal with future changes.

Soil and nutrient management in cropland

Many subsistence crop production systems in Africa are depleted and have low nutrient content especially nitrogen and phosphorus. Improving soil fertility is therefore essential to increase yields. Application of inorganic fertilizers enhances crop production and C inputs, with many experiments showing a positive effect of N addition on the soil C balance (Glendining and Powlson, 1995; Vlek, 1990). Given the low rate of N fertilizer application per ha in Africa, there is an opportunity to increase the rate of application to improve crop production and at the same time reduce the need to clear native ecosystems for new croplands (the second largest source of CO_2 emissions). If increasing fertilizer can double productivity then each improved hectare of agriculture will reduce the need to clear forest with their associated emissions. However, to maintain crop production means that fertilizer will need to be added continually through time and there will likely be a point where the cumulative N_2O emissions will outweigh the advantages of stopping the clearing of native ecosystems.

Improving N-use efficiency can reduce N_2O emissions and indirectly reduce GHG emissions from N fertilizer manufacture (Schlesinger, 1999). Practices that improve N-use efficiency include adjusting application rates based on precise estimation of crop needs (e.g., precision farming); improved timing; placing the N more precisely into the soil to make it more accessible to crops roots; or avoiding N applications in excess of immediate plant requirements (Robertson, 2004; Dalal *et al.*, 2003; Cole *et al.*, 1997).

The use of compost and manure reduces the need for synthetic fertilizers which through their production and transport, contribute to GHG emissions. A problem in many countries is the decreasing source of such amendments, linked to animal husbandry. There is competition for plant residues or plant cover — to be used for feeding animals or for returning to the soil. Therefore, adopting the strategy of integrated nutrient management (INM) is crucial to sustainable use of soil and water resources. Several research results have shown large crop yield increases with the improvement of soil fertility using organic and inorganic fertilizers (Abdullahi and Lombin 1978; Bationo *et al.*, 1993; Bationo *et al.*, 1998; Pieri, 1989). The application of 4 t of crop residue per hectare for example maintained soil organic carbon at the same level as that in an adjacent fallow field in the top soil but continuous cultivation without mulching resulted in drastic reduction of organic carbon.

The use of legumes as green manure, planted in intercropping systems, or as part of a scheme of crop rotation or in agroforestry systems can contribute to soil fertility and production and reduce reliance on external N inputs. Field experiments at several sites in West Africa have shown cereal yield increases in cereal/legume rotations of between 15% and 79% compared with continuous cereal systems. Rotation of cereals with legumes increases N use efficiency. For example, Bationo and Vlek (1998) have shown that nitrogen use efficiency increased from 20% in continuous pearl millet cultivation to 28% when pearl millet was rotated with cowpea.

Water harvesting and irrigation

Improved water harvesting and retention systems such as terraces, tied ridges, stone bunds, vegetative barriers, *zai*, and *half-moons* are used in many semi-arid areas of Africa and are fundamental for increasing production and addressing increasing variability of rainfall patterns. Taonda *et al.* (2003) found that water harvesting alone with stone bunds did not improve yields but the combination of water harvesting with stone bunds or *zai* plus manure more than doubled sorghum yields when compared to the control (Table 2). Similarly, Zougmore (2003a) showed that combining *Zai* phosphorus and significantly increased sorghum yield compared to *Zai* without fertilizer or manure and therefore, can reduce risks of crop failure in erratic rainfall years (Table 3). In Niger, water-use efficiency of millet during the rainy seasons (1999–2000) varied according to the planting technique and soil amendment (Table 4). Water use efficiency using the *Zai* technique was higher than in the flat planting. Combining the *Zai* technique with manure significantly improved water use efficiency in all two sites. The *Zai* concentrates both nutrients and water and facilitate water infiltration and retention.

Table 2. Effect of water harvesting technique and manure application on sorghum grain yield.

Treatment	Grain yield (kg ha^{-1})			
	2000	2001	2002	Average
Control	353	393	215	331
Stone Bunds	394	574	504	397
Stone Bunds + Manure	1026	1168	1072	789
Zai + Manure	1188	176	1267	805

Source: Taonda *et al.*, 2003.

Table 3. Effect of *Zai* and fertilizer on sorghum grain and stover yield in Burkina Faso.

Treatment	Yields (kg ha^{-1})	
	Grain	Stover
Zai	285	807
Zai + 100% WP + Manure	1439	2774
Zai + 100% PR + Manure	710	1409
Zai + 75% PR + 25% WP + Manure	1085	2278
LSD (5%)	429	586

Source: Zougmore *et al.*, (2003); WP = Water soluble phosphate; PR = Rock phosphate.

Table 4.   Water productivity of millet as affected by planting technique and soil amendment during the 1999–2000 seasons at Damari and Kakassi in Niger.

| | | Amendment based on grain yield (kg mm^{-1}) | | | |
| | | Damari | | Kakassi | |
Sowing technique		1999	2000	1999	2000
Zai	Millet straw	0.34	0.43	1.19	1.07
	Manure	2.35	2.28	2.67	2.80
Flat	Millet straw	0.26	0.38	0.88	0.79
	Manure	1.43	1.25	1.41	1.38
	Sed	0.41	0.19	0.32	0.58

Source: Fatondji (2002).

Breeding for adaptation to climate change and variability

The predicted increases in climate variability associated with expected climate changes require more versatile systems and flexible germplasm that perform satisfactorily under a wider range of climatic conditions rather than under a narrow subset of conditions. This is particularly relevant because variability remains unpredictable for smallholder systems (Nelson *et al.*, 2009).

Breeding for enhanced tolerance to drought and heat

The development of new crop varieties including types, cultivars and hybrids, has the potential to provide crop choices that are better suited to temperature, moisture and other conditions associated with climate change. This involves the development of plant varieties that are more tolerant to climatic conditions such as heat or drought through conventional breeding and genetic engineering (Haussmann *et al.*, 2012).

Integrated genetic and natural resource management

The development of new improved and climate-proof cultivars must go hand in hand with sustainable soil fertility management and water conservation and drainage techniques (Haussmann *et al.*, 2012). It is important for breeders to develop cultivars with specific adaptation to natural resource management techniques. Integrated genetic and natural resource management strategies for adaptation to climate change may include (i) developing genotypes suitable for mixed cropping systems e.g. pearl millet and cowpea cultivars with complimentary maturities, plant and root characteristics that maximize resource utilization; (ii) breeding for enhanced resource-use efficiency e.g. high water-, nitrogen- or phosphorus-use efficiency.

Systems diversification

Enhancing population buffering via systems diversification is another adaptation strategy to climate change. Farmers often cultivate several different varieties with contrasting maturity in the same season. Farmer's strategy of using varietal diversity aims to optimize their whole farm production given their predictions of local climatic conditions and their various production objectives and conditions. Thus enhancing precision of seasonal climate forecast and would aid farmers in their tactical choices of cultivars to maximize production in each cropping season (Haussmann *et al.*, 2012; Smith and Lenhart, 1996). Efficient seed production systems and strengthening seed availability of diverse varieties, are required to ensure rapid access of farmers to varieties adapted to their new agro-ecological conditions.

Various types of intercropping of different species are also practiced by farmers that can provide a type of population buffering. Well-designed intercropping systems can possibly enhance water and nutrient-use efficiency, protect the soil from erosion after heavy rainfall, and help in weed management, thereby reducing vulnerability and enhancing system stability. Use of varietal diversity as well as intercropping options also leads to production systems diversification on farm and landscape levels required for adaptation to more severe climatic change (Howden *et al.*, 2007).

Preservation of genetic resources

The preservation of genetic resources of crops and breeds and their wild relatives is therefore fundamental in developing resilience to shocks, improving the efficient use of resources, shortening production cycles and generating higher yields per area of land. Farmers have developed and practiced an array of strategies for adaptation to variable climate, e.g. creating a diversity of land race varieties, using early and late varieties in their production systems, staggering planting dates and intercropping. It is important that researchers tap into this knowledge possessed by farmers to pursue the development of new varieties and crop management practices that will diversify options available to farmers.

Mitigation through soil carbon sequestration

The largest potential of soil carbon sequestration lies in restoration of degraded soils and ecosystems (Vagen *et al.*, 2005; Lal, 2004). These soils especially those affected by erosion have lost a large fraction of the original soil carbon pool. Restoration of degraded soils through improved farming systems could lead to a considerable increase in soil carbon. Some options for increased carbon sequestration in the soil are highlighted below.

Conservation tillage and crop residue management

Conservation tillage is a viable practice for soils of the humid and sub-humid areas of Africa (Ike, 1986; Lal, 1989). Experiments in the sub-humid and humid tropics of Africa have demonstrated the potential for no-till systems to maintain higher soil C levels compared to conventional cultivation (Juo and Lal, 1979; Agboola, 1981). Agboola (1981) reported organic matter losses of less than 10% with no-till compared with 19–33% losses in tilled treatments, after four years of continuous maize. Juo and Lal (1979) reported nearly double C contents in no-till versus ploughed treatments in the top 10 cm under maize. No-till may also contribute to a more effective use of increased C inputs (e.g. from crop residue or mulches) in sustaining soil C levels. Under no-till, Juo and Kang (1989) reported that soil organic carbon (SOC) levels were up to two times higher where residues were added compared to where residues were removed. Brown et al. (2012) report significant mitigation potential when converting rainfed cropland from full tillage to reduced tillage, for nine African countries. The change in soil carbon varies between 0.4 t CO_2e ha^{-1} yr^{-1} in Senegal's dry climate (with a shift from low to medium input), up to 5.3 t CO_2e ha^{-1} yr^{-1} in Uganda's moist climate, with a shift from low to high inputs with manure.

Crop residue management is another important method of sequestering C in soil and increasing the soil organic matter (OM) content. The positive effects of using crop residues to induce C sequestration have been estimated by Lal (1997) at 0.2 Pg C yr-1 with transformation of 15 percent of the total C (1.5 Pg C globally). Under conservation agriculture, 0.2–0.5 t C ha-1 yr-1 can be sequestered in the humid tropics and 0.1–0.2 in semi-arid zones (Lal, 1999).

Cover crops and improved fallows

The use of improved (planted) fallows and cover crops within cropping sequences and woody species in agroforestry systems have the potential to reverse SOC loss and help restore degraded land (Lal et al., 1995). Attainable rates of C sequestration through the establishment of natural or improved fallows is in the range of 0.1 to 5.3 Mg C ha^{-1} yr^{-1} (Vagen et al., 2005). Kang et al. (1981, 1991) showed for instance, that alley cropping systems on Alfisols an Psamments in sub-humid and humid regions of tropical Africa were effective in maintaining higher levels of SOC and nutrients than under conventional cropping. Studies with planted fallows, comparing three different species of grass and five species of legumes, showed that SOC (0–10 cm) increased on average by 25% after two years, on an eroded Alfisol in Nigeria (Lal et al., 1979; Wilson et al., 1982). Swaine and Hall (1983) reported substantially higher SOC stocks after four years of fallow ($\sim$83..3 Mg C ha^{-1}) than in natural forest ($\sim$63.5 Mg C ha^{-1}) in humid forests in Ghana.

Table 5.   Total Grassland, overgrazed grassland and percentage of grassland that is overgrazed, by continent.

Continent	Grassland 10^6 ha	Overgrazed Grassland 10^6 ha	Percentage Overgrazed
Africa	838.2	87.7	10.4
Australia/Pacific	437.1	40.1	11.2
Eurasia	1385.9	85.6	6.2
North America	353.7	14.0	4.0
South America	402.2	26.2	6.5
Total	3417.1	262.5	7.7

Source: Conant and Paustian (2002).

Table 6.   Total potential C sequestration (Tg C yr^{-1}) within each overgrazing severity class, by continent.

Continent	Light	Moderate	Strong	Extreme	Total
Africa	1.9	8.6	6.1	0.1	16.7
Australia/Pacific	4.5	−0.1	0.0		4.4
Eurasia	0,8	3.2	0.0	0.3	4.3
North America	0.0	1.6	0.6		2.2
South America	6.1	11.3	0.7		18.1
Total	13.3	24.6	7.4	0.4	45.7

Source: Conant and Paustian (2002).

Improving management of grasslands and pastures

Grassland/grazing lands in Africa occupy much larger areas (838 Mha) compared to other continents, out of which 88 Mha (10%) are overgrazed grassland (Table 5). Many grassland areas in the semiarid and humid zones of Africa are badly managed and degraded; these offer a range of carbon sequestration possibilities. Restoring overgrazed grassland and improving forage species has a high potential to sequester carbon (16.7 Tg C yr^{-1}) (Table 6). Management options for improving grasslands and pastures include judicious use of fertilizers, controlled grazing, sowing legumes and grasses or other species adapted to the environment. Conant and Paustian (2001) reported rates of SOC sequestration through pasture improvement ranging from 0.11 to 3.04 Mg C ha^{-1} yr^{-1} with a mean of 0.54 Mg C ha^{-1} yr^{-1}. Establishing deep-rooted grasses in savannahs has been reported to yield very high rates of carbon accrual (Fisher *et al.*, 1994). Reducing the frequency or intensity of bush fires through more effective fire suppression and vegetation management to reduce fuel load should lead to increased tree and shrub cover, resulting in a CO_2 sink in soil and biomass (Scholes and van der Merwe, 1996). All the above interventions are technically feasible but may be more difficult to apply because of social aspects. Economic input and policy improvement can be determining factors.

Restoration of degraded lands

A large proportion of agricultural land in Africa has been degraded by excessive disturbance, erosion, organic matter loss, salinization, acidification, or other processes that curtail productivity (Lal, 2001a, 2003, 2004b; Vlek *et al.*, 2010). Conant and Paustian (2002), estimated that much of the land in Africa in degraded categories is savannah and grassland, reflecting the role of livestock and grazing in land degradation processes. Restoration of degraded soils is a development strategy to reduce desertification, soil erosion and environmental degradation, and alleviate chronic food shortages with great potential in sub-Saharan Africa (SSA; Vagen *et al.*, 2005). Further, it has the potential to provide terrestrial sinks of carbon (C) and reduce the rate of enrichment of atmospheric CO_2. The scope for soil organic carbon gains from improved management and restoration within degraded and non-degraded croplands and grasslands in Africa is estimated at 20–43 Tg C year^{-1}, assuming that 'best' management practices can be introduced on 20% of croplands and 10% of grasslands (Batjes, 2004). Under the assumption that new steady state levels will be reached after 25 years of sustained management, this would correspond with a mitigation potential of 4–9% of annual CO_2 emissions in Africa (Batjes, 2004). Often, carbon storage in these soils can be partly restored by practices that reclaim productivity including re-vegetation (e.g., planting grasses); improving fertility by nutrient amendments; applying organic substrates such as manures and composts; reducing tillage and retaining crop residues; and conserving water (Lal, 2001b; 2004b; Bruce *et al.*, 1999; Batjes, 2001).

Land cover/use change

One of the most effective methods of reducing emissions is often to allow or encourage the reversion of cropland to another land cover, typically one similar to the native vegetation. Such land cover change often increases carbon storage. For example, converting arable cropland to grassland typically results in the accrual of soil carbon because of lower soil disturbance and reduced carbon removal in harvested products. Vagen *et al.* (2005) summarized data on cumulative changes and rate of change in SOC stocks for a range of land use conversions for different regions of Sub-Saharan Africa (SSA). In cultivated areas, sequestration in the range of 0.05–0.36 Mg C ha^{-1} yr^{-1} may be attainable through conversion from traditional cultivation techniques (with tillage) to no-till systems or systems where a combination of animal manure and fertilizers are applied. Conversion from cultivated land to fallow (agroforestry) in semiarid and arid savanna areas increased SOC between

0.1 and 5.3 Mg ha^{-1} yr^{-1}. Attainable rates of SOC sequestration on permanent crop-land in SSA under improved cultivation systems (e.g. no-till) range from 0.2 to 1.5 Tg C yr^{-1}, while attainable rates under fallow systems are 0.4 to 18.5 Tg C yr^{-1}. Fallow systems generally have the highest potential for SOC sequestration in SSA with rates up to 28.5 Tg C yr^{-1}. These rates may be achieved through improved management on degraded cultivated land and establishment of improved fallows. Brown *et al.* (2012) reported that the range of mitigation potential in soil alone from converting rainfed cropland with reduced tillage and low inputs to native ecosystem (i.e. abandon cultivation and allow for system to return to native system) is from 1.0 t CO$_2$e ha^{-1} yr^{-1} in Senegal's tropical dry climate to 4.1 t CO$_2$e ha^{-1} yr^{-1} in Uganda's tropical moist climate. Given the vast areas of tropical savanna in SSA, there is thus a high potential for sequestering C in soils through improved management of cultivated land and introduction of agroforestry systems such as mixed fallows and natural fallow systems.

Afforestation of marginal agricultural soils or degraded soils has a large potential for SOC sequestration. In Uzbekistan, Khamzina *et al.* (2009) showed as much as 120 metric tons of carbon sequestered over four years by trees planted on highly salinized agricultural land. Vlek *et al.* (2004) show that the sequestration of carbon through afforestation far outweighs the emissions that are associated with the production of the extra fertilizer needed to maintain agricultural output. Mitigating GHG emissions through reforestation, should aim to set aside marginal land for the purpose of sequestering carbon such as to minimize the threat to food security in the region. Tree species adapted to many ecosystems in Africa that can be planted include tamarisk (*Tamarix spp.*), gum tree (*Eucalyptus spp.*), Leucaena (*Leucaena spp.*), cypress (*Cupressus spp.*), neem (*Azadirachta spp.*), acasia (*Acacia spp.*), teak (*Tectona grandis*), cassia (*Casia siamea*) and many others (Lal *et al.*, 1999). In addition to biofuel production, tree plantations also enhance SOC content.

Agroforestry, which is the association of trees with crops or pastures, represents a sustainable alternative to deforestation and shifting cultivation (Sanchez *et al.*, 1999; Schroeder, 1994; Sanchez, 1995). Agroforestry systems can be established on unproductive croplands with low levels of OM and nutrients that are widespread in sub-humid areas of tropical Africa. The promotion of agroforestry systems and practices can help to tackle the triple challenge of food security, mitigation and reducing the vulnerability and increasing the adaptability of agricultural systems to climate change. Nitrogen fixing leguminous trees and shrubs contributes to soil fertility and increased agricultural productivity. The carbon uptake rate can be very high because of sequestration by both trees and crops ranging from 2 to 9 t C yr^{-1} depending on the duration. The conversion to agroforestry would permit tripling the C stocks, from 23 to 70 t/ha over a 25-year period. In sub-humid tropical Africa alone, the benefit would be 0.04–0.19 Gt C yr^{-1}.

Planting trees on agricultural lands provides other co-benefits important for improved farm family livelihoods and climate change adaptation. The application of the Kyoto protocol will be a good opportunity to promote such initiatives, including tree planting, assuming that adequate socio-economic incentives can be provided under Clean Development Mechanism (CDM). In Uganda, community-based efforts are attempting to design Clean Development Mechanism-Agroforestry/Reforestation (CDM-AR) projects, which include hundreds to thousands of small farmers adopting agroforestry, and increasing carbon within the larger mixed farming landscape. Likewise, on-going work in Western Kenya funded by Global Environmental Facility (GEF) (World Bank, 2005) attempts to quantify the potential carbon sequestration benefits of improved farming and increased soil organic matter on small holder farms, in addition to the inclusion of trees in the farming system and the landscape.

Reducing emissions from livestock production

Livestock are a major component of the farming system in Africa, providing food, income, drought power for land preparation and transport. Despite the importance of livestock, the sector is a major cause of overgrazing and land degradation and an important driver of deforestation. It is also responsible for methane and nitrous oxide emissions from ruminant digestion and manure management (Table 1). It is estimated that livestock population would increase by 68% by 2030 from the 2000 base period (FAO, 2006). It is important to improve productivity of the sector to meet growing food security while minimizing resource use and GHG emissions. The extension and adoption of advances in feeding and nutrition technologies and improvements in animal husbandry can play a key role in mitigation and building resilience to climate change. The substitution of manures for inorganic fertilizers can also lower emissions and improve soil conditions and productivity. The composting of solid manure can lower emissions and produce useful organic amendments for soil improvement.

Institutional and policy reforms

A major transformation of the agricultural sector requires institutional and policy reforms to meet the challenges of ensuring food security and development under climate change while maintaining low emissions trajectories. Better aligned policy approaches across agricultural, environmental and financial boundaries and innovative institutional arrangements to promote their implementation are required. This section covers some institutional and policy adjustments required to support the agricultural transformation.

Mainstreaming climate change adaptation and mitigation into national and regional development

At the regional level, agricultural development and investment strategies are being developed under the Comprehensive African Agricultural Development Programme (CAAD). Policies are generally focused on improving productivity and returns to small scale agriculture and generally include some emphasis on sustainable land management and soil restoration (FAO, 2010). At the national levels, climate change policies are expressed through the National Action Plan for Adaptation (NAPAs) and the Nationally Appropriate Mitigation Actions (NAMAs). Policy statements made for NAPAs and NAMAs focus on increasing reliance of the agricultural sector by better management of land and soil resources. It is important that climate change adaptation and mitigation strategies are mainstreamed in the regional and national development. It is important that governments put in place institutional and macro-economic conditions that support and facilitate adaptation and resilience to climate change at local, national and transnational level.

Strengthen early warning systems

There is the need for the development of information systems capable of forecasting weather and climate conditions associated with climate change. Weather predictions over days or weeks have relevance to the timing of operations such as planting, spraying, or harvesting. Seasonal forecasts such as estimates of the likelihood of conditions associated with El Nino, have the potential to aid risk assessment and production decisions over several months (Smit and Skinner, 2002). Food production can be improved dramatically in dry areas when governments and/or organizations use climate forecasts and prepare accordingly by potentially distributing drought-tolerant seeds (Patt *et al.*, 2005). Farmers can also take advantage of climate forecasts by planting less drought-tolerant and higher-yield, long season maize when wetter than usual growing seasons are forecast (Patt *et al.*, 2005).

Enhancing the capacity of extension services

Agricultural extension systems are the main conduits for disseminating the information required to make such changes. There is a need to improve understanding of extension and other service providers on issues related to climate variability and on the range of possible responses that farmers should consider when making management decisions through proper use and application of climate information including forecast information.

Institutions to support financing and insurance

Financial incentives are crucial for farmers to adopt or implement some adaptation and mitigation activities e.g. for crop and grassland restoration projects (which often takes land out of production for a long time); reducing cultivated or grazing land; and incorporating crop residues that are expected to increase soil fertility and water retention capacity, thereby increasing yields over the medium-long term. Government programs and insurance are institutional responses to the economic risks associated with climate change and have the potential to influence farm level risk management strategies (Smit and Skinner, 2002). Index insurance program are one potential response to the insurance gap in developing country agriculture. Safety nets are a form of social insurance comprising program supported by the public sector or non-governmental organizations (NGOs) that provide transfers to prevent the poor from falling below a certain poverty level. These programs include cash transfers, food distribution, seeds and tools distribution. The continent should explore and expand the above opportunities to help people adapt to some of the unavoidable risks associated with climate.

Strengthening institutions at the grassroots

There is the need for institutions to improve coordination and collective action. For instance input supply e.g. fertilizer and seeds, is one activity that requires coordination beyond the farm. Many of the biophysical improvements to increase resilience in smallholder agricultural production systems discussed above require action and coordination amongst many stakeholders in the rural areas (FAO, 2010). Certain adaptation modalities may exist that can only be implemented at higher tiers of societal structure, beyond the capacity of individual and household ability and skills. For example, restoration of degraded areas to improve soil quality, improved management of communal resources, informal seed systems and the excavation of silted dams and rivers as potential adaptation options cannot be implemented solely by an individual or a few members of a household. It certainly needs the involvement of actors at higher tiers, with higher levels of skills, knowledge and techno-financial means. Integration of such adaptation modalities in an institutional framework is considered necessary.

Payment for environmental services

Payments for environmental services are one potential source of alternative financing for agricultural transitions (FAO, 2007b). Emerging carbon markets and payments for emissions removal or reduction have attracted much interest and anticipation of such financing as a source of income for some agricultural activities and producers.

However, carbon crediting schemes that pay projects to reduce greenhouse gas emissions and sequester carbon have largely ignored agriculture. A system for assessing the contribution of improved grassland management could help herders earn money by selling carbon credits. The main problem has been finding ways to measure how much carbon is actually being trapped in grazing-based livelihood systems. With the development of a methodology for sustainable grassland management by the FAO, herders who are investing in restoring grasslands can prove they are sequestering measurable amounts of carbon and earn carbon credits.

Conclusion

Agriculture in Africa must undergo a significant transformation in order to meet the related challenges of food security and climate change. Effective climate-smart practices already exist that could be implemented to lift millions of people out of poverty and hunger by increasing the productivity and profitability of small scale farmers in Africa. There exists a high potential for Africa to contribute to mitigating greenhouse gases emissions through SOC sequestration in cropping and grazing lands, restoration of degraded soils and land-use and cover changes. Although the potential of SOC sequestration in degraded soils and ecosystems of Africa is high, the realization of this potential is challenging. Yet the need to restore degraded soils and ecosystems is also urgent and of high priority.

Considerable investment is required in filling data and knowledge gaps and in research and development of technologies and methodologies. Strengthened institutional capacity will be required to improve the dissemination of climate information. Adaptation will involve improved risk management through insurance schemes and Information Communication Technologies (ICTs), like cell phones, which provide farmers with rapid access to information such as weather forecasts or facilitate their insurance payments. Institutional reforms and financial support will be required to enable small scale farmers in Africa make the transition to climate smart agriculture. Climate finance could provide a major stimulus to improving agriculture and making farmers more climate-resilient. The mechanisms that are being put in place to implement the Kyoto Protocol — through carbon emission trading — and prevailing agricultural policies will largely determine whether farmers can engage in activities that enhance C sequestration in Africa. Currently, Africa gets more benefit from capacity building on CDM rather than from the actual CDM projects (UNEP, 2006). Thus, it is important to make efforts to create a more equitable distribution of CDM projects. Developed nations should help Africa in integrating adaptation goals into sustainable development strategies; increasing the use of renewable energy; and improving the capacity for environmental and climatic research. Where the policy environment is conducive and where the incentives are enabling, then can farmers rise to the challenge.

References

Abdullahi A. and G. Lombin. 1978. Long-term fertility studies at Samaru, Nigeria: Comparative effectiveness of separate and combined applications of mineralizers and farmyard manure in maintaining soil productivity under continuous cultivation in the Savanna, Samaru, Samaru Miscellaneous Publication. No. 75, Zaria, Nigeria: Ahmadu Bello University.

Agboola, A.A. 1981. The effects of different soil tillage and management practices on the physical and chemical properties of soils and maize yield in a rainforest zone of western Nigeria. *Agronomy Journal*, 73:247–251.

Bationo, A., B.C. Christianson and M.C. Klaij. 1993. The effect of crop residue and fertilizer use on pearl millet yields in Niger. *Fertilizer Research* 34:251–258.

Bationo, A., F. Lompo and S. Koala. 1998. Research on nutrient flows and balances in West Africa: state-of-the art. *Agric. Ecosyst. Environ.* 71:19–35.

Bationo, A. and P.L.G. Vlek. 1998. The role of nitrogen fertilizers applied to food crops in the Sudano-Sahelian zone of West Africa. In: Renard G Neef A., Becker, K. and von Oppen, M. (Eds.). Soil Fertility Management in West African Land Use Systems. Margraf Verlag, Weikersheim, Germany. pp. 41–51.

Batjes, N.H. 2001. Options for increasing carbon sequestration in West African soils: An exploratory study with special focus on Senegal. *Land Degradation and Development*, 12:131–142.

Batjes, N.H. 2004. Estimation of soil carbon gains Upon Improved Management Within Croplands and Grasslands of Africa. *Environment, Development and Sustainability*, 6(1–2):133–143.

Bellassen, V., R.J. Manley, J.P. Chery, V. Gitz, A. Toure, M. Bernoux and J.-L. Chotte. 2010. Multicriteria spatialization of soil organic carbon sequestration potential from agricultural intensification in Senegal. *Climate Change*, 98, No. 1–2, 213–243.

Brown, S., A. Grais, S. Ambagis and T. Pearson. 2012. Baseline GHG Emissions from the Agricultural Sector and Mitigation Potential in Countries in East and West Africa. CCAFS Working Paper no. 13. CGIAR Research Program on Climate Change, Agriculture and Food Security (CCAFS). Copenhagen, Denmark. Available at: www.ccafs.cgiar.org

Bruce, J.P., M. Frome, E. Haites, H. Janzen, R. Lal and K. Paustian. 1999. Carbon sequestration in soils. *Journal of Soil and Water Conservation*, 54:382–389.

Burney, J.A., S.J. Davis and D.B. Lobell. 2010. Greenhouse gas mitigation by agricultural intensification. *Proceedings of the National Academy of Sciences*, 107(26):12052–12057.

Canadell, J.G., M.R. Raupach and R.A. Houghton. 2009. Anthropogenic CO_2 emissions in Africa. *Biogeosciences*, 6:463–468.

Cole, C.V., J. Duxbury, J. Freney, O. Heinemeyer, K. Minami, A. Mosier, K. Paustian, N. Rosenberg, N. Sampson, D. Sauerbeck and Q. Zhao. 1997. Global estimates of potential mitigation of greenhouse gas emissions by agriculture. *Nutrient Cycling in Agroecosystems*, 49: 221–228.

Conant, R.T. and K. Paustian. 2002. Potential soil carbon sequestration in overgrazed grassland ecosystems. *Global Biogeochemical Cycles*, 16(4):1143 pp., doi:10.1029/2001GB001661.

Conant, R.T., K. Paustian and E.T. Elliott. 2001. Grassland management and conversion into grassland: Effects on soil carbon. *Ecological Applications*, 11:343–355.

Dalal, R.C., W. Wang, G.P. Robertson and W.J. Parton. 2003. Nitrous oxide emission from Australian agricultural lands and mitigation options: a review. *Australian Journal of Soil Research*, 41:165–195.

Darwin, R., T. Marinos, L. Jan and R. Anton. 1995. World Agriculture and Climate Change. Agricultural Economic Report 703. US Department of Agriculture, Washington D.C.

Elasha, B., *et al.*, 2006. Background paper on impacts, vulnerability and adaptation to climate change in Africa. Paper presented at the African Workshop on Adaptation Implementation of Decision 1/CP.10 of the UNFCC convention, Accra, Ghana, 21–23 September, 2006.

FAO, 2006. World Agriculture: towards 2030/2050, Interim Report, Food and Agricultural Organization, Rome.

FAO, 2007b. The State of Food and Agriculture: Paying Farmers for Environmental Services, Food and Agricultural Organization, Rome.

FAO, 2010. Climate Smart Agriculture: Policies, Practices and Financing for Food Security, Adaptation and Mitigation, Rome, 41 pp.

Fatondji, D. 2002. Organic amendment decomposition, nutrient release and nutrient uptake by millet (Pennisetum glaucum (L.) R.Br.) in a traditional land rehabilitation technique (Zaï) in the Sahel. *Ecology and Development Series* No. 1 140 p. Germany: University of Bonn.

Fisher, M.J., I.M. Rao, M.A. Ayarza, C.E. Lascano, J.I. Sanz, R.J. Thomas and R.R. Vera. 1994. Carbon storage by introduced deep-rooted grasses in the South American savannas. *Nature*, 371:236–238.

Glendiningm, J. and Powlson, D.S. 1995. The effects of long continued applications of inorganic nitrogen fertilizer on soil organic nitrogen: a review. In: R. Lal and B.A. Stewart (eds.), *Soil Management — Experimental Basis for Sustainability and Environmental Quality. Advances in Soil Science*. CRC Lewis Publishers, Boca Raton, Florida, pp. 385–446.

Haussmann, B.I.G., H. Fred Ratunde, E. Weltzien Ratunde, P.S.C. Traore, K. vom Brocke and H.K. Parzies. 2012. Breeding Strategies for Adaptation of Pearl Millet and Sorghum to Climate Variability and Change in West Africa. *Journal of Agronomy and Crop Science*, DOI: 10.1111/j.1439-037X.2012.00526.X

Held, I.M., T.L. Delworth, J. Lu, K.L. Findell and T.R. Knutson. 2005. Simulation of Sahel Drought in the 20th and 21st Century. *PNAS*, 102(50):17891–17896.

Howden, S.M., J.-F. Soussana, F.N. Tubiollo, N. Chetri, M. Dunlop and H. Meinke. 2007. Adapting Agriculture to Climate Change. *PNAS*, 104(50):19691–19696.

Ike, L.F. 1986. Soil and crop responses to different cultural practices in a ferruginous soil in the Nigerian Savanna. *Soil and Tillage Research*, 6:261–272.

IPCC (Intergovernmental Panel on Climate Change), 2001a. *Climate Change 2001: The Scientific Basis. Contribution of Working Group I to the Third Assessment Report of the Intergovernmental Panel on Climate Change* [Houghton, J.T., Y. Ding, D.J. Griggs, M. Noguer, P.J. van der Linden, X. Dai, K. Maskell and C.A. Johnson, (eds.)], Cambridge University Press, 881 pp.

IPCC (Inter-governmental Panel on Climate Change), 2001b. *Climate Change 2001: Mitigation: Contribution of Working Group III to the Third Assessment Report of the Intergovernmental Panel on Climate Change* [Metz, B., O. Davidson, R. Swart and J. Pan, (eds.)], Cambridge University Press, 752 pp.

IPCC, 2007b. *Climate Change 2007: Impacts, adaptation and vulnerability. Contribution of Working Group II to the Fourth Assessment Report of the IPCC*. Cambridge, UK, Cambridge University Press.

Jarvis, A., C. Lau, S. Cook, E. Wollenberg, J. Hansen, O. Bonilla and A. Challinor. 2011. An integrated adaptation and mitigation framework for developing agricultural research: Synergies and Trade-offs. *Experimental Agriculture*, 47(2):185–203.

Jones, P.G. and P.K. Thornton. 2003. The potential impacts of climate change on maize production in Africa and Latin America in 2055. *Global Environment Change*, 13:51–59.

Juo, A.S.R. and R. Lal. 1979. Nutrient profile in a tropical Afisol under conventional and no-till systems. *Soil Science*, 127:168–13.

Juo, A.S.R. and B.T. Kang. 1989. Nutrient effects of modification of shifting cultivation in West Africa. In: J. Proctor (ed), *Mineral Nutrients in Tropical Forest and Savanna Ecosystems*. Blackwell Scientific Publications, Oxford, pp. 289–300.

Kamga, A., G. Jenkins, A. Gaye, A. Garba, A. Sarr and A. Adedoyin. 2005. Evaluating the National Center for Atmospheric Research Climate System Model over West Africa: Present-day and the 21st century A1 Scenario. *J. Geophysical Research*, 110(D3): doi:10.1029/2004JD004689.issn: 0148-0227.

Kang, B.T., G.E. Wilson and L. Sipkens. 1981. Alley cropping maize *(Zea mays L.)* and leucaena *(Leucaena leucocephala* L.) in Southern Nigeria. *Plant and Soil*, 63:165–179.

Kang, B.T., M. Gichuru, N. Hulugalle and M.J. Swift. 1991. Soil constraints for sustainable upland crop production in humid and sub-humid West Africa. In: *Soil Constraints on Sustainable Plant Production in the Tropics. Proceedings of the 24th International Symposium on Tropical Agriculture Research*. Tropical Agriculture Research Series 24, Kyoto, Japan, August 4–16, 1990, pp. 101–112.

Khamzina A., R. Sommer, J.P.A. Lamers and P.L.G. Vlek. 2009. Transpiration and early growth of tree plantations established on degraded cropland over shallow saline groundwater table in northwest Uzbekistan. *Agricultural and Forest Meteorology*, 149:1865–1874.

Knox, J.W., T.M. Hess, A. Daccache and M. Perez Ortola. 2011. What are the projected impacts of climate change on crop productivity in Africa and South Asia? *DFID Systematic Review Final Report*. Cranfield University, 77 p.

Lal, R.. G.F. Wilson and B.N. Okigbo. 1979. Changes in properties of an Alfisol produced by various crop covers. *Soil Science*, 127:377–380.

Lal, R. 1989. Conservation tillage for sustainable agriculture: Tropic vs. temperate environments. *Advances in Agronomy*, 42:84–191.

Lal, R., J.M. Kimble, E. Levine and C. Whitman. 1995. World soils and greenhouse effect: an overview, pp. 1–7, In: R. Lal, J.M. Kimble, E. Levine and B.A. Steward (eds.), Soils and Global Change. CRC Press, Boca Raton, Florida, MI.

Lal, R.. 1997. Degradation and resilience of soils. *Phil. Trans. Royal Soc. London B.*, 352:987–1010.

Lal, R. 1999. Soil management and restoration for C sequestration to mitigate the accelerated greenhouse effect. *Progress in Environmental Science*, 1:307–326.

Lal, R., H.M. Hassan and J. Dumanski. 1999. Desertification control to sequester carbon and mitigate the greenhouse effect. In: Rosenberg, N., R.C. Izaurralde and E.L. Columbus, (eds.), *Carbon sequestration in soils: science, monitoring and beyond*. Malone Battelle Press, Ohio.

Lal, R., 2001a. World cropland soils as a source or sink for atmospheric carbon. *Advances in Agronomy*, 71:145–191.

Lal, R.. 2001a. Fate of eroded soil carbon: emission or sequestration. pp. 173–181. *In*: Lal, R. (Ed.), *Soil Carbon Sequestration and the Greenhouse Effect*. Soil Science Society of America Special Publication, Vol. 57. Madison, WI.

Lal, R. 2001b. Potential of desertification control to sequester carbon and mitigate the greenhouse effect. *Climate Change*, 15:35–72.

Lal, R. 2004. Soil carbon sequestration to mitigate climate change. *Geoderma*, 123:1–22.

Lal, R. 2004b. Soil carbon sequestration impacts on global climate change and food security. *Science*, 304(5677):1623–1627.

Lal, R. 2003. Global potential of soil carbon sequestration to mitigate the greenhouse effect. *Critical Reviews in Plant Sciences*, 22:151–184.

Nelson. G.C., M.W. Rosegrant, J. Koo, R. Robertson, T. Sulser, T. Zhu, C. Ringler, S. Msangi, A. Pallazo, M. Batka, M. Magalhaes, R. Valmonte-Santos, M. Ewing and D. Lee. 2009. *Climate Change impact on agriculture and cost of adaptation*. International Food Policy Research Institute, Food Policy Report no. XX. ISBN: 978-0-89629-535-4, 30 p. Washington DC.

Patt, A., P. Suarez and C. Gwata. 2005. Effects of seasonal climate forecasts and participatory *workshops among subsistence farmers in Zimbabwe. Proceedings of the National Academy of Sciences of the United States of America*, 102:12623–12628.

Pieri, C. 1989. *Fertilite de terres de savane. Bilan de trente ans de recherché et de developpmenet agricoles au sud du Sahara.* Ministere de la cooperation. CIRAD. Paris, France. 444 p.

Robertson, G.P. 2004. Abatement of nitrous oxide, methane and other non-CO_2 greenhouse gases: the need for a systems approach. In: C.B. Field and M.R. Raupach (eds.), *The global carbon cycle: Integrating Humans, Climate, and the Natural World*, SCOPE 62, Island Press, Washington D.C., pp. 493–506.

Roudier, P., B. Sultan, P. Quirion and A. Berg. 2011. The impact of future climate change on West African crop yields: What does the recent literature say? *Global Environmental Change.* 21(3):1073–1083. ISSN 0959 3780.

Sanchez, P.A. 1995. Science in agroforestry. *Agroforestry Systems*, 30:5–55.

Sanchez, P.A. 2002. Soil fertility and hunger in Africa. *Science*, 295:2019–2020.

Sanchez, P.A., R.J. Buresh and R.R.B. Leakey. 1999. Trees, soils and food security. *Philosophical Transactions of the Royal Society of London, Series B*, 353:949–961.

Schlesinger, W.H. 1999. Carbon sequestration in soils. *Science*, 284:2095.

Schroeder, P. 1994. Carbon storage benefits of agroforestry systems. *Agroforestry Systems*, 27: 89–97.

Scholes, R.J. and M.R. van der Merwe, 1996. Sequestration of carbon in savannas and woodlands. *The Environmental Professional*, 18:96–103.

Scholes, R.J. and R. Biggs. 2004. *Ecosystem services in southern Africa: a regional assessment.* CSIR, Pretoria, South Africa.

Smith, J.B. and S.S. Lenhart. 1996. Climate Change Adaptation Policy Options. *Climate Research*, 6:193–201.

Smit, B. and M.W. Skinner. 2002. Adaptation options in agriculture to climate change: A typology. *Mitigation and Adaptation Strategies for Global Change*, 7:85–114.

Swaine, M.D. and J.B. Hall. 1983. Early succession in cleared forest land in Ghana. *J. Ecology*, 71:601–627.

Taonda S.J.B., A. Barro, R. Zoumgore, B. Yelemou and B. Ilboudo. 2003. Review article of the Inter CRSP activities of Burkina Faso, 24 pp.

UNECA (United Nations Economic Commission of Africa), 2002. *Harnessing Technologies for Sustainable Development. Economic Commission of Africa Report*, pp. 33.

UNEP (United Nations Environment Programme). 2006. African regional implementation review for 14th session on the commission on sustainable development (SCD-14): Report on climate change. Available online: http://www.un.org/esa/sustdev/csd/csd14/ecaRIM_bp1.pdf

UN (United Nations). 2007. The Millennium Development Goals 2007. Available online: http://www.un.org/millenniu/pdg/mdg2007.pdf

Vagen, T.-G, R. Lal and B.R. Singh. 2005. Soil carbon sequestration in Sub-Saharan Africa: A Review. *Land Degradation and Development*, 16:53–71.

Vlek, P.L.G. 1990. The role of fertilizers in sustaining agriculture in sub-Saharan Africa. *Fertilizer Research*, 26:327–339.

Vlek, P.L.G., G. Rodríguez-Kuhl and R. Sommer. 2004. Energy use and CO_2 production in tropical agriculture and means and strategies for reduction or mitigation. *Environment, Development and Sustainability*, 6(1–2):213–233.

Vlek, P.L.G., Q.B. Le and L. Tamene. 2010. Assessment of Land Degradation, Its Possible Causes and Threat to Food Security in Sub-Saharan Africa. In: Rattan Lal and B.A. Stewards, (eds.), *Advances in Soil Science — Food Security and Soil Quality*. CRC Press, Boca Raton, pp. 57–86.

Wilson, G.F., R. Lal and B.N. Okigbo. 1982. Effects of cover crops on soil structure and on yield of subsequent arable crops grown under strip tillage on an eroded Alfisol. *Soil and Tillage Research*, 2:233–250.

World Bank/GEF, 2005. *Western Kenya Integrated Ecosystems Management Project*. Project Document.

Zougmore R, A. Mando, J. Ringersma and L. Stroosnidjer 2003a. Effect of combined water and nutrient management on runoff and sorghum yield in semiarid Burkina Faso. *Soil Use and Management*, 19(3):257–264.

Chapter 7

Australia and New Zealand Perspectives on Climate Change and Agriculture

Peter J. Thorburn[1], Michael J. Robertson[2], Brent E. Clothier[3], Val O. Snow[4],
Ed Charmley[5], Jon Sanderman[6], Edmar Teixeira[7], Robyn A. Dynes[4],
Alistair Hall[3], Hamish Brown[7], S. Mark Howden[8], and Michael Battaglia[9]

[1] *CSIRO Climate Adaptation and Sustainable Agriculture Flagships,
GPO Box 2583, Brisbane Qld 4001, Australia*

[2] *CSIRO Climate Adaptation and Sustainable Agriculture Flagships,
PMB5, Wembley PO, WA 6913, Australia*

[3] *Plant & Food Research, Private Bag 11-600, Manawatu Mail Centre,
Palmerston North, 4442, New Zealand*

[4] *AgResearch — Lincoln Research Centre, Private Bag 4749,
Christchurch 8140, New Zealand*

[5] *CSIRO Sustainable Agriculture Flagship,
PMB Post Office Aitkenvale Qld 4814, Australia*

[6] *CSIRO Sustainable Agriculture Flagship, PMB 2,
Urrbrae SA 5034, Australia*

[7] *Plant & Food Research, Private Bag 4704,
Christchurch Mail Centre, Christchurch 8140, New Zealand*

[8] *CSIRO Climate Adaptation Flagship, GPO Box GPO Box 284,
Canberra, ACT 2601, Australia*

[9] *CSIRO Sustainable Agriculture Flagship, College Road,
Sandy Bay Tas 7005, Australia*

[1] *peter.thorburn@csiro.au*

Introduction

Australia and New Zealand both make important contributions to world food trade through the high (>70%) proportions of their food production that is exported (Robertson, 2010). Further, the two countries are unusual amongst developed nations, as agriculture is a major part of their greenhouse gas (GHG) profile as

well as their economies. In developed countries, agricultural GHG typically contribute <10% of the national GHG inventory (Ministry for the Environment, 2011b).
However, the contribution is higher in Australia (14.6%; DCCEE, 2012a) and particularly in New Zealand (46.5%; Ministry for the Environment, 2011b), which has
the highest agricultural emissions of any developed country (Ministry for the Environment, 2011b). The corollary of this is that the agricultural sectors potentially play
an important role in mitigating GHG is these countries. This importance has been
recognised in government mitigation policies. Australia has the 'Carbon Farming
Initiative' (DCCEE, 2012b) which provides a voluntary mechanism for GHG abatement achieved in farming enterprises to be sold to other sectors in order to offset
their liabilities under the nation's carbon pricing mechanism (which did not initially
include agriculture) that comes into effect in mid-2012 (Clean Energy Regulator,
2012). New Zealand has an emissions trading scheme that will include agriculture
in 2015 (Emissions Trading Scheme Review Panel, 2011). Thus, the climate change
impacts on agriculture and the potential for the agricultural sector to adapt to climate change and mitigate GHG emissions from the sector are important in these
countries.

On top of the export-orientation of the agricultural sectors and their importance
in National Greenhouse Gas Inventory (NGGI), the two countries also share other
similarities in the agricultural sector. Production in both countries is dominated
by family-owned and operated farming businesses (Robertson, 2010), that operate
under very low levels (e.g. <5%) of government support (OECD, 2010). Despite
being an important contributor to global food trade, the agriculture share in the gross
domestic product of both countries is relatively small — 5% and 3% in New Zealand
and Australia respectively (Robertson, 2010). Also, agriculture productivity and
profitability in both nations has always been sensitive to variation in climatic factors.
Consequently, there has been considerable evolution of both policy and practice in
dealing with climate variability. The climate is anticipated to change considerably in
both nations over the next decades (as outlined below). Thus, policies and practices
will need to adapt to the changing environment so as to reduce emergent risks and
take advantage of opportunities (Easterling et al., 2007).

Despite the similarities, there are significant differences in the agricultural sector between the two countries (Robertson, 2010). The agricultural sector dominates the landscape in New Zealand (37% of the land area), whereas it occupies
a small fraction (6%, excluding extensive grazing) of Australia. The proportion of
New Zealand's exports coming from agriculture (48%) is much higher than Australia's (14%). The profile of agricultural exports is also different in the two countries.
The main products in Australia are grains and meat, each accounting for 25% of
agricultural exports, with other products such as dairy, wool, cotton, wine, sugar etc.,
contributing to <10% of exports. In contrast, dairy (60%) and meat (30%) account

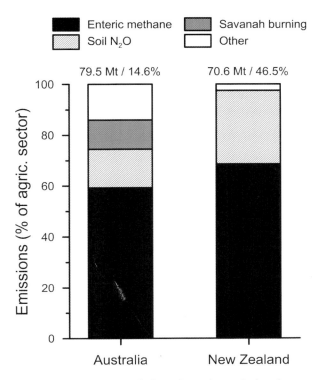

Fig. 1. The profile of greenhouse gas emissions from the agricultural sector in Australia and New Zealand (DCCEE, 2012a; Ministry for the Environment, 2011). The total emissions from agriculture and the proportion of the country's total greenhouse gas emissions coming from agriculture is also shown.

for the bulk of New Zealand's agricultural exports. Some of these differences are reflected in the GHG emission profiles of the agricultural sector in the two countries. Both are dominated by enteric methane. However, the proportionate contribution of nitrous oxide (N_2O) emissions from agricultural soils in New Zealand is approximately double that in Australia (Fig. 1). These emissions, being driven largely by animal excreta deposited onto pastures during grazing or from the indirect effects after nitrogen (N), are lost from excretal patches in the pastures through leaching or runoff (de Klein *et al.*, 2001). Another difference is the GHG emission from savannah burning (Fig. 1), which is important in Australia due to the vast expanse of dry tropical rangelands in the north of the continent, a biome absent in New Zealand.

Thus, Australia and New Zealand occupy a somewhat unusual place in the world with respect to agriculture in that they have developed economies but a GHG emissions profile more akin to those found in developing countries. In occupying that place, they have some similarities in their agricultural sectors, as well as some notable differences. There is potentially much to learn about both the opportunities for, and obstacles to, climate change adaptation and mitigation from considering these two countries. In this chapter, we outline adaptation and mitigation opportunities for agriculture in Australia and New Zealand and identify obstacles to and synergies between adaptation and mitigation in order to better understand the potential for reducing the impacts of climate change on agriculture.

Impacts of Climate Change on Agriculture

Climate change will have impacts on Australia's and New Zealand's agriculture through rising atmospheric CO_2 concentrations, changing amounts and patterns of rainfall, and higher temperatures. In Australia, median temperature rises by 2070 are predicted to be 1.8°C with low emissions scenarios and 3.4°C with high emissions scenarios (Hennessy et al., 2010). Temperature increases will be greater in inland regions. Rainfall is likely to decrease by up to 20% in southern Australia, although there is no clear trend for northern parts of the country. The number of droughts and intense rainfall events are likely to increase. Stokes and Howden (2010) have recently collated the impacts of climate change on Australian agriculture. That collation suggests that the general drying trend in southern Australia will reduce yields of rain-fed crops and productivity of pasture-based animal systems, and possibly reduce the availability of irrigation. Also, higher temperatures are likely to negatively impact crop and animal production (e.g. poorer grain filling, increased animal heat stress and lower milk yields). However, higher temperatures may reduce low temperature limitations (e.g. frost risk) in some areas and create new opportunities for agriculture through adaptations like earlier or later planting or changes in the geographic limits for some crops. In the rangelands of northern Australia, where rainfall is expected to increase, low soil fertility will constrain the growth benefit of higher atmospheric CO_2 concentrations and thus only modest increases in plant production are likely (Stokes and Howden, 2010).

Climate change impacts in New Zealand also include higher temperatures and changed rainfall patterns (Ministry for the Environment, 2001; Reisinger et al., 2010). These effects will not be uniform, with temperature increases greater in the North Island than the South Island, and rainfall higher in western parts compared to the eastern parts of the islands. New Zealand's pastoral farming systems are more sensitive to changes in the summer rainfall than winter. A case study by Dynes et al. (2010) on the likely impacts in dairy farming in regions of the lower west coast of the North Island found that pasture growth in the medium term (2030) would likely rise by about 12% due to increased winter temperature and rainfall, which would arise primarily through an earlier spring flush. Similar changes in pasture growth have been identified in Australian intensive livestock industries (Miller et al., 2010). However, the positive effects of further increases in temperature will eventually be constrained by other limitations such as nutrient availability and, in Australia, water availability. In New Zealand, there will be a decrease in pasture quality as higher temperatures favour the dominance of C_4 plants in the pasture.

For cropping in New Zealand, cereals and winter forages (e.g., brassicas and winter cereals) are likely to benefit through reduced current low temperature limitations on growth, particularly during autumn and early spring. Also, the CO_2-fertilization effect is expected to increase photosynthesis rates and increase water-use efficiency, particularly for temperate-zone crops (Jamieson and Cloughley, 1998). On the other hand, warmer temperatures may accelerate plant development and shorten growth cycle lengths in other species such as maize, potato and pea crops, partly offsetting the benefits from CO_2 fertilization.

Perennial horticultural crops in both countries will be most affected by changing temperatures due to effects on chilling requirements, floral abundance and fruit ripening. Reduced winter precipitation for the key horticultural regions on the east-coast of New Zealand could reduce groundwater recharge and irrigation availability, especially in summer months when irrigation is critical to alleviate water stress (Reisinger *et al.*, 2010). The impacts of reduced irrigation availability could be magnified by the other effects of climate change by 2050, which will result in increased crop water requirements. Similar impacts have been projected for Australian perennial horticultural crops (Webb and Whetton, 2010).

Adaptations to Climate Change

Adaptation to climate change is undertaken across the whole continuum of farm management decisions. Many management decisions in agriculture are tactical (Stafford-Smith *et al.*, 2011), in that they are made with short lead times, their consequences are felt for relatively short durations, and the management decisions can be revisited/adjusted frequently. Other decisions are more strategic, take considerable planning and have impacts over longer (e.g. decadal) time scales (Park *et al.*, 2012; Rickards and Howden, 2012). These different adaption pathways have been classified on the basis of whether they are incremental or transformational (Howden *et al.*, 2010; Rickards and Howden, 2012). It has been argued that where there is urgency for agriculture to adapt to climate change, early transformative action can sometimes be rewarded (Howden *et al.*, 2007). The need for transformative change may be driven by increasing rates of climate change that will increase the likelihood that environmental thresholds for current production systems, such as maximum temperatures or water availability, will be breached and the current system becomes increasingly less viable. This is in contrast to incremental change where relatively minor adjustments are made more frequently, allowing for current farming objectives to be met under changed conditions. If climate change in Australia and New Zealand remains small relative to season-to-season variations, the impacts of climate change will be modest and gradual in the minds of most farmers in comparison to the other 'shocks' agriculture has experienced in the recent past (Robertson, 2010). In this scenario, adaptations will be made incrementally, within normal responsive management paradigms. This is especially the case given that changes in varieties and breeds, production technologies and farming systems have combined to increase many measures of agricultural production by 1–3% per year for the last 30 or so years (Robertson, 2010). These improvements are expected to offset many of the predicted negative climate change impacts over the coming decades. In reality, though there is a continuum between the two types of adaptation and farmers' decision, pathways often cycle between incremental and transformational phases (Ash *et al.*, 2008; Park *et al.*, 2012).

Below, we highlight some examples of both incremental and transformational adaptations in Australia and New Zealand agriculture.

Incremental adaptation

Incremental and transformative adaptation responses have also been classified in terms of their type and extent (Stafford-Smith *et al.*, 2011). The class of response named 'same type and extent' includes management options currently being used by farmers and the appropriate adaptation response is to draw on them at a similar frequency and extent as present. Thus, these are a suite of practices that are currently considered by farmers, and may be adopted even in the absence of climate change. In many farming systems, such practices have been adopted in response to recent droughts, periods of increased climate variability, or changed temperatures.

In broadacre rainfed cropping in low rainfall zones, which include much of the cereal producing regions of Australia, 'same type and extent' responses include the broad suite of standard tools used by farmers to manage seasonal variability, such as moving to early maturing crop varieties, earlier sowing, residue retention and minimum tillage (Howden *et al.*, 2010). These practices will potentially allow farmers to adapt to climate change associated with ∼2°C increase in temperature.

Perennial horticulture is already an intensive and highly managed system of primary production. Through pruning, both in summer and winter, and by manipulating flower numbers, growers seek a balance between the vegetative and reproductive parts of the plant to achieve the right fruit sizing and fruit quality characteristics. The projected consequence of climate change scenarios by 2050 will primarily be a rise in vegetative vigour that will simply require more pruning, especially in winter, to establish the desired balance between vegetative and floral growth, provided that adequate irrigation is available. It is predicted that there will only be a very small, if any, change in fruit or berry mass as a result of climate change, even under high climate-change scenarios by 2050 (Clothier *et al.*, 2012).

'Same type same extent' responses may also occur in irrigated crops. Adopting well-established practices to improve water use efficiency would be a valuable response in many situations (Gaydon *et al.*, 2010; Bange *et al.*, 2010). For example, deficit irrigation strategies, currently practiced in many horticultural crops to limit vegetative vigour and enhance fruit and berry quality, could be fine-tuned to help meet the irrigation needs of these crops under future climates where water resources could be more limiting (Mpelasoka *et al.*, 2001).

In semi-intensive grazing systems in Australia, there has been a move towards the use of perennial plants (e.g. saltbush, lucerne) to provide valuable out-of-season forage for livestock (John *et al.*, 2005), a trend that is likely to continue as the climate changes. Likewise, in New Zealand's intensive pastoral farming systems, adapting to the changed timing and quality of pasture production may require calving earlier, increasing stocking rates and increasing the amount of feed-conserved and fed-out. These management changes might be viewed by many farmers as 'business as usual', but could be profitable adaptions under climate change projections to 2080 (Dynes *et al.*, 2010).

In addition to the 'same type and extent' responses, there is a class of responses called 'same type with different extent' (Stafford-Smith *et al.*, 2011). These responses include management options currently being used by farmers that may increase in scale, intensity or frequency, in response to climate change.

One example of this adaptation response is the use of fallowing by dryland grain growers in Australia. Fallowing is practiced sporadically under current conditions, but may increase in frequency and scale as rainfall declines in future climates and the farming systems change (e.g., declining sheep numbers) in response (Oliver *et al.*, 2010). Another example is in the fully-irrigated rice-based cropping systems in inland southern Australia. In response to severe droughts and substantially reduced irrigation water availability over 2002–2008, these farmers moved away from cropping systems requiring high amounts of irrigation, (e.g. traditional flooded rice) to systems that had a lower irrigation requirement, such as increased fallow land, rainfed or partially-irrigated crops (e.g. wheat), or rice systems based on aerobic or wet-dry conditions (Gaydon *et al.*, 2011). Many of these changes are part of the conventional management strategies employed by these farmers, but their frequency is likely to increase under projected climate change (Connor *et al.*, 2008).

In horticultural crops, there is a trend to erect shade netting to limit hail damage. Netting can also protect some crops from high temperatures. So, with the occurrence of extreme temperatures predicted to rise, it is likely that the installation of shade netting and overhead irrigation for fruit cooling (where cost-effective water supplies are available) will increase as a climate change adaptation (Webb and Whetton, 2010).

The strategy of New Zealand's horticulture industries of regularly introducing novel new cultivars to maintain a competitive edge in the market place also provides an important pathway for incremental adaptation to climate change. In the kiwifruit industry, many cultivars currently in the breeding pipeline produce much higher flower numbers following warm winters than does the dominant 'Hayward' cultivar, and some of these cultivars may well incrementally come to dominate the industry in the future (Clothier *et al.*, 2012).

Transformational adaptation

There are many factors that cause transformation in agriculture; some examples being the development of an irrigation scheme or processing infrastructure (like a cotton gin or sugar factory) that allows new crops to grow in particular regions. In these cases, a strategic decision is made, and capital is invested in the scheme or infrastructure, leading to new farming systems, transport and other components of the value chain. Substantial climate change may well be another cause for transformation of farming systems, and thus an additional adaptation strategy available to farmers (Howden *et al.*, 2007; 2010; Rickards and Howden, 2012). Transformation adaptation was less well studied than incremental adaptation as there are fewer clear examples of farmers 'transforming' their agricultural systems in response to climate change. However, now there are examples, both real and anticipated, starting to occur.

One example is efforts by the Australian peanut industry to establish a 'greenfield' production base in Katherine, Northern Territory, to protect against climate-driven poor production in traditional areas. Decades of below-average rainfall in Kingaroy, the 'traditional home' of peanut production in Australia, resulted in declining production and, particularly, increased contamination with the highly carcinogenic and immunosuppressant chemical aflatoxin (Chauhan *et al.*, 2010). In the belief that the climate change rather than climate fluctuations was causing the decline in production, the main peanut processing company (the Peanut Company of Australia, PCA) purchased land and established farms in the Katherine region where there were suitable soils and adequate irrigation water supplies. Unfortunately, production costs were higher than anticipated, and yields lower, so PCA withdrew from Katherine after ~5 years. The 'failure' of this venture illustrates the risks inherent in transformational adaptation. Other than the financial difficulties encountered, there may have been other problems with developing intensive cropping in the Katherine region (Thorburn *et al.*, 2012b): there was potential for substantial N leaching to groundwaters from peanut-based crop rotations, which sustain dry season flows and valuable ecosystems in the regions' rivers, unless N fertiliser inputs were carefully managed. Also, future climates in the Katherine region may be less favourable to agricultural production than they are now, so the long-term advantages sought by PCA may not have been as great in future decades as hoped. Accessing labour for an expanded production base in Katherine may also have been difficult as traditional peanut farmers in Kingaroy are unlikely to be mobile because of their attachment to place and occupation. Clearly, this example illustrates that many factors are involved in transformation, many of which are beyond the control, or outside the experience, of an individual farmer or industry body. Thus, government policy and community support will be an important component in establishing new agricultural industries.

Another currently more successful example of transformation is in the Australia wine industry (Park *et al.*, 2012). Both wine companies and grape growers are investing in vineyards in currently cool areas, in anticipation of more favourable climatic conditions in coming decades and less suitable temperatures in current grape growing regions. Also, some wineries are becoming reluctant to purchase grapes from regions with increasingly unfavourable climatic conditions.

While there are only two examples given above, there are many indications that the need for transformational adaptation may increase in Australia and New Zealand. Deciduous fruit trees and vines require a certain amount of accumulated chilling, or vernalisation, to break winter dormancy. In New Zealand, rising temperatures may lead to inadequate chilling, which would result in prolonged dormancy, poorer fruit quality, lower yield, and potentially higher costs, especially in kiwifruit (Kenny, 2008). While new cultivars may provide the solution to this problem, if production of the currently dominant 'Hayward' kiwifruit variety were to continue, then kiwifruit growers may also need to transform; that is, move 'Hayward' production to more suitable regions, such as the inland basins of both islands. Concomitantly, there would be a need to establish different agricultural systems in the existing 'Hayward' kiwifruit-producing areas.

In Australia, broadacre irrigated industries may be highly sensitive to climate change because of reduced water availability, both in terms of the amount of water availability and the higher price for water as it is traded to higher value industries. In these circumstances, broadacre cropping may be replaced by (Cullen *et al.*, 2010):

- High value irrigated crops (e.g. pome fruit and citrus), that can 'justify' higher cost of irrigation water.
- Resilient irrigated crops (e.g. olives, jojoba, pomegranates, capers, Australian native bush foods or dates), that have lower water requirements and/or can tolerate periods when irrigation is not available.
- Rainfed farming systems, based on both traditional (e.g. cereals, grazing) and new or emerging crops (e.g. crambe, quinoa, tepary bean, Australian native grass crops etc.).

While these examples represent 'pure' transformational strategies, adaptation often follows cycles of incremental and transformational change (Ash *et al.*, 2008). This relationship can be considered as two concentric action-learning cycles; the inner cycle focusing on incremental adaption and the outer cycle on transformative adaptation (Park *et al.*, 2012). Some external driver or opportunity triggers the move from one cycle to the other, as exemplified by the decision to establish a new peanut production base in Katherine. Some transformation may not be the result of strategic decisions, but the unintended (although possibly predictable) consequence. This situation is illustrated by the case of the decline of sheep numbers as part of mixed farming enterprises in the low rainfall regions of Western Australia. Droughts trigger de-stocking, which is followed by reduced investment in livestock infrastructure (fences, watering points, shearing sheds) and loss of specialist livestock management skills. These changes are exacerbated by the need (or desire) to invest in other farm activities, such as cropping. These changes introduce inertia to re-stocking when good seasonal conditions return, and the inertia can be reinforced by high livestock prices due to high demand for stock in this circumstance. Thus, a series of small incremental changes eventually leads to a transformation from one state to another.

Mitigation Potential

GHG emissions from the agricultural sector in both countries are dominated by enteric methane, with soils N_2O emissions being the second largest contributor (Fig. 1). As described above, emissions from savannah burning are important in Australia as are the aggregate of emissions from other agricultural sources (e.g. manure management, rice cultivation, field burning of agricultural residues). In this section, we will focus on enteric methane and soil N_2O emissions because of their dominance. We will also consider the potential for sequestration of carbon (C) in soils because of the possible role of that process in GHG abatement.

N_2O from soils

Nitrous oxide emissions from soils in Australia's crop lands vary widely (Fig. 2). Emissions (relative to fertiliser N applications) are substantially lower than IPCC defaults in cereal and oilseed grain crops in southern Australia (i.e. Wheat-Vic, Wheat-WA, Canola-WA; Fig. 2). Conversely, emissions are substantially more than default values in sugarcane production systems. The highest emissions come from organic soils (i.e. soil organic C ~10%; Denmead *et al.*, 2010). However, these soils only account for a small area of sugarcane production, and overall emissions from sugarcane systems are likely to be closer to those measured at other sites

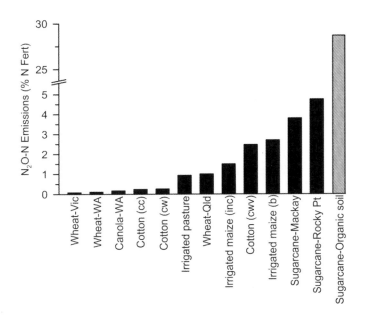

Fig. 2. Nitrous oxide emissions (relative to N fertiliser additions) from cropped soils in Australia. NB: The hatched bar is plotted against the broken scale. [Data sources: Wheat-Vic, Galbally *et al.* (2005); Wheat-WA, Barton *et al.* (2008); Canola-WA, Barton *et al.* (2010); Cotton-cotton (cc), Galbally *et al.* (2005); Cotton-wheat (cw), Galbally *et al.* (2005); Irrigated pasture, Galbally *et al.* (2005); Wheat-Qld, Wang *et al* (2011); Irrigated maize (inc), Galbally *et al.* (2005); Cotton-wheat-vetch (cwv), Galbally *et al.* (2005); Irrigated maize (b), Galbally *et al.* (2005); Sugarcane-Mackay, Denmead *et al.* (2010); Sugarcane-Rocky Pt, Allen *et al.* (2010); Sugarcane-Organic soil, Denmead *et al.* (2010).]

(Thorburn *et al.*, 2010). The high emissions from sugarcane result from the crop having high applications of N fertiliser in combination with the crop growing in warm and wet conditions that are conducive to rapid soil N and C cycling (Thorburn *et al.*, 2010). The low emissions in southern Australia arise from a combination of factors. In some soils, water contents rarely get above 60% water-filled pore space, so nitrification is the main process producing N_2O (Barton *et al.*, 2008; Galbally *et al.*, 2010). Low soil N and C, and winter being the main time of rainfall and crop growth, may well be additional contributing factors. As water supplies become more plentiful, as with irrigated crops and crops grown in more humid climates (e.g. Wheat-Qld and sugarcane; Fig. 2), emissions are higher, although soil organic C can constrain emissions. For example, in different irrigated cotton rotations, continuous cotton and cotton-wheat rotations had lower emissions than cotton-wheat-vetch. Presumably, the green manure vetch crop, which is a legume, provides a source of C and N to stimulate emissions in the rotation (P. Grace, pers. comm.). Including legumes in cereal crop rotations is also likely to increase N_2O emissions (Huth *et al.*, 2010).

The potential benefits of mitigation are likely to be the greatest in the farming systems with the highest emissions, (i.e. sugarcane). Approximately 980 kt of fertiliser N was applied in Australia in 2009–2010, which included ~79 kt applied to sugarcane (DCCEE, 2011). Assuming (from Fig. 2) that N_2O emissions average 4% of N fertiliser applications in sugarcane and 0.5% in other crops, emissions from sugarcane probably accounted for ~33% of total N_2O emissions driven by fertiliser application. Conventionally, large amounts of N fertiliser have been applied to sugarcane to ensure maximum production (Thorburn et al., 2011), a situation that has resulted in large N surpluses (the difference between N fertiliser applied and N removed in crop off-take) and the potential for losses of N to the environment (Thorburn and Wilkinson, 2012). Not surprisingly, N_2O emissions are strongly related to N fertiliser applications (Thorburn et al., 2010). Also, there is considerable scope (e.g. >30%) for reducing N fertiliser applications through adoption of 'agri-environmental' management practices (Thorburn et al., 2011). Simplistically, the reduction in N fertiliser applications that would result from universal adoption of these practices could reduce N_2O emissions by up to 20%. Thus, there would be considerable N_2O mitigation benefit from improving N fertiliser management in Australian sugarcane production systems. Interestingly, there are currently government policies aimed at improving water quality that are encouraging farmers to improve N management practices over the majority of sugarcane producing areas (Thorburn and Wilkinson, 2012). There are similar water quality policies in New Zealand (Ministry for the Environment, 2011a) that may incidentally also produce an N_2O mitigation benefits.

There are other crops in Australia with high N surpluses, which are amenable to management practices that reduce N fertiliser applications. Cotton is an example (Rochester, 2010; Thorburn and Wilkinson, 2012). Adoption of improved N fertiliser management practices in these crops will help mitigate N_2O emissions. However, the extent of the mitigation will be less than in sugarcane, because both N_2O emissions (relative to fertiliser applications) and the scope to reduce N fertiliser applications are lower than in sugarcane production.

There is less information about N_2O emissions from soils in New Zealand's arable lands. A New Zealand-specific emission factor of 1% of N fertiliser applications is used in the country's NGGI (Ministry for the Environment, 2011b). Given that N fertiliser applications can be quite high on some crops, (e.g. ~200 kg ha^{-1} applied to wheat in the Canterbury region (MAF, 2011)), emissions from arable systems can be significant. However, the small proportion (i.e. 3.2%; WorldStat, 2007) of New Zealand's agricultural lands devoted to arable cropping means that a limited share of total N_2O emissions comes from these systems and so their overall mitigation potential is limited.

Emissions of N_2O from soils in intensive pastoral systems can be substantial (de Klein *et al.*, 2001; de Klein and Eckard, 2008). In New Zealand, about 75% of the N_2O emitted directly or indirectly from soils comes from animal excreta deposited onto pastures during grazing or from the indirect effects after N is lost from excretal patches in the pastures through leaching or runoff (de Klein *et al.*, 2001). The relative importance of N_2O emissions from intensive pastoral systems in Australia is lower because of the smaller proportion of Australia's agricultural sector occupied by these systems. The emissions profiles from intensive pastoral farms in Australia and New Zealand mean that the mitigation options often used for arable or housed-livestock farms (e.g. Table 9 of de Klein *et al.*, 2001) will have limited effectiveness. The mitigations that will be most effective involve manipulation of N_2O losses from urine patches.

The primary mitigation option of N_2O losses from urine patches in New Zealand is nitrification inhibitors (de Klein *et al.*, 2001; Di and Cameron, 2002). The inhibitor dicyandiamide (DCD) is available in several commercial formulations for application to soil and this practice is included in New Zealand's NGGI (Clough *et al.*, 2008). At a farm scale, its price and application practicalities restrict its use to intensive dairy farming systems. Nitrification inhibitors work by slowing the biological oxidation of ammonium to nitrate. This can have several effects: Firstly, the retention of the N in the NH_4^+ form means that the cation can be retained on the particle surfaces (Di and Cameron, 2002) and so can reduce leaching provided there is sufficient surface charge to retain the high concentrations (Bryant *et al.*, 2007). Secondly, the reduction in N leaching will reduce indirect N_2O emissions. Also, retention of the N in the NH_4^+ form will reduce the NO_3^- concentration and therefore the source for N_2O production. However, the retained N will increase the N availability to the pasture roots and so increase growth in these generally N-deficient pastures (Di and Cameron, 2002). A likely management response to the increased growth would be to increase animal intake/silage production to use the additional pasture and maintain good pasture quality. The potential outcome of such a management response will be an increase in the urinary N return to the paddock which may increase total N_2O emissions (Bryant *et al.*, 2007), although the amount of N_2O emitted per unit of production (i.e. the emissions intensity) would be reduced. Increased stocking in this scenario would also increase total methane GHG emissions (even at reduced emissions intensity), further offsetting the benefits sought from application of nitrification inhibitors. Thus, N_2O mitigation options should consider the whole N cycle and would generally aim to increase the N efficiency of the farms (de Klein *et al.*, 2001; Luo *et al.*, 2010b).

Other N_2O mitigation options that have been considered for pastoral systems include: manipulation of the soil pH to favour the production of N_2 rather than N_2O

from denitrification (van der Weerden *et al.*, 1999); selectively removing the grazing ruminants from the pasture (e.g., to feed pads where the effluent can be managed) at times of the year when the potential denitrification losses from urine patches is high (de Klein *et al.*, 2006); and manipulating the feeding of animals with low-N feeds (de Klein *et al.*, 2001) or supplements such as condensed tannins (Burggraaf and Snow, 2010) to reduce the urine patch N concentration and the denitrification potential. Some of these suggested options have positive co-benefits, but none are currently cost effective at the farm scale or are included in the NGGI.

Soil carbon

In Australia, conversion of native land to agriculture has resulted in decreases in soil organic C (SOC) in the order of 40–60% from pre-clearing levels (Luo *et al.*, 2010a). Recapturing even a small fraction of these legacy emissions through improved land management would represent a significant climate benefit with multiple potential co-benefits to farm productivity and profitability (Lal, 2004). For example, increasing Australia's current SOC stock of 19 Pg C (0–0.3 m; Grace *et al.*, 2006) by a mere 0.08% per annum would effectively mitigate Australia's annual GHG emissions, although there are substantial practical barriers to capturing this soil C.

Australian soils may impose unique constraints on the mitigation potential of SOC. Many soils in Australia are much older than those in the northern hemisphere. Thus, in many soils, weathering-derived nutrients have been exhausted (Chen *et al.*, 2009), salts have accumulated (Rengasamy, 2006), surface soils are coarse-textured, and there are pedogenic features (such as hardpans) that impose substantial limitations to agriculture (Chan *et al.*, 2003). These edaphic constraints, especially when combined with a generally less favourable climate, may act to reduce SOC sequestration potential in Australian soils.

In general, any management shift that increases C inputs to the soil should increase SOC or, at least decrease the rate of SOC loss (Table 1). While there is a strong theoretical basis for SOC stock increases under a number of potential management options, field evidence is generally sparse at best. The majority of the evidence for SOC stock changes in Australia comes from agricultural field trials, summarized in detail by Sanderman *et al.* (2010) and Luo *et al.* (2010a). These trials primarily focused on tillage practices, stubble and fertilizer management, fallowing, rotation complexity and a number of pasture improvements. Improved management of cropland and pastures generally resulted in relative gains between 0.1 and 0.5 t C ha^{-1} yr^{-1}, while conversion of cultivated land to permanent pasture resulted in relative gains of 0.3–0.6 t C ha^{-1} yr^{-1} (Fig. 3). These values are consistent with, but generally lower than those found for similar management improvements in the northern hemisphere (Hutchinson *et al.*, 2007; Smith *et al.*, 2008; West and Post, 2002).

Table 1.   Summary of major management options for sequestering carbon in agricultural soils (after Sanderman *et al.*, 2010).

Management	SOC benefit[a]	Conf.[b]	Justification
1. Shifts within an existing cropping/mixed system			
a. Maximizing efficiencies – (1) water-use (2) nutrient-use	0/+	L	Yield and efficiency increases do not necessarily translate to increased C return to soil
b. Increased productivity – (1) irrigation (2) fertilization	0/+	L	Potential trade-off between increased C return to soil and increased decomposition rates
c. Stubble management – (1) Eliminate burning/grazing	+	M	Greater C return to the soil should increase SOC stocks
d. Tillage – (1) Reduced tillage (2) Direct drilling	0 / 0/+	M / M	(1) Reduced till has shown little SOC benefit; (2) Direct drill reduces erosion and destruction of soil structure thus slowing decomposition rates; however, surface residues decompose with only minor contribution to SOC pool
e. Rotation – (1) Eliminate fallow with cover crop (2) Inc. proportion of pasture to crops (3) Pasture cropping	+ / +/++ / ++	M / H / M	(1) Losses continue during fallow without any new C inputs — cover crops mitigate this; (2) Pastures generally return more C to soil than crops; (3) Pasture cropping increases C return with the benefits of perennial grasses (listed below) but studies lacking
f. Organic matter and other offsite additions	++/+++	H	Direct input of C, often in a more stable form, into the soil; additional stimulation of plant productivity (see above)
2. Shifts within an existing pastoral system			
a. Increased productivity – (1) irrigation (2) fertilization	0/+	L	Potential trade-off between increased C return to soil and increased decomposition rates
b. Rotational grazing	+	L	Increased productivity, inc. root turnover and incorporation of residues by trampling but lacking field evidence
c. Shift to perennial species	++	M	Plants can utilize water throughout year, increased belowground allocation but few studies to date
3. Shift to different system			
a. Conventional to organic farming system	0/+/++	L	Likely highly variable depending on the specifics of the organic system (i.e. manuring, cover crops, etc...)

(Continued)

Table 1.   (*Continued*)

Management	SOC benefit[a]	Conf.[b]	Justification
b. Cropping to pasture system	+/++	M	Generally greater C return to soil in pasture systems; will likely depend greatly upon the specifics of the switch
c. Retirement of land and restoration of degraded land	++ +++	H	Annual production, minus natural loss, is now returned to soil; active management to replant native species often results in large C gains

[a]Qualitative assessment of the SOC sequestration potential of a given management practice (0 = nil, + = low, ++ = moderate, +++ = high).
[b]Qualitative assessment of the confidence in this estimate of sequestration potential based on both theoretical and evidentiary lines (L = low, M = medium, H = high).

Importantly, nearly all of the trials that had time series data (i.e. approximately 50% of those reviewed by Sanderman *et al.*, 2010) indicated that the relative gains reported in Fig. 3 were actually due to a reduction or cessation in the rate of SOC loss. The exception was the conversion to permanent pastures, which did show absolute gains in SOC over time (Sanderman and Baldock, 2010). While a reduction in the rate of loss of SOC is a real mitigation benefit, accounting for this type of mitigation in a national accounting system is difficult because the 'business as usual' scenario also needs to be tracked (Sanderman and Baldock, 2010).

Little information exists on management impacts on SOC stocks in the extensive rangelands in Australia. Due to a combination of grazing pressure from livestock and feral animals, a drying climate and shifting fire regimes, most of Australia's extensive rangelands are likely losing moderate amounts of SOC (Hill *et al.*, 2006). Halting or reversing this trend would represent a significant abatement potential; however, limited management options exist on these lands. GHG emissions are correlated with stocking rates (Howden *et al.*, 2003) and Hill *et al.* (2006) estimated that stocking densities would have to decrease to 20% or less of 1997 levels to obtain positive SOC outcomes.

Some of the results described above possibly indicate that SOC sequestration can be an attractive mitigation option. However, there are a number of practical considerations that greatly limit the real potential. While the agricultural sector covers approximately 60% of the land area of Australia, greater than 90% of this area (419 Mha) is used for low-density grazing of natural vegetation. Of the remainder, only 50 Mha of land that is actively managed, primarily for grazing of modified pastures (25 Mha) and various cropping systems (25 Mha) with less than 2 Mha under irrigation (BRS 2001/2002 National Land Use Summary; http://adl.brs.gov.au/

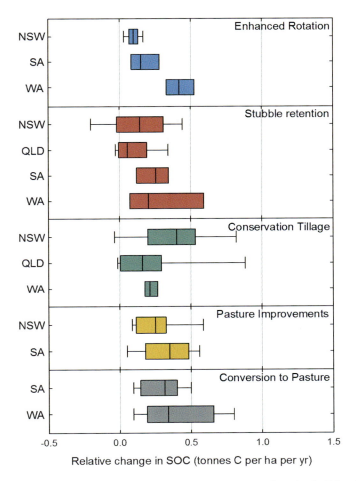

Fig. 3. Relative changes in SOC normalized to the median sampling depth (15 cm) in response to improved agricultural management (see Sanderman *et al.*, 2010, for details). Boxes represent the interquartile range, lines indicate median values and whiskers incorporate the 95% confidence intervals.

mapserv/landuse/docs/2001_02_Nat_Luse_Summary_Stats.pdf). This distribution of land use places an effective cap on the total mitigation potential of most of the management practices considered above. Assuming that improved management practices with a mean (± 1 S.E. mean) relative sequestration rate of 0.25 ± 0.11 t C ha^{-1} yr^{-1} can be adopted on all 50 Mha of actively managed land, it would result in a mitigation potential between 26 and 65 Mt CO$_2$-e yr^{-1} for the next few decades. However, this simple calculation is unrealistic and produces a very misleading picture of the real SOC mitigation potential, because there are biophysical, climatic, economic and sociopolitical constraints to SOC mitigation (Smith *et al.*, 2005) operating to varying degrees across the continent (Battaglia, 2011). Thus, the real sequestration potential is much lower.

Unlike the situation in Australia, there has not been a general assessment of the potential for SOC sequestration in New Zealand. However, the current Land Use and Carbon Analysis System program (http://www.mfe.govt.nz/issues/climate/lucas/) being undertaken by New Zealand's Ministry for the Environment aims to provide that information. Some of the most general information on SOC comes from a study by Schipper *et al.* (2010) who re-sampled 31 pasture soils down to 1 m. In general, New Zealand soils, especially in contrast to Australian soils, have high levels of soil C thus there is potential for gains and losses of soil C and N in New Zealand's pasture soils over decades depending on land-use. For example, over 27 years, the 0–30 cm zone of soils in flat, dairy pastures lost 0.73 ± 0.16 t C ha^{-1} y^{-1}, and 57 ± 16 kg N ha^{-1} y^{-1} (Schipper *et al.*, 2010). In contrast, no significant change was observed under flat dryland pasture grazed by sheeps and cattles. Soils (0–30 cm) on grazed hill country, however, had a rise in C and N of 0.53 ± 0.18 t C ha^{-1} y^{-1} and 66 ± 18 kg N ha^{-1} y^{-1}. Furthermore the pattern of losses and gains were shown to extend below the IPCC depth of 30 cm.

In horticultural production systems on andic soils in New Zealand, there can be substantial SOC at depth under deep-rooted trees and vines. For example, under kiwifruit vines only 65% of SOC in the top metre of soil was stored in the 0–0.3 m layer (Deurer *et al.*, 2010). Also, there can be considerable sequestration at depth. By comparing SOC under neighbouring 10 and 25 year old kiwifruit vines, SOC was found to be sequestered at 0.4 t C ha^{-1} yr^{-1}, and this sequestration occurred below 0.5 m soil depth (Deurer *et al.*, 2010). Whether these results are unique to andic soils with high SOC storage capacity or are more generally applicable to horticultural systems regardless of soil type has yet to be demonstrated. Management can also affect SOC in the topsoils of orchards. In a paired study of an organically- and conventionally-managed apple orchard by Deurer *et al.*, (2009), SOC (0–0.1 m) of the organic orchard was 3.8 kg C m^{-2} compared with 2.6 kg C m^{-2} in the conventionally-managed orchard. In addition, the organic orchard had greater and more inter-connected macroporosity, which would enhance soil gaseous exchange and potentially lower soil N$_2$O emissions, providing an additional measure of climate change mitigation. Surface soil properties and SOC were also affected by wheel trafficking, a common feature in orchards. In the organically-managed apple orchard mentioned above, SOC under the wheel track were higher than under the row as the moderate compaction by trafficking provided a physical protection mechanism for SOC (Deurer *et al.*, 2012).

In New Zealand's cropped soils, SOC stocks can be increased relative to conventional cultivation by implementing conservation agriculture techniques such as reduced-tillage (Aslam *et al.*, 2000). These techniques increase soil cover and minimise soil loss through erosion. In addition, land use change is relatively a major contributor of GHG emissions in arable cropping systems (MAF, 2011). By reducing the frequency of conversion of land, for example from long-term pasture into arable, it is therefore possible to mitigate emissions and maintain higher C stocks in soils.

Methane from livestock

Enteric methane from ruminant livestock represents 9.7 and 34.8% of national GHG emissions in Australia and New Zealand, respectively (Table 2). However, the total enteric emissions in Australia are approximately double those of New Zealand (DCCEE, 2012a; Ministry for the Environment, 2011b). In both countries, these emissions are significant and major research efforts are being devoted to mitigating these emissions.

There are marked contrasts between the livestock systems of Australia and New Zealand and these affect the strategies most likely to result in significant mitigation of methane emissions. In New Zealand, intensive dairy, sheep, beef and deer production occupy a large proportion of the landscape (Ministry for the Environment, 2011b), but a small total area compared to Australia. On the other hand, Australian livestock agriculture is dominated by extensive beef production (ABS, 2012) that occupies a large area. The sheep population in Australia, although large, has been reduced markedly in recent decades. The dairy industry accounts for only 5% of cattle in Australia (ABS, 2012), while New Zealand's 4.5 M dairy cows (LIC, 2011) account for over half of all cattle. In addition, dairy-industry-bred cattles contribute

Table 2.   Livestock numbers and enteric methane emissions in Australia and New Zealand, 2009.

	Australia	New Zealand[1]
Total dairy cattle (,000)	1,604[2]	5,861
Total beef cattle (,000)	28,809[2]	4,101
Total sheep (,000)	74,282[2]	32,384
Total GHG emissions (Gg CO_2 equiv.)	564,542[3]	70,564
Livestock enteric emissions (Gg CO_2 equiv.)	54,736[3]	24,650
Dairy cattle (Gg CO_2 equiv.)	6,129[3]	10,364
Beef cattle (Gg CO_2 equiv.)	37,765[3]	5,299
Sheep (Gg CO_2 equiv.)	10,544[3]	8,341
Deer (Gg CO_2 equiv.)	11.6[3]	588
Livestock enteric emissions (% of total)	9.7	34.8

[1] Ministry for the environment (2011).
[2] ABS (2012).
[3] DCCEE (2011).

an estimated 55% of beef production comprising bull beef, dairy-beef cross animals and cull cows (calculated from B + LNZ statistics 2011). The climate of New Zealand supports temperate grassland production and the focus of New Zealand research has been on mitigation of CH_4 from animals grazing such pastures *in situ* year round (Clark *et al.*, 2011).

Methane is produced by ruminants as an end product of methanogenic fermentation in the rumen and is eructed from the digestive tract via the mouth. Most methane production is attributed to the fermentative activity of archea rather than bacteria. Protozoa themselves are not methanogenic, but host methanogenic micro-organisms (McAllister and Newbold, 2008). The rumen is a strongly reducing environment and produces hydrogen (H_2) which is converted to methane by reaction with CO_2:

$$4H_2 + CO_2 \rightarrow CH_4 + 2H_2O.$$

Typically methane accounts for between 4 to 8% of ingested energy and therefore represents a loss of productive energy that otherwise could have gone into production of meat, milk, wool or a foetus. While early research focused on understanding the relationship between diet and methane to increase productivity (e.g. Moe and Tyrrell, 1979), current research activity in Australia and New Zealand is driven by the need to reduce total greenhouse gas emissions and the intensity of these emissions in cost-effective ways.

Mitigation options fall into four broad categories depending on their mode of action. These are, dietary additives, best management practices, genetic selection and rumen manipulation. Figure 4 demonstrates that some of these options can be realized in the short term, but would be expected to have quite modest impacts on methane mitigation. Others are unlikely to have an impact on methane emissions in

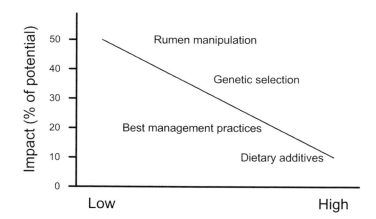

Fig. 4.   The relationship between the liklihood of a mitigation effect by 2030 and the magnitude of that mitigation effect.

the next 20 years or so, but future research developments can be expected to yield significant emissions abatement.

In Australia and New Zealand, research into dietary additives has been ongoing for several decades and there is already a range of products and compounds known to inhibit methanogenesis in the rumen (Grainger and Beauchemin, 2011). Australian research has focused on fat supplements which are known to reduce methane production by inhibiting methanogenic micro-organisms (Beauchemin *et al.*, 2008). They can be included in the diet as oilseed supplements or as extracts from various sources. Moate *et al.* (2011) reported on an ongoing research program investigating locally available fats for dairy cattle. These included canola, brewers' grains and hominy meal and they showed that for each increase of $10 \, \text{g} \, \text{kg}^{-1}$ dry matter (DM) in dietary fat concentration, methane emissions were reduced by $0.79 \, \text{g} \, \text{kg}^{-1}$ DM intake. Similar work feeding cottonseed to beef cattle is also ongoing (Grainger *et al.*, 2008; A. Klieve, pers. comm.).

Many plant species contain secondary compounds, such as tannins and saponins, that have anti-methanogenic effects, particularly when studied *in vitro* (Kamra *et al.*, 2008; Ramírez-Restrepo *et al.*, 2011). In Australia, research on tropical legumes has demonstrated the antimethanogenic activity of several species using animals in methane chambers (Kennedy and Charmley, 2012). They found that leucaena (*Leucaena leucocephala*), a widely grown tree legume in northern Australia, was the most effective of these and reduced methane emissions in cattle in proportion to the inclusion rate of leucaena in the diet. New Zealand research into tannin containing plants (e.g. Ramírez-Restrepo and Barry, 2005) has focussed on *Lotus coniculatus*, as well as more unusual sources such as willow (*Salix* spp.) trees (Ramírez-Restrepo *et al.*, 2010): however, no commercially viable forages have been identified for New Zealand farming systems. Long-term solutions to this challenge may come from plant breeding, for example cultivars with enhanced lipid content or genetically modified high tannin content white clover (Clark *et al.*, 2011). Current research in Australia is focusing on the antimethanogenic activity of some of the native eucalypt and shrub (*Atriplex* spp.) species which are rich in secondary plant compounds.

Management practices to mitigate methane emissions cover a range of options that increase productivity (e.g. Ulyatt *et al.*, 2002) and in so doing reduce the amount of methane emitted per unit of animal product; i.e. the emissions intensity (Howden and Reyenga, 1999; Charmley *et al.*, 2008). In the beef industry, for example, reducing days to market and increasing reproductive performance can have marked improvements in methane intensity (Hunter and Neithe, 2009). Likewise, in the dairy industry, management systems designed to take advantage of increased pasture quality can substantially decrease (e.g. by 19%; Eckard *et al.*, 2010) the emissions intensity of milk production. There are additional management strategies that could lead to a significant and sustainable reduction in emissions from dairy systems, including reducing reproductive wastage, increased genetic merit, better pasture quality and improved energy yield of supplements and stand-off pads (Beukes *et al.*, 2010; Buddle *et al.*, 2011). However, reducing methane emissions intensity often leads to increased total methane output in both beef (Rolfe, 2010) and dairy production (Eckard *et al.*, 2010) as producers increase animal numbers in response to increased efficiency of production. Globally, this is positive as methane emissions intensity of food production, worldwide, is reduced. However in countries where there is, or may be in the future (e.g. Australia and New Zealand), an impost associated with agricultural emissions, there is a potential disincentive towards increased production.

In extensive grazing systems, another way to reduce GHG emissions is to diversify land area from ruminant production into C sequestration enterprises such as forestry (Charmley *et al.*, 2011). In Australia, for example, retention of strips of native vegetation regrowth may be financially viable at C prices of AU$10 $(t\,CO_2\text{-}e)^{-1}$ (Donaghy *et al.*, 2010). However, undertaking C sequestration may be less attractive in more intensive livestock enterprises and/or where there are no external financial incentives for C sequestration and alternative approaches will be necessary. For example, collaborative modelling with farmers in intensive livestock enterprises in New Zealand (Dynes *et al.*, 2011) has shown that farmers can increase profitability, through higher total milk and/or meat production, and decrease the intensity of GHG emissions per unit of product. At the scale of the individual farmer, this is a win-win strategy, although total GHG emissions increase.

Genetic selection for low emissions phenotypes represents a permanent and long term mitigation strategy (Wall *et al.*, 2010). Research efforts in Australia and New Zealand to select cattle and sheep with high net feed efficiency (also known as low residual feed intake) has resulted in divergent lines for these traits (Hegarty *et al.*, 2007). Incidentally, selection for this trait also results in selection for high and low methane emissions intensity, with more efficient individuals having lower emission per unit of product. However, research has also attempted to select for lower methane emissions independent of animal efficiency. This approach implies that the host elicits an effect on the rumen microbial population through changed conditions in the rumen. Results to date are equivocal (Waghorn and Hegarty, 2011). Nevertheless, heritability values for methane production of between 20 and 30% have been reported for sheep (Pinares-Patino *et al.*, 2011).

Manipulation of the rumen ecosystem represents the most direct and potentially the most effective manner in which methane production could be reduced in the rumen, and active research is ongoing in both countries (e.g. Wright and Klieve, 2011). There are several options to consider that either inhibit the methanogenic micro-organisms or provide an alternate hydrogen sink to methane (McAllister and Newbold, 2008). Options to inhibit methanogens include genomic manipulation of existing mircobiota (Attwood and McSweeney, 2008), microbial production of anti-microbials such as bacteriocins and bacteriophages (McAllister and Newbold, 2008), and immunization against methanogens (Wright *et al.*, 2004). Although these techniques offer the potential to markedly reduce methane production in the rumen, their commercial application is well into the future.

The ruminant industry has a range of options for reducing methane emissions. These options will increase into the future, such that a significant impact on ruminant methanogenesis can be expected in the next 20 years. Depending on the production enterprise, some methods are more applicable than others. In intensive systems, such as dairying and feedlots, there are opportunities for intervention through diet, inoculation and dosing which are simply impractical in low cost (e.g. New Zealand sheep, beef and deer) and extensive systems, such as the pastoral industries of Australia. For the latter industries, plant breeding for novel anti-methanogenic forages, improvements in production efficiency and genetic selection will be the main options for mitigating enteric methane emissions from extensive livestock systems.

Challenges in Adaptation and Mitigation

There are a number of challenges in successfully adapting to climate change or mitigating GHG emissions in agriculture. Adaptation, particularly incremental adaptation, is well studied. Many of the strategies needed for adaptation are those that may well be already applied as part of adaptive management strategies in agricultural enterprises (Howden *et al.*, 2007; Meinke *et al.*, 2009; Conant, 2010; Stafford-Smith *et al.*, 2011). These studies also show how success in adapting to climate changes can be enhanced by appropriate institutional settings to support and include adaptive management in every day decision making. With climate change likely to be manifested as greater climate variability as well as changes in climate averages, adaptation will focus on risk management at paddock, farm and regional scales. Retrospective studies of how farmers and farm businesses have coped with extended dry periods will be instructive in understanding how they will cope with increased climate variability in the future. In particular, the mix of agronomic and farm business strategies that induce resilience will be important. As the impacts of climate change become more severe, the role of traditional production-based agricultural science in providing incremental adaptation options will decline and the importance of business, financial and social adaptation strategies will increase.

As noted above, examples of transformational adaptation are only just starting to emerge in Australia and New Zealand, as is probably the situation elsewhere. Thus, the requirements to enhance success in transformation are not as clear as they are for incremental adaptation, although some insights into the attributes that have to be considered in transformation are starting to emerge (Thorburn *et al.*, 2012b). However, as with incremental adaptation, the institutional settings and support are likely to be important (Park *et al.*, 2012).

One potential issue to emerge with transformation is the potential mal-adaptation that may result from intensification associated with transformation. In the peanut industry case study described above, establishing intensive crop production on a site previously occupied by native vegetation is likely to increase emissions of GHG from soils in that landscape (Thorburn *et al.*, 2012b). Whether these emissions are off-set to any extent by potentially lower GHG intensity and/or lower emitting agricultural activities in the area affected by climate change would be interesting to consider. Transformational adaptation is not the only response to climate change that may increase GHG emissions from the landscape. The potential for increased methane emission (per unit area) from livestock production to result from activities (i.e. higher quality feed and increased stocking) aiming to reduce emissions intensities was described above (Rolfe, 2010; Eckard *et al.*, 2010). What was not considered was the possible source of the higher quality feed. If the feed came from an expansion of forage cropping that was associated with N fertilisation and irrigation, the emissions may be further increased. These complex interactions in capturing mitigation benefits need to be placed in the context of the farm business and year-to-year climate variability experienced by Australian and New Zealand farmers. For example, different management options to reduce greenhouse gas emissions from a 'hill country' sheep and cattle farm in New Zealand's North Island were predicted to only have a small impact on whole-farm GHG emissions, ranging from a -3% to $+8\%$ (Dynes *et al.*, 2011). Most of the management systems considered in that study exposed the business to greater risk from market and climate variability.

Maintaining a balance between adaptation measures to climate change and other environmental and health concerns may also prove to be a challenge. For example, the New Zealand kiwifruit industry is currently dependent on application of hydrogen cyanamide (HC) to ensure adequate and synchronous flowering, particularly for the dominant 'Hayward' cultivar. With warmer winters following climate change, the need for such treatment will increase, but concerns about the human health impacts of HC (Schep *et al.*, 2009) may well lead to it being banned. A search for alternatives technologies is currently under way, but none has yet proved to perform as well as HC.

Not all interactions between adaptation and mitigation are negative (Smith *et al.*, 2007). As noted above, sequestering C in agricultural soils could potentially make a substantial contribution to mitigating Australia's GHG emissions, and the Australian Government has introduced policy to provide a mechanism whereby farmers can be paid for GHG abatement arising from sequestering C in soils (DCCEE, 2012b). Aside from providing GHG mitigation benefits, many of the management practices that will increase C sequestration in soils (e.g. Fig. 3) are also positive adaptations to climate change. Crop residue conservation and reduced-tillage or no-tillage are good examples because of the soil moisture conservation and soil fertility benefits associated with these practices that will help overcome the impacts of reduced rainfall projected in many parts of Australia (Howden *et al.*, 2010).

The benefits of crop residue conservation and reduced-tillage or no-tillage have long been recognised in Australia (e.g., Thomas *et al.*, 2007). Hence, these practices are already widely adopted where it is practical and profitable to do so, and the capacity to sequester more soil C in response to pricing C may be limited. In south-eastern Australia for example, it has been predicted that only a fraction (<15%) of the available additional soil C sequestration potential would be realised at a C price of AU\$25 $(t\, CO_2\text{-}e)^{-1}$, and even at AU\$200 $(t\, CO_2\text{-}e)^{-1}$, most regions would achieve <33% of potential sequestration (Grace *et al.*, 2010). These prices compare with the C price of AU\$23 $(t\, CO_2\text{-}e)^{-1}$ in Australia from mid-2012 (DCCEE, 2012b). In addition, because these practices are widely used, they may not meet tests of 'additionality' in terms of emissions trading.

In other regions, the desire to maximise C sequestration will mean a shift towards more pastures (e.g. Fig. 3) and hence more livestock in the farming system. The shift from cropping to livestock systems raises many issues. One is the relative economic performance of cropping versus livestock production. If livestock production is less profitable than cropping, as is currently the case in some regions of Australia, the question arises as to what C price is needed to support greater amounts of pastures and livestock production on farms. A recent case study from the Western Australian wheatbelt found that a price of AD\$80 $(t\, CO_2\text{-}e)^{-1}$ would be required to support the mix of enterprises (i.e. including longer pasture phases) and maximise soil C sequestration (Kragt *et al.*, 2012), a price substantially higher than the C price in Australia in 2012 (i.e. AU\$23 $(t\, CO_2\text{-}e)^{-1}$). Apart from the economic considerations of a shift from cropping to pastures and livestock, the GHG emissions associated with livestock would have to be considered to estimate the net GHG mitigation achieved by such enterprise shifts.

The foregoing analyses ignore the interactions between practices to increase soil C and their potential effects on soil N_2O emissions. Crop residue and tillage management have complex interactions with soil N_2O emissions, although the combination of residue retention and no-tillage systems may reduce N_2O emissions from grains production systems (Wang *et al.*, 2011) and so increase the combined GHG mitigation benefits of this management system. Another way to increase soil C in cropping is to increase crop growth and hence the return of C to the soil through residues and roots (Option 1.b.1, Table 1). N is one of the major limitations to cereal production in Australia (Hochman *et al.*, 2009), thus improved N fertilisation has the potential to boost growth and help sequester soil C. However, increasing

N fertilisation will stimulate N_2O emissions from the soil (Stehfest and Bouwman, 2006). In cereal production systems in southern Australia, soil N_2O emissions are so small that the impact of higher N fertilisation maybe insignificant (Barton *et al.*, 2008). In north-eastern Australia however, conditions are more conducive to denitrification and N_2O emissions are higher (Wang *et al.*, 2011). In this situation, applying agronomically-optimum rates of N fertiliser may not only increase crop growth and soil C sequestration, it may also increase soils GHG emissions through N_2O to such an extent that net soil GHG emissions are increased even though more C is sequestered in soils (Huth *et al.*, 2010).

The complex interactions between soil C and N_2O are expressed in other cropping systems as well. Retaining crop residues in sugarcane production systems increases soil C (Thorburn *et al.*, 2012a), but also stimulates N_2O emissions (Thorburn *et al.*, 2010). Residue retention has been previously adopted in most sugarcane producing regions in Australian for a range of practical reasons, not strongly related to soil C. Thus, management of GHG emissions in these regions can concentrate on minimising N_2O emission. However, where residues are still dominantly burnt, e.g. the Burdekin region which produces ~30% of Australia's sugarcane, there will need to be careful consideration of the GHG implications of any change in residue management. Potentially increased N_2O emissions will need to be balanced against higher soil C and reduced emissions of methane that would otherwise have occurred when residues were burnt. This issue illustrates the complex systems nature of GHG emissions.

The issues described above illustrate the complexity in understanding or predicting the likely GHG abatement arising from the different management options. As C pricing is introduced to the Australian and New Zealand agricultural sectors, farmers will be seeking management options to deliver least-cost abatement. They will want to weight returns, risks and trade-offs, especially given that C prices will vary over time and some management options will reduce flexibility in managing their farm in the future. There are decision support systems currently available that allow farmers to explore GHG mitigation options of different management practices. They vary in complexity, in the range of greenhouse gasses and gas sources considered, and of the underlying methodological complexity behind the system, e.g. the 'emission factor' approach to mechanistic simulation models (Paustian *et al.*, 2011). Clearly, further development and testing of decision support systems will be needed in Australia and New Zealand to allow farmers to make informed choices about how they manage their farms to maximise benefits under C pricing regimes, and adapt to the changing climate. There is a wealth of experience in Australia on decision support systems for helping farmers manage the impact of climatic variation on crop production (Jakku and Thorburn, 2010; Hochman and Carberry, 2011) and this experience will provide a good basis for that development.

The evolution of farm management to adapt to climate change and mitigate GHG emissions in Australia and New Zealand will need to take place within the broader context of increasing demand for food production, likely scarcer water supplies and increasing demands on land use (e.g. expansion of urban areas, GHG mitigation through afforestation). These constraints pose additional challenges, which span across disciplines and government departments, that will need to be considered in the search for optimal outcomes for greenhouse gas abatement, food production, and the environment (Battaglia, 2011).

Conclusions

Australia and New Zealand represent a microcosm of the challenges facing agriculture under climate change. Unlike other developed countries, agriculture is economically important in the two countries and contributes a greater share of GHG emissions to the national GHG inventory than in most developed countries, so climate change in agriculture has a high 'profile' in both countries. While the composition of the agricultural sector in each country is contrasting, i.e. intensive livestock production dominating in New Zealand whereas grains and extensive livestock production dominate in Australia, there are substantial similarities in the issues surrounding climate change and agriculture in each country. Impacts will be felt by farmers not only through a drier and warmer climate, but also through a regulatory environment that will soon impose costs upon their greenhouse gas generating activities.

Agriculture in Australia and New Zealand has always been affected by substantial variation in climate, so the agricultural sector has considerable experience in dealing with climate variability, which will help adaptation. Pathways for adaptation in the short and medium term seem clear based on existing technologies and anticipated productivity improvements. However, changes in climate beyond 2°C pose some particular issues for the geographic location of agricultural industries, supply of irrigation water, viability of enterprises and resilience of rural communities. These issues will potentially see some industries ceasing to be viable in their traditional locations, creating an impetus for industries to move to new locations and new, more resilient agricultural systems emerge in traditional production areas. There are already real examples of such transformation in Australia.

Within the context of the foregoing analyses, the most promising opportunities for GHG mitigation in agriculture appear to lie in reduction of methane emissions by livestock, although there may be some complex interactions in the fodder-livestock system that limit the mitigation potential achieved. For various reasons, substantial reductions in emissions of N_2O seem unlikely, with the possible exception of promoting the adoption of optimal N fertilisation practices in intensive crops that are currently being over-fertilised. Such a practice change is likely to be consistent with current initiatives to reduce other environmental impacts of cropping in these regions (e.g. N pollution of water), and so may happen independent of, or be accelerated by, climate change mitigation initiatives. Soil C sequestration is also likely to be problematical. Slowing the decline of soil C (in Australia) or building upon existing levels (in New Zealand) will require large changes to farming practices that will be potentially profit-losing under current economic and policy circumstances. As both countries move to pricing C in the agricultural sector, farmers will need support in assessing the most cost-effective management systems for GHG mitigation against potential uncertainties in C prices, the degree of abatement achieved and impacts on farm management into the future.

Acknowledgments

The authors would like to thank Dr Elizabeth Meier for editorial assistance in compiling the chapter.

References

ABS. 2012. Principal Agricultural Commodities, Australia, preliminary. Australian Bureau of Statistics Publication 7111.0. Available at http://www.ausstats.abs.gov.au/ausstats/subscriber.nsf/0/F36E81BC6A9A2F26CA2579590011A055/$File/71110_2010-11.pdf (Accessed 25th March, 2012).

Allen, D., G. Kingston, H. Rennenberg, R. Dalal and S. Schmidt. 2010. Effect of nitrogen fertiliser management and waterlogging on nitrous oxide emission from subtropical sugarcane soils. *Agric. Ecosyst. Environ.* 136:209–217.

Ash, A., R. Nelson, M. Howden and S. Crimp. 2008. Australian agriculture adapting to climate change: Balancing incremental innovation and transformational change. ABARE Outlook 2008 Conference, Canberra. Available at http://www.abare.gov.au/interactive/Outlook08/files/day_1/Ash_ClimateChange.pdf Accessed 25 Jan. 2010.

Aslam, T., M.A. Choudhary and S. Saggar. 2000. Influence of land-use management on CO_2 emissions from a silt loam soil in New Zealand. *Agric. Ecosyst. Environ.* 77:257–262.

Attwood, G. and C. McSweeney. 2008. Methanogen genomics to discover targets for methane mitigation technologies and options for alternative H_2 utilisation in the rumen. *Aust. J. Exp. Agric.* 48:28–37.

Bange, M.P., G.A. Constable, D. McRae and G. Roth. 2010. Cotton, pp. 49–66, *In* C. Stokes and S.M. Howden (eds.), *Adapting agriculture to climate change: Preparing Australian agriculture, forestry and fisheries for the future*, CSIRO Publishing, Melbourne.

Barton, L., R. Kiese, D. Gatter, K. Butterbach-Bahl, R. Buck, *et al.* 2008. Nitrous oxide emissions from a cropped soil in a semi-arid climate. *Glob. Change Biol.* 14:177–192.

Barton, L., D.V. Murphy, R. Kiese and K. Butterbach-Bahl. 2010. Soil nitrous oxide and methane fluxes are low from a bioenergy crop (canola) grown in a semi-arid climate. *Glob. Change Biol. Bioenergy* 2:1–15.

Battaglia M. 2011. Greenhouse gas mitigation: Sources and sinks in agriculture and forestry, pp. 97–108. In: H. Cleugh, M. Stafford Smith, M. Battaglia and P. Graham P. (eds.), *Climate change: Science and solutions for Australia*. CSIRO Publishing, Collingwood.

Beauchemin, K.A., M. Kreuzer, F. O'Mara and T.A. McAllister. 2008. Nutritional management for enteric methane abatement: A review. *Aust. J. Exp. Agric.* 48:21–27.

Bryant, J.R., C.J. Hoogendoorn and V.O. Snow. 2007. Simulation of mitigation strategies to reduce nitrogen leaching from grazed pasture. *Proc. New Zeal. Grassland Assoc.* 69:145–151.

Buddle, B.M., M. Denis, G.T. Attwood, E. Altermann, P.H. Janssen, R.S. Ronimus, C.S. Pinares-Patino, S.Muetzel and D.N. Wedlock. 2011. Strategies to reduce methane emissions from farmed ruminants grazing on pasture. *The Veterinary J.* 188:11–17.

Beuekes, P.C., P. Gregorini, A.J. Romera, G. Levy and G.C. Waghorn. 2009. Improving production efficiency as a strategy to mitigate greenhouse gas emissions on pastoral dairy farms in New Zealand. *Agric. Ecosys. Environ.* 136:358–365.

Burggraaf, V.T. and V.O. Snow. 2010. Effects of condensed tannins on nitrogen dynamics in grazed temperate agricultural systems, pp. 317–326. *In* G.K. Petridis (ed.), *Tannins: Types, Foods Containing, and Nutrition*, Nova Science Publishers Inc., New York.

Chan, K.Y., D.P. Heenan and H.B. So. 2003. Sequestration of carbon and changes in soil quality under conservation tillage on light-textured soils in Australia: A review. *Aust. J. Exp. Agric.* 43:325–334.

Charmley, E., S. Eady and C. McSweeney. 2011. Strategies for measuring and reducing methane emissions from beef cattle in northern Australia, pp. 73–80. *In* R. Holroyd (ed.), *Proceedings of the Northern beef Research Update Conference*, Darwin, Australia. Northern Australia Beef Research Council, Park ridge, Australia.

Charmley, E., M.L. Stephens and P.M. Kennedy. 2008. Predicting livestock productivity and methane emissions in northern Australia: Development of a bio-economic modelling approach. *Aust. J. Exp. Agric.* 48:109–113.

Chauhan, Y.S., G.C. Wright, R.C.N. Rachaputi, D. Holzworth, A. Broome, *et al.* 2010. Application of a model to assess aflatoxin risk in peanuts. *J. Agric. Sci.* 148:341–351.

Chen, W., R.W. Bell, R.F. Brennan, J.W. Bowden, A. Dobermann, *et al.* 2009. Key crop nutrient management issues in the Western Australia grains industry: A review. *Aust. J. Soil Res.* 47:1–18.

Clark, H., F. Kelliher and C. Pinares-Patino. 2011. Reducing CH_4 emissions from grazing ruminants in New Zealand: Challenges and opportunities. *Asian-Aust. J. Anim. Sci.* 24:295–302.

Clean Energy Regulator, 2012. Guide to carbon price liability under the Clean Energy Act 2011. Available at (http://www.cleanenergyregulator.gov.au/Carbon-Pricing-Mechanism/Reports-and-publications/Pages/default.aspx Accessed 2/4/12.)

Clothier, B.E., A. Hall and S.R. Green. 2012. Horticulture. *In* A. Clark and R. Nottage (eds.), *Enhanced climate change impact and adaptation evaluation for New Zealand's primary sectors*, Technical Report to MAF, NIWA, Wellington (in press).

Clough, T.J., F.M. Kelliher, H. Clark and T.J. van der Weerden. 2008. Incorporation of the Nitrification Inhibitor DCD into New Zealand's 2009 National Inventory. Report to the Ministry of Agriculture and Forestry Ministry of Agriculture and Forestry, Wellington.

Conant, R.T. 2010. Challenges and opportunities for carbon sequestration in grassland systems: A technical report on grassland management and climate change mitigation. Integrated Crop Management Vol. 9-2010. FAO, Rome.

Connor, J., M. Kirby, K. Schwabe, A. Lukasiewicz and D. Kaczan. 2008. Estimating impacts of climate change on lower Murray irrigation, Australia. Australian Agricultural and Resource Economics Society 52nd Annual Conference, Canberra. Available at http://econpapers.repec.org/paper/agsaare08/5974.htm Accessed 26 Apr. 2012.

Cullen, B., P. Thorburn, E. Meier, S. Barlow and M. Howden. 2010. New rural industries for future climates. Rural Industries Research and Development Corporation Publication No. 10/010, Rural Industries Research and Development Corporation, Canberra.

DCCEE. 2011. National Greenhouse Gas Inventory. Department of Climate Change and Energy Efficiency. http://ageis.climatechange.gov.au. Accessed 24 Mar. 2012.

DCCEE. 2012a. Australian National Greenhouse Accounts. National Inventory Report 2010, Volume 1, The Australian Government Submission to the United Nations Framework Convention on Climate Change, April 2012. Department of Climate Change and Energy Efficiency, Canberra.

DCCEE. 2012b. The Carbon Farming Initiative Handbook. Version 1.0. April 2012. Department of Climate Change and Energy Efficiency, Canberra.

de Klein, C.A.M. and R.J. Eckard. 2008. Targeted technologies for nitrous oxide abatement from animal agriculture. *Aust. J. Exp. Agric.* 48:14–20.

de Klein, C.A.M., R.R. Sherlock, K.C. Cameron and T.J. van der Weerden. 2001. Nitrous oxide emissions from agricultural soils in New Zealand — a review of current knowledge and directions for future research. *J. Roy. Soc. New Zeal.* 31:543–574.

de Klein, C.A.M., L.C. Smith and R.M. Monaghan. 2006. Restricted autumn grazing to reduce nitrous oxide emissions from dairy pastures in Southland, New Zealand. *Agric. Ecosyst. Environ.* 112:192–199.

Denmead, O.T., B.C.T. Macdonald, G. Bryant, T. Naylor, S. Wilson, D.W.T. Griffith, *et al.* 2010. Emissions of methane and nitrous oxide from Australian sugarcane soils. *Agric. Forest Meteorol.* 150:748–756.

Deurer, M., D. Grinev, I. Young, B.E. Clothier and K. Müller. 2009. The impact of soil carbon management on soil macro-pore structure: A comparison of two apple orchard systems in New Zealand. *Eur. J. Soil Sci.* 60:945–955.

Deurer, M., K. Müller, I. Kim, I. Young, G.-I. Jun and B.E. Clothier. 2012. Can moderate compaction increase soil carbon sequestration? A case study in a soil under a wheel track. *Geoderma* (in press).

Deurer, M., H. Rahman, A. Holmes, S. Saunder, B. Clothier and A. Mowat. 2010. Quantifying soil carbon sequestration in kiwifruit orchards. Development of a sampling strategy, pp. 445–459. *In Proceedings of the Workshop Farming's Future: Minimising footprints and maximising margins,* 10–11 February 2010. Fertiliser & Lime Research Centre, Massey University.

Di, H.J. and K.C. Cameron. 2002. The use of a nitrification inhibitor, dicyandiamide (DCD), to decrease nitrate leaching and nitrous oxide emissions in a simulated grazed and irrigated grassland. *Soil Use Manage.* 18(4):395–403.

Donaghy, P., S. Bray, R. Gowen, J. Rolfe, M. Stephens, M. Hoffman and A. Stunzer. 2010. The bioeconomic potential for agroforestry in Australia's northern grazing systems. *Small-scale Forestry* 9:463–484.

Dynes R.A., T. Payn, H.E. Brown, J.R. Bryant, P. Newton, V.O. Snow, M. Lieffering, D.R. Wilson and P. Beets. 2010. New Zealand's land-based primary industries & climate change: Assessing adaptation through scenario-based modelling, pp. 44–55. *In* R.A.C. Nottage, D.S. Wratt, J.F. Bornman and K. Jones (eds.), *Climate change adaptation in New Zealand. Future scenarios and some sectoral perspectives,* New Zealand Climate Change Centre, Wellington.

Dynes, R.A., D.C. Smeaton, A.P. Rhodes, T.J. Fraser and M.A. Brown. 2011. Modelling farm management scenarios that illustrate opportunities farmers have to reduce greenhouse gas emissions while maintaining profitability. *Proc. New Zeal. Anim. Prod. Soc.* 71:161–171.

Easterling, W.E., P.K. Aggarwal, P. Batima, K.M. Brander, L. Erda, S.M. Howden, A. Kirilenko, J. Morton, J.-F. Soussana, J. Schmidhuber and F.N. Tubiello. 2007. Food, fibre and forest products, pp. 273–313 *In* M.L. Parry, O.F. Canziani, J.P. Palutikof, P.J. van der Linden and C.E. Hanson (eds.), *Climate Change 2007: Impacts, Adaptation and Vulnerability. Contribution of Working Group II to the Fourth Assessment Report of the Intergovernmental Panel on Climate Change,* Cambridge University Press, Cambridge, UK.

Eckard, R.J., C. Grainger and C.A.M. de Klein. 2010. Options for the abatement of methane and nitrous oxide from ruminant production: A review. *Livest. Sci.* 130:47–56.

Emissions Trading Scheme Review Panel. 2011 Doing New Zealand's fair share. Emissions trading scheme review 2011: Final Report. Ministry for the Environment, Wellington. Available at http://www.climatechange.govt.nz//emissions-trading-scheme/ets-review-2011/review-report.pdf. Accessed 2/4/2012.

Galbally, I., M. Meyer, S. Bentley, I. Weeks, R. Leuning, *et al.* 2005. A study of environmental and management drivers of non-CO_2 greenhouse gas emissions in Australian agro-ecosystems. *Environ. Sci.* 2:133–142.

Galbally, I., C.P. Meyer, Y.-P. Wang and W. Kirstine. 2010. Soil–atmosphere exchange of CH_4, CO, N_2O and NO_x and the effects of land-use change in the semiarid Mallee system in Southeastern Australia. *Glob. Change Biol.* 16:2407–2419.

Gaydon, D.S., H.C. Beecher, R. Reinke, S. Crimp and S.M. Howden. 2010. Rice, pp. 49–66. *In* C. Stokes and S.M. Howden (eds.), *Adapting Agriculture to Climate Change: Preparing Australian Agriculture, Forestry and Fisheries for the Future*, CSIRO Publishing, Melbourne.

Gaydon, D.S., H. Meinke and D. Rodriguez. 2011. The best farm-level irrigation strategy changes seasonally with fluctuating water availability. *Agric. Water Manage.* 103:33–42.

Grace, P.R., J. Antle, S. Ogle, K. Paustian and B. Basso. 2010. Soil carbon sequestration rates and associated economic costs for farming systems of south-eastern Australia. *Aust. J. Soil Res.* 48:1–10.

Grace, P.R., J.N. Ladd, G.P. Robertson and S.H. Gage. 2006. SOCRATES — A simple model for predicting long-term changes in soil organic carbon in terrestrial ecosystems. *Soil Biol. Biochem.* 38:1172–1176.

Grainger, C. and K.A. Beauchemin. 2011. Can enteric methane emissions from ruminants be lowered without lowering their production? *Anim. Feed Sci. Tech.* 166–167:308–320.

Grainger, C., T. Clarke, K.A. Beauchemin, S.M. McGinn and R.J. Eckard. 2008. Supplementation with whole cottonseed reduces methane emissions and can profitably increase milk production of dairy cows offered a forage and cereal grain diet. *Aust. J. Exp. Agric.* 48:73–76.

Hegarty, R.S., J.P. Goopy, R.M. Herd and B. McCorkell. 2007. Cattle selected for lower residual feed intake have reduced daily methane production. *J. Anim. Sci.* 85:1479–1486.

Hennessy, K.J., P.H. Whetton and B. Preston. 2010. Climate projections, pp. 13–20. *In* C. Stokes and S.M. Howden (eds.), *Adapting Agriculture to Climate Change: Preparing Australian Agriculture, Forestry and Fisheries for the Future*, CSIRO Publishing, Melbourne.

Hill, M.J., S.H. Roxburgh, G.M. McKeon, J.O. Carter and D.J. Barrett. 2006. Analysis of soil carbon outcomes from interaction between climate and grazing pressure in Australian rangelands using Range-ASSESS. *Environ. Modell. Softw.* 21:779–801.

Hochman, Z. and Carberry, P.S. 2011. Emerging consensus on desirable characteristics of tools to support farmers' management of climate risk in Australia. *Agric. Syst.* 104:441–450.

Hochman, Z., D.P. Holzworth and J.R. Hunt. 2009. Potential to improve on-farm wheat yield and water-use efficiency in Australia. *Crop Pasture Sci.* 60:708–716.

Howden, S.M. and P.J. Reyenga. 1999. Methane emissions from Australian livestock: Implications of the Kyoto Protocol. *Aust. J. Agric. Res.* 50:1285–91.

Howden, S.M., R.G. Gifford and H. Meinke. 2010. Grains, pp. 21–40. *In* C. Stokes and S.M. Howden (eds.), *Adapting Agriculture to Climate Change: Preparing Australian Agriculture, Forestry and Fisheries for the Future*, CSIRO Publishing, Melbourne.

Howden, S.M., C. Stokes, A.J. Ash and N.D. MacLeod. 2003. Reducing net greenhouse gas emissions from a tropical rangeland in Australia, pp. 1080–1082. *In* N. Allsopp, A.R. Palmer, S.J. Milton, K.P. Kirkman, G.I.H. Kerley, C.R. Hurt and C.J. Brown (eds.), *Rangelands in the New Millennium, Proceedings of the VIIth International Rangelands Congress*, 26 July–1 August 2003, Rangelands Society, Durban.

Howden, S.M., J.F. Soussana, F.N. Tubiello, N. Chhetri, M. Dunlop and H. Meinke. 2007. Adapting agriculture to climate change. *P. Natl. Acad. Sci. USA* 104:19691–19696.

Hunter, R. A. and G. E. Niethe. 2009. Efficiency of feed utilisation and methane emissions for various cattle breeding and finishing systems. *Rec. Adv. Anim. Nutr. Aust.* 17:75–79.

Hutchinson, J.J., C.A. Campbell and R.L. Desjardins. 2007. Some perspectives on carbon sequestration in agriculture. *Agric. Forest Meteorol.* 142:288–302.

Huth, N.I., P.J. Thorburn, B.J. Radford and C.M. Thornton. 2010. Impacts of fertilisers and legumes on N_2O and CO_2 emissions from soils in subtropical agricultural systems: A simulation study. *Agric. Ecosyst. Environ.* 136:351–357.

Jamieson, P.D. and C.G. Cloughley. 1998. Chapter 5. Impacts of Climate Change on Wheat Production. *In* CLIMPACTS Assessment Report. Available at www.waikato.ac.nz/igci/climpacts/ Linked%20documents/Chapter_5.pdf. Accessed Jan. 2012.

Jakku, E. and P.J. Thorburn. 2010. A conceptual framework for guiding the participatory development of agricultural decision support systems. *Agric. Syst.* 103:675–682.

John, M., D. Pannell and R. Kingwell. 2005. Climate change and the economics of farm Management in the face of land degradation: Dryland salinity in Western Australia. *Can. J. Agric. Econ.* 53:443–459.

Kamra D. N., A.K. Patra, P.N. Chatterjee, R. Kumar, N. Agarwal and L.C. Chaudhary. 2008. Effect of plant extracts on methanogenesis and microbial profile of the rumen of buffalo: A brief overview. *Aust. J. Exp. Agric.* 48:175–178.

Kennedy, P.M. and E. Charmley. 2012. Methane yields from Brahman cattle fed tropical grasses and legumes. *Anim. Prod. Sci.* 52:225–239.

Kenny, G. 2008. Adapting to climate change in the kiwifruit industry. Prepared for MAF Policy Climate Change — 'Plan of Action' Research Programme 2007–08. http://maxa.maf.govt.nz/ climatechange/slm/grants/research/2007-08/pdf/2008-25-kiwifruit-adaptation.pdf. (accessed 26 Apr. 2012).

Kragt. M.E., D.J. Pannell, M.J. Robertson and T. Thamoa. 2012. Assessing costs of soil carbon sequestration by crop-livestock farmers in Western Australia. *Agric. Syst.* 112:27–37.

Lal, R. 2004. Soil carbon sequestration impacts on global climate change and food security. *Science* 304:1623–1627.

LIC. 2011. 2010/2011 New Zealand Dairy Statistics. Available at http://www.lic.co.nz/lic_News_ Archive.cfm?nid=371 Accessed 26/3/2012.

Luo, Z., E. Wang and O.J. Sun. 2010a. Soil carbon change and its responses to agricultural practices in Australian agro-ecosystems: A review and synthesis. *Geoderma* 155:211–223.

Luo, J., C.A.M. de Klein, S.F. Ledgard and S. Saggar. 2010b. Management options to reduce nitrous oxide emissions from intensively grazed pastures: A review. *Agric. Ecosyst. Environ.* 136:282–291.

MAF. 2011. Carbon Footprint of New Zealand Arable Production — Wheat, Maize Silage, Maize Grain and Ryegrass Seed. MAF Technical Paper No: 2011/97. Available at http://www.fedfarm.org.nz/f3304,130326/130326_2011_-_Carbon_Footprint_of_ New_Zealand_ Arable_Production.pdf Accessed Mar. 2012.

McAllister, T. A. and C. J. Newbold. 2008. Redirecting rumen fermentation to reduce methanogenesis. *Aust. J. Exp. Agric.* 48:7–13.

Meinke, H., S.M. Howden, P.C. Struik, R. Nelson, D. Rodriguez and S.C. Chapman. 2009. Adaptation science for agriculture and natural resource management — urgency and theoretical basis. *Curr. Opin. Env. Sus.* 1:69–76.

Miller, C.J., S.M. Howden and R.N. Jones. 2010. Intensive livestock industries. pp. 171–185, *In* C. Stokes and S.M. Howden (eds.), *Adapting Agriculture to Climate Change: Preparing Australian Agriculture, Forestry and Fisheries for the Future*, CSIRO Publishing, Melbourne.

Ministry for the Environment. 2001. Climate change impacts on New Zealand. Publication number: ME 396, Ministry for the Environment, Wellington. Available at http://www.mfe.govt.nz/ publications/climate/impacts-report/index.html. Accessed 16 Apr. 2012.

Ministry for the Environment. 2011a. National Policy Statement for Freshwater Management 2011: Implementation Guide. Ministry for the Environment, Wellington. Available at http://www.mfe. govt.nz/publications/rma/nps-freshwater-guide-2011/nps-freshwater-management-guide.pdf. Accessed 25 Mar. 2012.

Ministry for the Environment. 2011b. New Zealand's Greenhouse Gas Inventory. Publication number ME 1045, Ministry for the Environment, Wellington. Available at http://www.mfe. govt.nz/publications/climate/greenhouse-gas-inventory-2009/. Accessed 25 Mar. 2012.

Moate, P.J., S.R.O. Williams, C. Grainger, M.C. Hannah and E.N. Ponnampalam. 2011. Influence of cold pressed canola, brewers grains and hominy meal as dietary supplements suitable for reducing enteric methane emissions from lactating dairy cows. *Anim. Feed Sci. Tech.* 166–167: 254–264.

Moe, P.W. and H.F. Tyrrell. 1979. Methane production in dairy cows, pp. 59–62. *In* L.E. Mount (ed.), *Energy metabolism*. Proceedings of the Eighth Symposium on energy metabolism, Cambridge. Butterworths, London.

Mpelasoka, B.S., M.H. Behboudian and T.M. Mills. 2001. Effects of deficit irrigation on fruit maturity and quality of 'Braeburn' apples. *Sci. Hortic.-Amsterdam* 90:279–290.

OECD. 2010. Agricultural Policies in OECD Countries: At a Glance 2010. Available at http://www.oecd.org/dataoecd/17/0/45539870.pdf. Accessed 26 Apr. 2012.

Oliver, Y., M.J. Robertson and C. Weeks. 2010. A new look at an old practice: Quantifying the benefits of long fallowing to wheat yield in a Mediterranean climate. *Agric. Water Manage.* 98:291–300.

Park, S.E., N.A. Marshall, E. Jakku, A.M. Dowd, S.M. Howden, *et al.* 2012. Informing adaptation responses to climate change through theories of transformation. *Global Environ. Chang.* 22:115–126.

Paustian K., K. Peterson, S. Archibeque, K. Brown, E. Campbell, K. Denef, *et al.* 2011. Farm-level accounting of greenhouse gas (GHG) emissions: Can farmer's be the new carbon accountants?. Abstract 169–4, ASA-CSSA-SSSA International Annual Meetings, San Antonio, Texas. Available at (http://a-c-s.confex.com/crops/2011am/webprogram/Paper64952.html)

Pinares-Patiño, C.S., J.C. McEwan, K.G. Dodds, E.A. Cardenás, R.S. Hegarty, *et al.* 2011. Repeatability of methane emissions from sheep. *Anim. Feed Sci. Tech.* 166–167:210–218.

Ramírez-Restrepo, C.A. and T.N. Barry. 2005. Alternative temperate forages containing secondary compounds for improving sustainable productivity in grazing ruminants: A review. *Anim. Feed Sci. Tech.* 120:179–201.

Ramírez-Restrepo, C.A., T.N. Barry, A. Marriner, N. López-Villalobos, E.L. McWilliam, *et al.* 2010. Effects of grazing willow fodder blocks upon methane production and blood composition in young sheep. *Anim. Feed Sci. Tech.* 155:33–43.

Ramírez -Restrepo, C., L. Xixi, Z. Durmic, P. Vercoe, C. Gardiner, *et al.* 2011. Assessment of tropical legumes to reduce methane emissions from pastoral systems using in vitro fermentation and near infrared reflectance spectroscopy methods, p. 87. *In* Greenhouse 2011 the Science of Climate Change, Cairns, Australia. Australian Government, Canberra.

Reisinger, A., B. Mullan, M. Manning, D. Wratt and R. Nottage, 2010. Global and local climate change scenarios to support adaptation in New Zealand, pp. 27–43. *In* R.A.C. Nottage, D.S. Wratt, J.F. Bornman, and K. Jones (eds.), *Climate Change Adaptation in New Zealand: Future Scenarios and Some Sectoral Perspectives*, New Zealand Climate Change Centre, Wellington.

Rengasamy, P. 2006. World salinization with emphasis on Australia. *J. Exp. Bot.* 57:1017–1023.

Rickards, L. and S.M. Howden. 2012. Transformational adaptation-agriculture and climate change. *Crop Pasture Sci.* 63:240–250.

Robertson, M.J. 2010. Agricultural productivity in Australia and New Zealand: Trends, constraints and opportunities. *In* H. Dove and R.A. Culvenor (eds.), *Food Security from Sustainable Agriculture*, Proceedings of 15th Agronomy Conference 2010, 15–18 November 2010, Lincoln, New Zealand. Australian Society of Agronomy.

Rochester, I.J. 2010. Assessing internal crop Nitrogen use efficiency in high-yielding irrigated cotton. *Nutr. Cycl. Agroecosys.* 90:147–156.

Rolfe, J. 2010. Economics of reducing methane emissions from beef cattle in extensive grazing systems in Queensland. *Rangeland J.* 32:197–204.

Sanderman, J. and J.A. Baldock. 2010. Accounting for soil carbon sequestration in national inventories: A soil scientist's perspective. *Environ. Res. Lett.* 5:034003.

Sanderman, J., R. Farquharson and J.A. Baldock. 2010. Soil carbon sequestration potential: A review for Australian agriculture. CSIRO Publishing, Canberra. Available at http://www. csiro.au/resources/Soil-Carbon-Sequestration-Potential-Report.html. Accessed 26 Apr. 2012.

Schep, L., W. Temple and M. Beasley. 2009. The adverse effects of hydrogen cyanamide on human health: An evaluation of inquiries to the New Zealand National Poisons Centre. *Clinical Toxicology* 47:58–60.

Schipper, L.A., R.L. Parfitt, C. Ross, W.T. Baisden, J.J. Claydon and S. Fraser. 2010. Gains and losses in C and N stocks of New Zealand pastures depend on land use. *Agric. Ecosyst. Environ.* 139:611–617.

Smith, P., O. Andren, T. Karlsson, P. Perala, K. Regina, *et al.* 2005. Carbon sequestration potential in European croplands has been overestimated. *Glob. Change Biol.* 11:2153–2163.

Smith, P., D. Martino, Z. Cai, D. Gwary, H. Janzen, *et al.* 2007. Policy and technological constraints to implementation of greenhouse gas mitigation options in agriculture. *Agric. Ecosyst. Environ.* 118:6–28.

Smith, P., D. Martino, Z. Cai, D. Gwary, H. Janzen, *et al.* 2008. Greenhouse gas mitigation in agriculture. *Philos. T. R. Soc. B* 363:789–813.

Stafford-Smith, M., L. Horrocks, A. Harvey and C. Hamilton. 2011. Rethinking adaptation for a 4°C world. *Phil. Trans. R. Soc. A.* 369:196–216.

Stehfest, E. and L. Bouwman. 2006. N_2O and NO emission from agricultural fields and soils under natural vegetation: Summarizing available measurement data and modeling of global annual emissions. *Nutri. Cycl. Agroecosys.* 74:207–228.

Stokes, C. and S.M. Howden, editors. 2010. Adapting agriculture to climate change: Preparing Australian agriculture, forestry and fisheries for the future. CSIRO Publishing, Melbourne.

Thomas, G.A., G.W. Titmarsh, D.M. Freebairn and B.J. Radford. 2007. No-tillage and conservation farming practices in grain growing areas of Queensland — a review of 40 years of development. *Aust. J. Exp. Agric.* 47:887–898.

Thorburn, P.J., J.S. Biggs, K. Collins and M.E. Probert. 2010. Using the APSIM model to estimate nitrous oxide emissions from diverse Australian sugarcane production systems. *Agric. Ecosyst. Environ.* 136:343–350.

Thorburn, P.J., J.S. Biggs, A.J. Webster and I.M. Biggs. 2011. An improved way to determine nitrogen fertiliser requirements of sugar cane crops to meet global environmental challenges. *Plant Soil* 339:51–67.

Thorburn, P.J., E.A. Meier, K. Collins and F.A. Robertson. 2012a. Changes in soil carbon sequestration, fractions and soil fertility in response to sugarcane residue retention are site-specific. *Soil Tillage Res.* 120:99–111.

Thorburn, P., N. Marshall, E. Jakku, C. Gambley, Y. Chauhan, *et al.* 2012. The basis for successful transformation of farming industries as an adaption to future climates. *In* 2012 Australian National Climate Change Adaptation Conference. National Climate Change Adaptation Research Facility, Melbourne.

Thorburn, P.J. and S.N. Wilkinson. 2012. Conceptual frameworks for estimating the water quality benefits of improved agricultural management practices in large catchments. *Agric. Ecosyst. Environ.* doi:10.1016/j.agee.2011.12.021 (in press).

Ulyatt, M.J., K.R. Lassey, I.D. Shelton and C.F. Walker. 2002. Methane emission from dairy cows and wether sheep fed subtropical grass-dominant pastures in mid-summer in New Zealand. *New Zeal. J. Agric. Res.* 45:227–234.

van der Weerden, T.J., R.R. Sherlock, P.H. Williams and K.C. Cameron. 1999. Nitrous oxide emissions and methane oxidation by soil following cultivation of two different leguminous pastures. *Biol. Fert. Soils* 30:52–60.

Waghorn, G.C. and R.S. Hegarty. 2011. Lowering ruminant methane emissions through improved feed conversion efficiency. *Anim. Feed Sci. Tech.* 166–167:291–301.

Wall, E., G. Simm and D. Moran. 2010. Developing breeding schemes to assist mitigation of green-house gas emissions. *Anim.* 4:366–376.

Wang, W.N., R.C. Dalal, S.H. Reeves, K. Butterbach-Bahl and R. Kiese. 2011. Greenhouse gas fluxes from an Australian subtropical cropland under long-term contrasting management regimes. *Glob. Change Biol.* 17:3089–3101.

Webb, L. and P.H. Whetton. 2010. Horticulture, pp. 119–136. In C. Stokes and S.M. Howden (eds.), *Adapting Agriculture to Climate Change: Preparing Australian Agriculture, Forestry and Fisheries for the Future*, CSIRO Publishing, Melbourne.

West, T.O. and W.M. Post. 2002. Soil organic carbon sequestration rates by tillage and crop rotation: A global data analysis. *Soil Sci. Soc. Am. J.* 66:1930–1946.

WorldStat. 2007. New Zealand Statistics. Available at http://en.worldstat.info/Oceania/NewZealand Accessed Mar. 2012.

Wright, A.D., P. Kennedy, C.J. O'Neill, A.F. Toovey, S. Popovski, *et al.* 2004. Reducing methane emissions in sheep by immunization against rumen methanogens. *Vaccine* 22:3976–3985.

Wright, A.-D. and A.V. Klieve. 2011. Does the complexity of the rumen microbial ecology preclude methane mitigation. *Anim. Feed Sci. Tech.* 166–167:248–253.

Chapter 8

Middle East and North Africa Perspectives on Climate Change and Agriculture: Adaptation Strategies

Shawki Barghouti, Shoaib Ismail*, and Rachael McDonnell

International Center for Biosaline Agriculture, Dubai, UAE.
*s.ismail@biosaline.org.ae

Introduction

Both global and regional climate change studies have highlighted that Arab countries are some of the most vulnerable, particularly as most of them may be described as naturally scarce of water (Giorgi, 2006, Solomon *et al.*, 2007; Evans, 2010). Results from climate models suggest that the region will largely become hotter and drier with greater inter-seasonal and inter-annual variability resulting in less water runoff (between 20–30%) (Milly *et al.*, 2005). This will have both direct and indirect impacts on the food security of the region's countries, agricultural production and on the livelihoods of those in rural areas. Food production and its availability, to a large extent, is dependent on climate, especially precipitation, and hence any chance in rainfall patterns will have both short- and long-term impacts.

In addition to projected high temperatures and lower precipitation affecting agricultural production, the latter is also a source of greenhouse gases — methane (CH_4) and nitrous oxides (N_2O) being the dominant ones (Fig. 1). This contributes to the greenhouse gas concentrations in the atmosphere and global warming (Verchot, 2007). Crop production and livestock release methane (from cattle and wetlands, especially rice paddies) and nitrous oxide (from fertilizer use). In addition, indirect impacts of unsustainable farming practices such as deforestation and soil degradation also contribute to the atmospheric carbon dioxide levels.

The impacts of climate change on agriculture may already be affecting global food production. The growth rates of the last thirty years appear to be slowing, even with continued improvement in agricultural practices, such as in varieties of crops and advanced irrigation technology. Compounding problems range from the loss

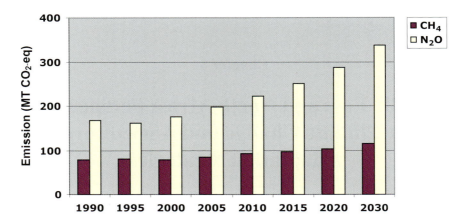

Fig. 1.   Trend in non-CO_2 emissions from agriculture in the MENA region.
Source: Adapted from Verchot, 2007.

of arable agricultural lands to natural and human-induced processes, urbanization, desertification, and the deterioration in quality of irrigation water (both surface and ground) from many causes (including seawater intrusion) in many parts of the world.

The Middle East and North Africa (MENA) region can be considered as the most water-scarce region of the world. As the MENA region's population continues to grow, projected to double over the next 40 years, per capita water availability is projected to fall by more than 50% by 2050. Moreover, climate change is likely to affect weather and precipitation patterns with the consequence that the MENA region may see more frequent and severe droughts.

Current Conditions of Food Security and Agricultural Production Systems

Food security may be observed from the perspectives of both food availability and the ability of local food production to meet demands. Schmidhuber and Tubiello (2007) described food security as linked to four factors:

Availability   The overall ability of the agricultural system (both home-grown and imported) to meet food demand. This is linked to the fundamentals of crop and pasture production as well as the factors that determine how farmers respond to different market conditions.

Stability   The ability of individuals to have access to food through various perturbations. Climate variability and savings are major factors affecting stability of access to food.

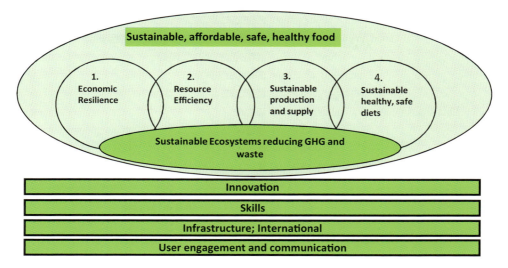

Fig. 2.   The Global food security themes.

Access   This represents the rights and resources of individuals (legal, political, financial and social) to acquire adequate and appropriate food. This is linked to real incomes and food prices.

Utilization   All food safety and quality aspects that ensure individuals, use of food.

Global food security relies on resources, production and supply, and ecosystem approaches complemented by innovation, capacity building and knowledge sharing (Fig. 2).

The issue of food security is related to the resources available to sustain agriculture, food consumption patterns in different countries, deficits in supplying of local food demands, dependence on external sources, and short- and long-term economic conditions. One of the major issues in terms of resources in the entire MENA region is the aridity and lack of availability of water resources to sustain agriculture.

Water scarcity and its use in agriculture

Food availability in the MENA region is more dependent on imports rather than on local production, mainly due to the non-availability of water. The Middle East and North Africa region can be considered as the most water-scarce region of the world (Fig. 3). The average consumption of internal fresh water resources is reported to be around $6800\,\mathrm{m}^3$ per capita, whereas, below $1000\,\mathrm{m}^3$ per capita is regarded as scarce. With the exception of a few countries (Sudan, Syria and Egypt), most countries in the MENA region currently have inadequate renewable water to

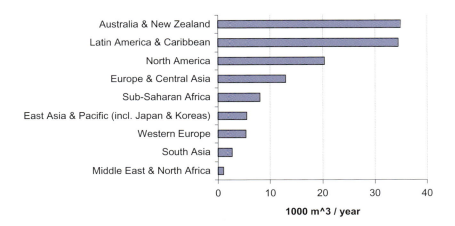

Fig. 3.   Renewable fresh water resources by region.
Source: The World Bank Report, 2007.

sustain agricultural production and, in many places, fossil or non-renewable water is extracted to continue present-day agriculture. Lebanon and Morocco have favorable topography, where coastal mountain ranges intercept moisture-laden weather systems to produce heavy winter precipitation. Both countries have renewable water resources around $1000 \, m^3$ per capita and receive no water from outside their boundaries. Water scarcity and variability will possibly be more felt in the poorest countries including Mauritania, Sudan, Somalia, Comoros, Djibouti, and Yemen, where most of the economically active population are engaged in agriculture.

The high evapotranspiration and soil percolation rates in arid regions reduce soil moisture content and consequently increase irrigation requirements, which typically surpass 80% of the total water withdrawals in most of the MENA countries.

The issue of climate change impacts on water has socio-economic implications (Table 1). With climate change, the predicted increases in evapotranspiration rates will likely lead to higher irrigation requirements. Even with pastoralism activities that currently have no irrigation input, Evans (2009) found that the lengthening of dry periods could reduce the grazing area available and consequently increase the need for irrigated fodder to maintain the same level of livestock. With increasing demand of good-quality water for agriculture and competition with other sectors, possible solutions to managing water demand will require agricultural activities to be considered within an integrated national socio-economic development strategy that involves other sectors. This will be particularly important given that the agriculture sector is the largest employer in many MENA countries and contributes significantly to provision of food supplies, yet it still does not meet the overall food requirements.

In addition to arable land and harsh climatic conditions, water scarcity and quality of water are dominating factors that affect agricultural production. The irrigated

Table 1. Socio-economic implications of climate change impacts on water resources in MENA region countries.

| | MENA region countries | | | | |
Impact	Iraq	Jordan	Lebanon	Palestinian authority	Syria
Increased industrial and domestic water demand	++	+	+	+	++
Increased agricultural water demand	+++	+	+++	+++	+++
Flood damage	+++	+	++	+	+
Reduced water quality	+++	+++	+++	+++	+++
Ecosystem damage and species loss	++	+	+++	++	++
GDP reduction (%)	3–6	1–2	2–5	2–5	4–7

+++ high; ++ moderate; + insignificant.
Source: *El-Fadel and Bou-Zeid, 2001.*

areas among the Mashreq countries (Iraq, Jordan, Lebanon, and Syria) range from 27–43% of total cropland. Maghreb countries (Algeria, Libya, Mauritania, Morocco, and Tunisia) are much less dependent on irrigation (7–18%), whereas farming in Egypt and Djibouti is almost entirely irrigated. Although water resources in Sudan are relatively less scarce, the proportion of irrigated land remains less than 10% (AOAD, 2007). The Gulf Cooperation Council states are totally dependent on irrigation due to extremely low levels of rainfall.

A recent World Bank (2011) study indicates that water shortage in the MENA region will be large in the next few decades. About 20% can be attributed to climate change and 80% to a steep increase in demand owing to strong population growth and fast economic development. The study clearly indicates that a mix of management needs to be formulated and that this must be country-specific.

The current annual flow of fresh water in the MENA region is close to 228 billion cm, which is 0.5% of total fresh water flows in the world. Per capita renewable internal fresh water resources amount to 757 cubic meters in the MENA region, representing only 11% of the world level (Slam, 2009). About 20% of the land in the MENA region is irrigated (which uses 85–90% of the total water), whereas the rainfed area is only half as large as the irrigated area (∼10% of the total land).

In most of the MENA countries, a large percentage of the overall water supply comes from groundwater, mostly non-renewable in nature. Furthermore, since there is little or no water pricing, over-pumping of this water resource has significantly depleted the aquifers. In addition, this has lead to intrusion of sea water in the coastal region, thus affecting water quality. This is particularly serious in Libya, Jordan, Yemen, Oman, Bahrain, the UAE and Qatar, where deterioration of groundwater quality has been observed and measured over the last few years (Barghouti, 2010).

The pressure to reallocate water among different users is likely to intensify in the next decade. Since irrigated agriculture is the main user of these scarce resources, pressure is mounting in several countries in the region to readjust water allocation to agriculture and other sectors. This issue is more serious in some countries where the relative contribution of agriculture to national economic growth is diminishing.

Agricultural production systems

The impacts of climate change on agriculture in a nation is dependent on (1) the biological effects on crop yields; (2) the resulting impacts and adaptations on the outcomes, including production, consumption and markets; and (3) the potential for meeting the deficit by importing foods. The biophysical effects of climate change on agriculture induce changes in production and prices, which is the economic component where farmers and other market participants play a role in altering crop mix, input use, production, food demand, food consumption and trade.

The agricultural sector is generally viewed as the driving force for the development of the MENA region, especially in non-oil producing countries, though the per capita arable land is 0.17 ha, 23% below the world average. However, agriculture and food production in the MENA region has also been associated with non-, under-, and mal-utilization of material and natural resources, resulting in low resource productivities and inadequate crop yield levels aggravated by a widening technological gap (Slam, 2009).

Climate change impacts agricultural production now and in years to come — from the perspectives of resources and consumption and of markets and policies. These systems will face even greater challenges in the future, in terms of preparing mitigation and adaptation strategies to respond to climate change. Furthermore, the efficiency of the irrigation systems and their contribution to depletion of water resources, versus the crop yields obtained also needs to be considered seriously. The main challenge in many countries is to improve the performance of the existing investments in the water sector, especially in regard to irrigation and crop yields. Some countries have also reported that moving from conventional flood to modern irrigation systems may have contributed to increases in yield but this did not have much impact on the quantity of water used, since the farmers respond by expanding the areas under irrigation.

Most agricultural production systems in the MENA region tend to be characterized by maximizing returns from irrigation through enhanced low-technology and low-skill levels, which are very much dependent on government subsidies for economic survival. The ability of smallholder farmers in particular to respond to and benefit from variability in food availability and associated increases in higher domestic prices was reported to be limited in a recent study by El-Dukheri *et al.* (2011).

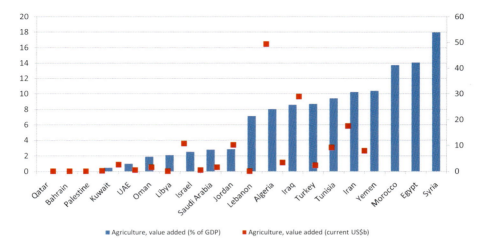

Fig. 4. Contribution of agriculture to GDP in MENA countries. Left axis represents % of GDP and right axis represents US$b in 2008 or latest available figures.
Source: http://data.worldbank.org/country and http://www.fao.org/nr/water/aquastat/dbase/index.stm. (Accessed on 25 July 2012)

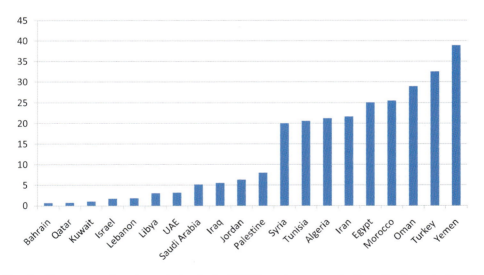

Fig. 5. Percentage of total population in the MENA region economically active in agriculture in 2010 or latest available.
Source: http://www.fao.org/nr/water/aquastat/dbase/index.stm. (Accessed on 25 July 2012)

The current rate of agricultural production and its contribution to GDP is variable in the different countries. Agriculture in the GCC countries contributes least to GDP (Fig. 4), while Oman, Morocco, Yemen and Egypt employ more than 25% of its population in the agriculture sector (Fig. 5).

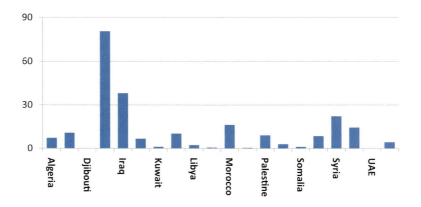

Fig. 6.   Arable land per capita (ha/person) for 2007 in the MENA region.

The total land under cultivation (relative to total area), particularly for cereal production, is relatively low, ranging from 0.05% to 22.22% for most of the countries. The exceptions are Iraq and Egypt, where cereal production exceeds 30% of the total agricultural land, with a maximum of 80.55% for Egypt (Fig. 6).

Current estimates suggest that real agricultural prices will increase from now through the next 40 years, in part because of the rising costs and negative productivity effects of climate change. Since June 2008, a reduction in world food prices has been observed due to the world financial crisis, the decline of world oil prices, and the appreciation of the international currencies. In addition, approximately 50% prices have retreated due to strong production gains in developed countries (FAO, 2008). The countries of the MENA region are vulnerable to trends in overall agricultural production, since they are dependent on imported foods, with net importation levels of 58.2%.

The yield of major cereals, including wheat, rice, maize, barley, oats, rye, millet, sorghum, buckwheat and mixed grains for the region varies between 1–2.5 tons/ha. Intensive irrigation practices in Kuwait, Oman, Qatar and Saudi Arabia have pushed the yield at maximum to 3 tons/ha. Egypt leads with the highest production, close to about 8 tons/ha. However, consumption of cereals far exceeds production, even in Egypt. This pushes the countries in the region to meet the deficits by importing cereals for domestic consumption, making them more vulnerable to food security issues (Fig. 7). The GCC countries are particularly dependent on food imports, followed by the Maghrib and Mashriq.

Climate change is projected to result in additional price increases for many important agricultural crops, including rice, wheat, maize and soybeans. Higher feed prices would increase the meat prices, thus potentially reducing meat consumption. This may indirectly result in increased production of cereals for feed.

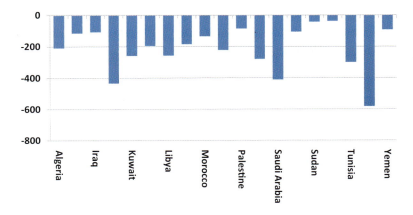

Fig. 7. Cereal deficit per capita (kg/person) for the year 2007 in the MENA region.

Impacts on agriculture

Using a regional climate change model and the A2 SRES scenario, Evans (2009) estimated that for the Middle East region alone, there would be a decrease of over 170 000 km^2 in viable rainfed agricultural land by the late-century. This would be exacerbated due to the projected increase in the number and length of periods of drought. In North Africa, maize yields were projected to fall by between 15 and 25% with an 3°C rise in temperature.

Rangeland areas are likely to be extremely vulnerable as they are located mostly in marginal areas with less available soil moisture, leading to degraded land (Elasha, 2010). Decline in available moisture would greatly affect nomadic systems over a wide area. Evans (2009) has predicted that at the end of the 21st century, due to an increase in the length of the dry season, there will a reduction in the length of time that the rangelands can be grazed. This is projected to be particularly pronounced in large parts of Syria and Iraq, where it is projected that there will be a significant increase in the length of the dry season by around two months. This would increase the need for supplementary water and feedstuffs or decreased herd sizes.

The most permanent and direct losses related to climate change will result from the inundation of agricultural land by sea level rise. The countries likely to be most affected under a temperature change of 1–3°C are Qatar, UAE and Kuwait. In terms of agricultural production and the impact on an agriculturally engaged population, the largest impacted area will be the Nile Delta in Egypt, a rich agricultural area. The extremely low elevation of the Delta arable cropland is particularly vulnerable as most of the 50 km-wide strip of land along the river is less than two metres above sea-level and is protected from flooding by only a 1–10 km-wide belt of coastal sand (El-Raey 2009; Elasha, 2010).

Salinization of soils and water is already a problem faced by the MENA region which leads to severe degradation of arable land, enhanced desertification, reduction in crops yields and increasing dependencies on food imports. Under the projected climate change conditions, increased evapotranspiration and irrigation, will like increase the saline content of both soils (reduced flushing and increased evapor-transpiration) and water (over-abstraction of water resources leading to increased salinization of groundwater supplies). Currently, approximately 397 m.ha of saline soils and 434 m.ha of sodic soils are reported to form globally (Taha *et al.*, 2005), with a rapid increase in affected areas in the Arab region.

These factors that directly or indirectly reduce agricultural yield (climate change, scarcity of fresh water, degradation of land, and salinization) demonstrate that both short- and long-term strategies are needed urgently. Furthermore, based on the extent of the climate change impacts and projected changes, both mitigation and adaptation strategies need to be formulated, keeping in mind the technological and economical constraints related to each of the MENA countries.

Adaptation to Climate Change

The expansion and unrestricted use of non-renewable groundwater supplies in many countries demonstrate the impact of inadequate policies and misguided investment in this sector. The absence of a strategic national water framework to protect non-renewable water supplies is driving many countries in the MENA region to use fresh water resources to grow low-value crops that do not significantly contribute to farmer income or the national GDP. Both economic and environmental assessments are needed.

Food production is dependent on local conditions and as these conditions change, agricultural practices, ranging from management adjustment to complete changes in cropping systems, need to be adapted. This has financial implications for farmers. Other adaptation options aim to maintain crop supplies in order to maintain market prices — with potential implications for government supporting policies and infras-tructure. Table 2 describes climate change scenarios in relation to impacts on water and agriculture in the Middle East.

The main adaptation strategies would revolve around the (i) water quality and quantity and (ii) agricultural production issues. The former will have to be considered from an integrated perspective of agricultural production, education and gender.

Integrated Water Resource Management (IWRM) is based on four principles as put forth by the International Conference on Water and the Environment in Dublin in 1992, and later adopted by the United Nations Conference on Environment and Development in Rio de Janeiro in 1992 (Agarwal *et al.*, 2000):

Table 2.　Climate change scenarios: Impacts on water and agriculture in the Middle East.

Scenario	Type of change	Impacts on human security	Affected areas
Water	1°C rise in mean earth temperature	Reduced water runoff in Ouergha watershed by 10%	Morocco
	1.2°C rise in mean earth temperature	Decreased water availability by 15%	Lebanon
	2°C rise in mean earth temperature	1 to 1.6 billion people affected by water shortages	Africa, the Middle East, Southern Europe, parts of South and Central America
	3°C rise in mean earth temperature	Increased water stress for additional 155 to 600 million people	North Africa
	3°C rise in mean earth temperature	Increased risks of coastal surges and flooding	Cairo
	Climate change	Repeated risk of drought known in recent years, with economic and political effects	Mauritania, Sudan and Somalia
	Climate change	Reduced average rainfall	Egypt, Jordan, Lebanon and OPT
	Climate change	50% decline in renewable water availability	Syria
	Climate change	Greater water shortages	Yemen
	Climate change	Reduced water flow by 40–60%	Nile River
	Rising sea levels	Risk of flooding and threats to coastal cities	Gulf coast of Arabian peninsula
Agriculture	2–3°C temperature rise in tropical regions	A drop by 25–35% in crop production (with weak carbon enrichment) and by 15–20% (with strong carbon enrichment)	Africa and West Africa (Arab countries included)
	3°C rise in mean earth temperature	Reduced agricultural productivity and unsustainable crops	North Africa
	1.5°C rise in mean earth temperature	70% drop in yields of sorghum	Sudan (Northern Kordofan)
	Climate Change	Flooding of 4,500 km^2 of farmland and displacement of 6 million people	Lower Egypt

Source: Adapted from UNDP 2009; Barghouti, 2010.

- Fresh water is a finite and vulnerable resource, essential to sustain life, development, and the environment;
- Water development and management should be based on a participatory approach, involving users, planners, and policy-makers at all levels;
- Women play a central part in the provision, management, and safeguarding of water, and;
- Water has an economic value in all its competing uses, and should be recognized as an economic good.

These principles emphasize the need to balance water demand management with supply management, ecosystem protection and social equity. IWRM also emphasizes the importance of water as an economic commodity that needs to be managed to reflect its scarcity and optimize its socio-economic and environmental services. The MENA countries could used IWRM as a foundation for climate change strategies.

On the other hand, key issues related to enhancing food security and agriculture in the region include:

1. Reducing food imports and raising cereal production
2. Gradually decreasing government subsidies and increasing investment to make agriculture sustainable
3. Strengthening weak marketing structures
4. Evaluating land tenure
5. Broadening farmer access to new investments
6. Augmenting farmer education

Adaptation measures will also include water conservation, exploration for other water resources, and increased efficiency in use of water, among others. Table 3 provides potential adaptation measures for both fresh and other water resources.

With growing acknowledgement that climate changes are already impacting the region, ideas of adaptation are beginning to be discussed and policies developed. However, few countries have initiated strategies because they are hampered by lack of data to support evidence-based decision-making. There will be many implications and these need to be carefully considered, especially in areas that will affect the most vulnerable. There is a need to further develop information through research and development for the conditions of the MENA region.

The adaptations required will need to be implemented at various levels. Since adaptation options will likely require farmers to alter their current practices, they may require resources that are currently used for other areas. Thus, there is a great need to consider programs at a number of different levels and support capabilities:

Table 3.   Technical adaptation measures and non-conventional water resources.

Adaptation measure	Potential benefits	Best uses
Conservation	Curbs water demand increase	Domestic, industrial, agricultural demand reduction
Use of surplus winter runoff	Total runoff, up to 10% of rainfall	Irrigation, aquifer recharge
Wastewater reclamation	Treated at different levels and use	Irrigation (landscaping), aquifer recharge
Seawater/brackish water reclamation	Unlimited water supply	Domestic, industrial
Seawater/brackish water use in agriculture	Seawater — unlimited but restricted use; Brackish water — mainly groundwater	Seawater for coastal agriculture; brackish groundwater used in agriculture in many countries

Source: Modified from El-Fadel and Bou-Zeid, 2001.

A. Local — capabilities and practices developed and implemented by societies to survive over time, particularly during extremes droughts/floods; local management practices.
B. National — governments need to have in place supporting policy and infrastructure (economic and physical), changes in management, structural changes.
C. Global trading regimes — to ensure that changes in comparative advantage translate into unimpeded trade flows to balance world supply and demand.

In terms of technical inputs for on-farm strategies, the adaptation of crops to high-temperature, low-water conditions is not new. Drought tolerant crops have been developed for many arid and semi-arid regions. However, the increasing temperatures would result in shorter growth period for crop maturity and thus affecting the overall crop yields. More crop varieties will need to be bred to meet these challenges that can give multiple yields in a given cropping seasons. In addition, priorities needs to be established as to whether growing crops (for food) or growing 'others' (feed, bioenergy, forests, etc) will be more feasible, based on environmental and economical perspectives. Crops diversification would be key to keep up with the changing climate. Crops grown and consumed in a country/region may not necessarily be suitable with the changing environment and food sharing through pricing or barter agreements may have to be strengthened.

With the projected availability of water resources, more food will need to be produced with less water (Cai, *et al.*, 2011; Cai and Sharma, 2010). Water productivity — WP (defined as the ratio of net benefits from crop, forestry, fishery, livestock and other mixed agricultural systems to the amount of water used in the production process) would have a serious role in agriculture for converting water (together with other resources) into products (Molden *et al.*, 2010). The outputs would be dependent in the form of biomass, the market value of the products and the nutritional value. This will have more impact in water-scarce regions where agricultural production is limited by the amount and quality of water and are mainly controlled by management strategies.

The future agriculture will have to be adapted to the changing scenario of climate change, water scarcity, more need of food, natural disasters and other factors. The 'efficient agriculture' will need to have:

1. Good knowledge of the changing physical resources.
2. The projected demands for the region and the countries.
3. The water efficiency in reference to quality, quantity, allocation and recovery
4. The rainfed versus irrigated farming.
5. Crops adapted to region, country, site (farm) — based on the prevailing situation.
6. Production to meet the local demands, it sustainability and efficiency.
7. The import of food and its costs — the physical and economic implications.
8. The regional and global market impact.
9. Policy framework for import and export of food commodities in the region.

The recent World Bank Report (2011) indicates that to overcome the future water shortage, countries have a range of options at their disposal to respond and adapt. These options can be divided into three broad categories: (i) increasing the productivity; (ii) expanding supply; and (iii) reducing demand. For each of these three categories typical options were explored in the study resulting in the following framework:

➢ Increasing the productivity:

   o Improved agricultural practice (including crop varieties)
   o Increased reuse of water from domestic and industry
   o Increased reuse of irrigated agriculture

➢ Expanding supply:

   o Expanding reservoir capacity (small scale)
   o Expanding reservoir capacity (large scale)
   o Desalinisation by means of solar energy
   o Desalination by means of reverse osmosis

➢ Reducing demand:

   o Reduce irrigated areas
   o Reduce domestic and industrial demand

The use of poor quality water and degraded lands, theoretically has a negative impact on agricultural yield. However, new technologies in genetic engineering, biotechnology and natural selection process, with appropriate management, will need to be exploited and introduced to identify genetic resources that can give a productive yield. *Biosaline Agriculture* (defined as crop production on saline soils where, in most cases, seawater and brackish/saline groundwater are the only sources of irrigation water) will have a big niche in many parts of the MENA region. It is most concerned with the development and propagation of sustainable production alternatives for salt-affected lands that are deemed unsuitable for conventional farming, including: (1) more effective soil/water management and improved crop salt-tolerance, and (2) the domestication of halophytes (salt loving plants) for commercial and/or environmental cultivation. The ultimate goal of this discipline is to help provide food and water security for future generations by conserving and rehabilitating scarce resources, augmenting them with more abundant saline ones in newly emerging agro-ecosystems (Taha and Ismail, 2010).

The question is what would be the impact if a business-as-usual (without adaptation) situation continues for the agricultural sector? Figure 8 shows a clear impact of improving the agricultural yield if strategies are adapted to combat climate change. Both for Middle East and Africa regions, the impacts are double on the positive end than the business-as-usual approach.

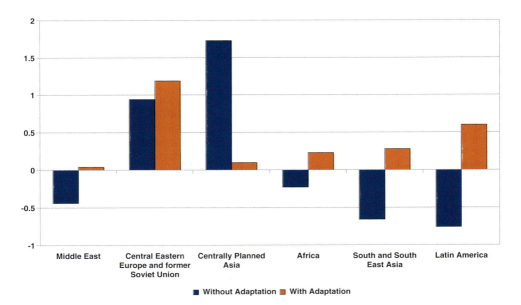

Fig. 8.   Climate change impact on agriculture. Based on 2.5°C increase in global mean temperature. Figures are expressed in % change from reference GDP projection.
Source: IFAD/GEF, 2008 (Adapted from Tol, 2002).

In general, with changing climate patterns, altering the timing or location of cropping activities will become very important in the region. Furthermore, in order to improve the livelihoods of poor farmers of the region, diversifying income through altering integration with other farming activities such as livestock raising will also play a key role.

Conclusions

Strategies to combat climate change needs to be prepared at different levels; i.e. from on-farm levels to the national and regional levels. This also needs to be supplemented with short- and long-term objectives keeping in mind the resources, projections and the economic implications to implement. As discussed earlier, the two key focuses will be on water management and the crops (and cropping systems) to improve the local/regional agricultural production systems.

In order to sustain agricultural systems, water will be needed and with increasing water scarcity projections, more efficient and/or less agricultural water consumption, including increased use of marginal waters will need to be part of the national strategies. The marginal water will mainly include the groundwater, in addition to the treated waste water in some countries (Jordan, Tunisia, Morocco, etc.).

Rainfed agriculture will be minimized and more irrigated agriculture will ensure production of many crops currently grown in the MENA region. In addition to the availability of water, irrigation technologies will need to be tested, introduced and adapted in the region. It will be equally important that these technologies are simple and cost-effective in consideration of the poor farmers of the region. Also, the government will need to assist the innovative agricultural technologies as part of their water/food security policies, targeted subsidy and investment plans.

References

Agarwal, A., M. S. S. Delos Angeles, R. Bhatia, I. Chéret, S. Davila-Poblete, M. Falkenmark, F. Gonzalez-Villarreal, T. Jonch-Clausen, M. Ait Kadi, J. Kindler, J. Rees, P. Roberts, P. Rogers, M. Solanes, and A. Wright, A. 2000. *Integrated Water Resources Management*. Global Water Partnership/Technical Advisory Committee (GWP/TAC) Background Papers, NO. 4. Global Water Partnership, Stockholm.

AOAD (Arab Organization for Agricultural Development). 2007. *Strategy for Sustainable Arab Agricultural*. League of Arab States AOAD, Cairo.

Barghouti, S., 2010. Water Sector Overview. *In* El-Ashry, M.; Saab, N. and Zeitoon, B., (eds.), *Arab Environment Water: Sustainable Management of a Scarce Resource*. Report of the Arab Forum for Environment and Development, Cairo.

Cai, X.L., D. Molden, M. Mainuddin, B.R. Sharma, M. Ahmad and P. Karimi, 2011. Producing more food with less water in a changing world: Assessment of water productivity in 10 major river basins. *Water International*, 36(1):42–62.

Cai, X.L. and B.R. Sharma. 2010. Integrating remote sensing, census and weather data for an assessment of rice yield, water consumption and water productivity in the Indo-Gangetic river basin. *Agricultural Water Management*, 97(2):309–316.

Elasha, B. 2010. Mapping climate change threats and human development impacts in the Arab Region. United Nations Development Program, Regional Bureau, Arab Human Development Report Research Series.

El-Fadel, M. and E. Bou-Zeid. 2001. Climate change and water resources in the Middle East: Vulnerability, socio-economic impacts and adaptation. *Nota dil Lavoro* 42: 2001. Foundazoine En Enrico Mattei. Available at http://www.feem.it/web/activ/_acctiv.html

El-Dukheri, I., N. Elamin and M. Kherallah. 2011. "Farmers' Response to Soaring Food Prices in the Arab Region". *Food Security*, 3(1):149–162. doi:10.1007/s12571-010-0098-9.

El-Raey, M. 2009. Impact of Climate Change: Vulnerability and Adaptation. *In* Tolba, M. and Najib S. (Eds.), *Arab Environment Climate Change: Impact of Climate Change on Arab Countries*. Arab Forum for Environment and Development, Beirut.

Evans, J. 2009. 21st Century Climate Change in the Middle East. *Climatic Change*, 92:417–432, doi:10/1007/s10584-008-9438-5.

Evans, J. 2010. Global warming impact on the dominant precipitation process in the Middle East. *Theoretical and Applied Climatology*, 99:389–402, doi:10.1007/s00704-009-0151-8.

FAO. 2008. *Climate Change and Food Security: A Framework Document*. Food and Agriculture Organization, Rome.

Giorgi, F. 2006. Climate change hot-spots. *Geophysical Research Letters*, 33: L08707. doi:10.1029/2006GL025734.

IFAD/GEF. 2008. *IFAD/GEF Partnership on Climate Change: Fighting a Global Challenge at the Local Level*. IFAD: Rome.

Molden, D., T. Oweis, P. Steduto, P. Bindraba, M.A. Hamjra and J. Kijne. 2010. Improving agricultural water productivity: Between optimism and caution. *Agricultural Water Management*, 97(4):528–535.

Milly, P.C.D., K.A. Dunne and A.V. Vecchia. 2005. Global pattern of trends in stream flow and water availability in a changing climate. *Nature*, 438(17):347–350.

Solomon, S., D. Qin, M. Manning, Z. Chen, M. Marquis, K.B. Averyt, M. Tignor and H.L. Miller, (eds.), 2007. *Contribution of Working Group I to the Fourth Assessment Report of the Intergovernmental Panel on Climate Change*. Cambridge University Press: Cambridge, United Kingdom and New York, NY, USA.

Schmidhuber, J. and F.N. Tubiello. 2007. Global food security under climate change. *PNAS*, 104(50):19703–19708.

Slam, G. 2009. Food supply crisis and the role of agriculture in the Middle East and North Africa (MENA) region. Economy and Territory, pp. 234–237.

Taha, F. and S. Ismail. 2010. Potential of marginal land and water resources: Challenges and opportunities. *Proceedings of the International Conference on Soils and Groundwater Salinization in Arid Countries*, pp. 99–104, Sultan Qaboos University.

Taha, F., S. Ismail and A. Dakheel. 2005. Biosaline Agriculture: An international perspective within a regional context of the Middle East and North Africa (MENA). Keynote Paper on 'Food Security and Use of Non-Conventional Water Resources' *Int. Conf. on "Water, Land and Food Security in Arid and Semi-Arid Regions*, September 6–11, 2005. pp. 255–278. Bari, Italy.

Tol, R.S.J. 2002. Estimates of the damage costs of climate change Part 1: benchmark estimates, *Environmental and Resource Economics*, 21: 47–73.

UNDP. 2009. *Arab Human Development Report: Challenges to Human Security in the Arab Countries*. United Nations Development Program, New York.

Verchot, L.V. 2007. *Opportunities for Climate Change Mitigation in Agriculture and Investment Requirements to take Advantage of these Opportunities. Report to the UNFCCC Secretariat, Financial and Technical Support Program.* Nairobi: World Agroforestry Centre.

World Bank. 2007. *Making the Most of Scarcity: Accountability for Better Water Management in the Middle East and North Africa.* World Bank, Washington, D.C.

World Bank. 2011. *Middle East and North Africa Water Outlook.* World Bank, Washington DC.

Chapter 9

Israeli Perspectives on Climate Change Influences on Semi-Arid Agriculture, Forestry and Soil Conservation

Shabtai Cohen*, Guy Levy[†], and Meni Ben-Hur[†]

*Department of Environmental Physics and Irrigation and
[†]Department of Soil Chemistry, Plant Nutrition and Microbiology
Institute of Soil, Water and Environmental Sciences
A.R.O. Volcani Center, POB 6
Bet Dagan, Israel 50250
*vwshep@agri.gov.il

Overview of Israeli Climate and Agriculture

Israel's climate is semi-arid to arid, potential evapo-transpiration greatly exceeds rainfall, and water resources are scarce. More than half of the country is technically desert. Soils are varied, ranging from sandy dunes to heavy clay soils. All are poor in organic matter and some are saline. These characteristics, which are common in the Middle East, can easily lead to land degradation and abandonment. However, taking advantage of favorable temperatures for crop production in most of the year has led to high productivity winter rain-fed agriculture and year-round irrigated agriculture.

Israel has traditionally given high priority to agricultural development. Israeli agriculture is today a high-tech industry, with an annual output exceeding $7 billion, and backed by an active local agricultural research community (IMS, 2011). Similar to other developed economies, it employs only 2% of the population, and its output is a small part of the gross national product. More than half of the land area is in the Negev desert, and therefore only productive if irrigated. As of 2007, Israeli agriculture occupied 13% of the land area and forests another 8%, both land uses covering much of the semi-arid northern part of the country, where annual rainfall exceeds 500 mm per year. Where possible, extensive rainfed agriculture is practiced in the winter. This includes much of the northern half of Israel. In the northern part of the Negev region, where rainfall is between 200 and 500 mm, extensive winter rainfed agriculture is maintained. In the dry summer season, almost all agriculture is irrigated.

Forest planting, a related land use, has also been given priority. Woodlands are strictly non-irrigated and have been found to thrive down to the southern areas where rainfall averages ~300 mm per year, which is the dry timberline for *Pinus halepensis*, the pioneer species used in most afforestation projects. Forests can be established if they are planted with two-year-old saplings after the first rains. At the lower end of the rainfall range no natural rejuvenation occurs. Currently, forests (natural and planted) occupy ~8% of the area of the state of Israel. However, per capita, the forested area in Israel is relatively low, only 0.2 Km^2 per 1000 inhabitants (CBS, 2008).

Since the 1980's, agricultural land area has been decreasing slowly (~1% per year, relative to 1970) while forest area has risen continuously since the 1960's (~3% per year). This trend reflects the main limit to agriculture, which is water availability. The fraction of intensive agriculture is increasing, which enables Israeli farmers to continuously increase production while using less land and water.

National priorities, infrastructure and food security

Most of the bread grain used in Israel is imported. Thus, Israeli agriculture is focused on cash crops and less on food security. This may seem surprising, given the political tension in the Middle East, which might encourage a greater focus on food security. But apparently the high cost of irrigation, relatively low cost of imported grains and free economy has over-ridden these concerns. In any event, the existence of an active agricultural industry does provide long term food security in that local agriculture can make the transition to food staples if necessary and if the price is right.

Although government controls on agricultural production are weak, Israeli agriculture is subsidized. These subsidies include (mainly) reduced water prices and public funded agricultural research. The former subsidy is currently being reduced, while the latter remains a main feature of Israeli agriculture (and a large portion of the Ministry of Agriculture's budget), which is constantly seeking new and improved crops, production systems, and enhanced water use efficiency. This is enabled by the high level of education of Israeli farmers, resulting in a knowledge-based system.

Israeli agriculture is thus highly dependent on well developed infrastructures, including education, energy, water, machinery and availability of capital[1]. These

---

[1] Availability of capital allows farmers to invest in expensive technology. In many countries, this depends on private ownership of agricultural land, which enables farmers to obtain mortgages. In Israel, almost all agricultural land is

dependencies make it vulnerable to international monetary and energy crises. Agricultural limitations as well as supply and demand of the local and international markets determine which crops will be preferentially grown each year.

Perturbations and Resilience in Response to Change

Easterling (2011) presented and discussed a conceptual framework for evaluation and assessment of adaptation to climate change. In his framework, based on that of the IPCC Third Assessment Report, a system is exposed to a change, initial effects are perceived and an impact is felt, then adaptation measures are applied, with residuals remaining from the impact. Policy responses, research, and planning feed back into the framework. We will discuss actual climate changes that have been observed in Israel. But first, we consider the resilience (or ability to adapt) of Israeli agriculture.

The rate of response of a system to a gradual change is much harder to quantify than its response to a step change or perturbation. Perturbations that require or result in adaptation can include catastrophic events, like droughts, floods or forest fires. Disastrous results of these often spur politicians to make populist policy changes, although changes in long term management may also result. The politically motivated responses can also confuse the analysis of system resilience. Analysis of the response to less publicized perturbations may offer a more appropriate opportunity to evaluate the resilience of agriculture. These can involve changes in agricultural policy or the introduction of innovations that transform the system.

Israeli agriculture has undergone a few large transformations that can help us in this evaluation. One is the transition from extensive irrigation, i.e. flood and sprinkler, to intensive and precisely focused irrigation (drip). That transformation was gradual and changed the face of agriculture over a period of several decades. Recently, an increasing amount of Israel's fresh water is being produced from seawater desalination. The high-quality desalinated water is allocated for domestic use, but farmers who irrigate with fresh water sometimes find it in their water supply, which can lead to nutrient deficiencies (Yermiyahu *et al.*, 2007). The step-change in water quality is recent and the response has not yet been played out. Another relatively abrupt perturbation was the transition of agriculture, for approximately half of its irrigation water, from fresh water to treated-effluent wastewater. It began in the late 1980's with a national program for wastewater treatment. There was no public outcry and no political response, and it was dealt with by the agricultural community quietly.

---

publically owned and cannot be mortgaged. Thus, Israeli agriculture focuses on cheaper and ever more intensive and innovative technologies.

An overview of the system response to that, with its implications for resilience and adaptive capacity, follows.

Resilience and adaptation — the case of the transition to effluent irrigation

Fuchs (2006) studied the response of Israeli agriculture and its research community to the supply of treated sewage effluents following the opening of the main wastewater reclamation project and treatment facility in the center of the country (Dan region) in 1987. The following is a synopsis of that study. In 1988, the National average irrigation rates (mm year^{-1}), which had decreased continuously for over 30 years, suddenly began to increase. The increase continued for the next ten years, after which the decrease resumed at an accelerated rate. The decreasing rate of irrigation is a normal feature of Israeli agriculture, as noted above, which is constantly introducing improved irrigation techniques and systems.

The increased rates of irrigation following 1988 can be attributed to the farmers' concerns about the use of effluents, with their relatively high mineral contents and high salinity ($\sim$290 mg Cl per liter), leading to the use of increased leaching fractions in irrigation volumes. Other possible reasons, e.g. changes in water prices, can be ruled out.

This initial reaction of farmers could only be reversed by improved knowledge and systems for appropriate application of effluents. Apparently the perception that 'we have a real problem' took several years to sink in. The number of research projects funded by the Ministry of Agriculture on effluent irrigation increased only four to five years later, in the early 1990's, especially from 1993–1994. Most research projects of this type are funded for three years. Permitting one year for implementation of research findings by the farmers, we get a good match to the date when the decreasing rates resumed.

The findings of Fuchs' (2006) study can teach us about the resilience of Israeli agriculture, which may be a paradigm for other Western agricultural systems. 'Exposure' (quotes are used for Easterling's (2010) conceptual framework) to the large influx of effluents began in 1988. 'Initial impacts and effects' could be felt within a few years as farmers increased irrigation rates. More advanced adaptive measures ('autonomous adaptation') awaited the results of research efforts, which only took off some five years after initial exposure. Thus, although crude initial responses involved increasing irrigation rates, real mitigation responses were only implemented after some 10 years. It is important to note that 'residuals of the net impacts' do exist and currently there is ongoing research in Israel on mitigation of the long-term detrimental effects of effluent irrigation (e.g. Bar Tal *et al.*, 2011).

Agriculture's Contribution to Climate Change: Evidence from Israel

Effects of land use changes

Agricultural land use can contribute to climate change through various mechanisms, their importance varying with the land use prior to conversion, as well as the type of agriculture practiced.

The most important direct effect is a modification of the radiation and energy balance of the land surface: Crops growing under the intensive irrigated agriculture practiced in Israel have surfaces which are darker and cooler than most other forms of land-use, reflecting less short wave solar radiation (Stanhill, 1966) and radiating less long wave terrestrial radiation than the land surfaces they replace; they also transpire more water, thereby reducing the sensible heat available to warm the air.

In Israel, these effects were triggered by the construction of the National Water Carrier in 1964, which led to a rapid expansion of the area of irrigated agriculture and its extension into Israel's arid southern regions, primarily during the rainless summer months. Direct measurements have confirmed the local cooling and moistening effects of irrigated agriculture (Stanhill, 1966; Stanhill *et al.*, 1973). Afforestation is another form of land use that strongly modifies the radiative and aerodynamic characteristics of land surfaces and results in micro-climate changes (Stanhill and Fuchs, 1968; Kalma and Stanhill, 1972), which is an important factor in the increasingly important recreational role of forested areas.

As noted above, land use changes have influenced surface albedo in Israel. Ben Gai *et al.* (1998) studied the changes that followed the introduction of the National Water Carrier aqueduct in 1964. This >200 km aqueduct transports water pumped from Lake Kinneret (i.e. the Sea of Galilee) in the Northern part of the country and 200 m below sea level, across lower Galilee and then southward to the central and southern Coastal Region. The aquaduct provides water for drinking and irrigation. Its introduction led to major increases in the area of irrigated agricultural land in central and southern Israel over a period of several decades. Ben Gai *et al.* (1998) present maps of land use for those parts of Israel for the 1930's, 1960's and 1990's, which were used with Aircraft-based measurements of albedo for the different land-use types to estimate the overall changes in albedo. In general, for most of the area mapped, albedo had changed by between 10 and 20% over that period. They conclude that these decreases in albedo have led to increased absorption of solar radiation at the surface. The influx of irrigation water has also increased latent heat flux and these together, in the first part of the rainy season, were advanced to provide an explanation for the observed changes in rainfall patterns. Those changes are not straightforward and are discussed below in the section 'Rainfall.'

Effects of intensive production systems

A secondary, indirect effect of agricultural land use on climate is through the effect of the fossil fuel consumed in providing the inputs used in intensive agriculture. The combustion of fossil fuels, by releasing CO_2 and other radiatively active gases into the atmosphere, is a major contributor to climate change (IPCC, 2007). In Israeli agriculture, major uses of the fossil fuel are in the imported animal-feed grains, the provision of water for irrigation, the agrochemicals for fertilization and plant protection, as well as the cultivation, harvesting and transportation of agricultural products (Stanhill, 1974).

Analyses have shown that the remarkable increases in the efficiency with which Israeli agriculture uses its limited resources of land, water and manpower have been achieved at the cost of an increased dependence on fossil fuel-based inputs. This has resulted in a negative energy balance in which the fossil fuel needed to sustain intensive irrigated agriculture exceeds the metabolic food energy which it produces. These trends, noted by Stanhill in 1974, have continued, i.e., agricultural land area has decreased while energy use by agriculture has increased. Data for electricity supplied to agriculture can give an idea of this trend. In 1970, agriculture used 200 million KWh (or GWh), which increased to 930 in 1990 and 1600 by 2000. However, the rate in 2010 (1610 GWh) may indicate that electrical energy use has been leveling off in the last decade. Some fraction of the large electrical energy requirements now being used for desalination may be attributed to agriculture in that this provides the necessary sink for domestic waste water.

Observed Climate Change in Israel

Goldreich (2003) reviewed what is known about changes in Israel's climate in the past and some of the work on expected changes in climate in the future. Up-to-date detailed analysis of temperature and rainfall changes in Israel and review of some of the simulations of future climate are available in the slide shows presented at a meeting on climate change in Israel in January, 2012, at the Israel Meteorological Society website (http://met-society.org.il/climate12.htm).

Here we will focus on several aspects that are especially relevant to agriculture and analyses from published papers.

Israel's climate is Mediterranean in the North and arid desert in the South. In this broad semi-arid transition zone rainfall is a major determinant of water availability for agriculture and forestry, but there are large fluctuations in rainfall from year to year. In addition, evaporation rates are very high, and altogether return three quarters of the rainfall back to the atmosphere. Hence for Israeli agriculture, as in other places where water availability is limited, the major concern over climate change is the possibility that rainfall will decrease and aridity will increase. This would imply increases in crop water requirements and decreasing water availability, while in forestry it might result in increased forest mortality from desiccation and related causes (McDowell *et al.*, 2011). In response to the latter, forest managers would need to reduce tree density in order to sustain the forest with less available soil water (e.g. Moreno-Gutiérrez *et al.*, 2011).

Reference evapo-transpiration, the main information necessary for irrigation scheduling, is best computed from a suite of climate variables, mainly air temperature and humidity, wind speed and solar radiation. Pan evaporation is also useful and has been used as the standard reference. Much of the climate data collected in Israel has been for this purpose and is normally made available to farmers from government and other public sources. Here, we review the changes that have occurred in these parameters and summarize their possible influence on the relevant issues above.

Temperature

The Israel Meteorological Service studied the temperature records from five stations with continuous data from 1951–2011. The stations were selected to represent a transect of the length of Israel. They found that temperature declined from the early 1950's relatively high temperatures to lower temperatures during the 1970's, leveled off in the 1980's, increased markedly during the 1990's and then leveled off at temperatures higher than in the 1950's during the most recent decade (Fig. 1). The past 15 years, i.e. 1996–2010, were the warmest period. 2010 was the single hottest year, but the warming trend is true even if that year is omitted from the analysis.

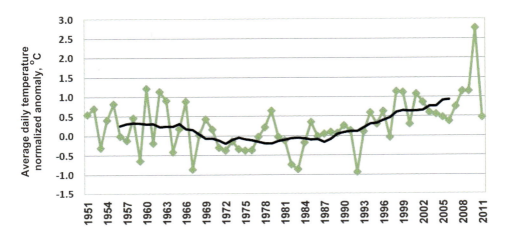

Fig. 1. Average temperatures for five representative stations in Israel. From IMS (2011).

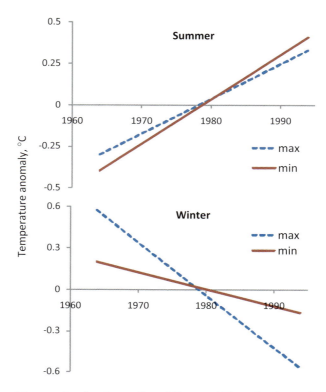

Fig. 2. Fitted serial regression for the median of T_{max} and T_{min}, for the cool and warm seasons 1964–1999. From Ben Gai *et al.*, 1999.

The main contribution to the warming trend is from the minimum temperature values, which have increased mostly in the summer months (June, July and August). For average temperatures the early decline was about 0.5°C, and the latter increase about 1°C, so that recent temperatures are about a half a degree higher than in the 1950's.

Ben Gai *et al.* (1999) studied temperatures measured at 40 stations over the period 1964–1994. They found that the summer has become warmer while the winter has become colder (Fig. 2). In the warm season (April–October), minimum temperatures (T_{min}) increased more than maximum temperatures (T_{max}), and thus the temperature range also decreased. The opposite pattern was observed in the cold season (November–March), when T_{max} decreased faster than T_{min}. Thus, a general decrease in daily temperature range was observed, while the annual temperature range increased. The summer and winter trends were in agreement with the 500–1000 hPa geopotential thickness patterns observed at the Israel Meteorological Service's main site at Bet Dagan in the coastal plain of central Israel where radiosonde measurements are made daily. There, in February and March, significant decreases in thickness were observed, while in August, September and October thickness increased. For all surface data, the average yearly changes for T_{max}, T_{min} and T_{range} were $-0.03°C$, $0.05°C$ and $-0.06°C$ per annum, respectively (statistically significant at $p<0.1$). The increase in mean temperature, $0.02°C$ per annum is approximately twice that of the global land-ocean temperature index for the same period.

Another change is the incidence and length of extreme temperature events. Extremely high temperatures and several-day periods of those temperatures have been found to occur more frequently (Saaroni, 2012). These can have important impacts on agriculture, since they can inflict severe damage to crops. An example is the heat wave in August 2010, when temperatures exceeded $40°C$ in much of Israel for several days. This resulted in sun-scald on much of the apple crop and reduced yields. It is interesting to note that there have been trials in Israel with deployment of 10–20% white or clear shade screens above apple orchards to reduce orchard water use and improve fruit quality (e.g. Tanny *et al.*, 2009). Under the screens during that heat wave sun-scald was minor. Thus a practical mitigation measure to protect from sun-scald during heat waves is already tested and available.

Humidity and wind speed

Humidity and wind speed are important climate factors affecting evaporation and aridity, but rarely studied in the context of climate change. Thus IMS (2011), Ben Gai *et al.* (1999), and Kliot *et al.* (2008) ignored humidity data when they analyzed climate changes in Israel. Actually, most studies in Israel focus on temperature and rainfall data only. Roderick and Farquhar (2002) discussed the causes for worldwide decreases in evaporation during the second half of the 20th century. Based on a study of data from the US, which found that vapor pressure deficit (VPD, a function of temperature and humidity) had not changed during that period, they proposed that the lack of change was due to the fact that minimum temperatures (T_{min}) increased faster than average temperatures during that period. As noted above, T_{min} in Israel increased faster than T_{max} (Ben Gai *et al.*, 1999), which may also have stabilized VPD. However, analysis of data from Bet Dagan in the central coastal region from 1964 to 1998 showed that VPD increased significantly during that period (Cohen *et al.*, 2002), and was a major factor in the increase of the aerodynamic component of reference evaporation. In addition, relative humidity for a number of Israeli stations has not changed from 1975–2010 (Saaroni, 2012) while temperatures have increased, so the increase in VPD observed at Bet Dagan for the early period may indicate a general trend that will continue henceforth. More work will be needed in order to determine whether the trends at Bet Dagan are indeed representative of the rest of the country.

Wind speed trends in Israel have only been studied for Bet Dagan, where they increased from 1964–1998 (Cohen *et al.*, 2002). There, predominant wind direction also changed, which may result from land use changes and the growth of the nearby city of Tel Aviv and its suburbs. McVicar *et al.* (2012 and references therein) found that worldwide wind speeds have decreased in the past half century and that this is the main reason for global decreases in reference evapo-transpiration.

Solar radiation

Solar radiation is the major factor determining evaporation rates from moist surfaces. Because accurate measurements of solar radiation are demanding and expensive few data are available. However some good measurements are available for Israel which show that in the central and northern parts of the country solar radiation decreased significantly between the 1950's and 1990's (Stanhill and Moreshet, 1992; Stanhill and Ianetz, 1997; Stanhill and Cohen, 2001) but subsequently recovered to a large extent, although the overall trend since the 1960's is negative (i.e. global dimming exceeds brightening). In the southern desert region no significant trends were found (David Faiman, personal comm.). The causes of the decreases are apparently increasing haziness and cloudiness, probably caused by increased aerosol loading of the atmosphere from local and/or upwind sources (e.g. Europe). The rate of decrease at Bet Dagan and Jerusalem was large, $\sim 0.9 \, \mathrm{W \, m^{-2} \, a^{-2}}$, equivalent to $> 10\%$ in 30 years.

Evaporation and evapo-transpiration

Analyzing the observed trends in temperature, humidity, solar radiation and wind speed, and data from Class A evaporation pan measurements, Cohen *et al.* (2002) showed that their total influence on evaporation at Bet Dagan was minor, evidently because the increases in the aerodynamic component of evaporation were offset by the decreasing radiative component. The increases in the aerodynamic component resulted from the combined increases in VPD and wind speed. However, this means that in conditions when the radiative component is more important, e.g. for evaporation from large open water surfaces, evaporation decreased significantly. Similarly, for situations where the aerodynamic component is more important, e.g. in shady areas or urban climates, evaporation increased. The validity of the latter conclusions may be limited because it is not clear to what extent land use changes in the vicinity of Bet Dagan influenced the changes observed in humidity and wind speed. This is especially important because worldwide wind speed has been decreasing together with pan evaporation (e.g. McVicar *et al.*, 2012), while at Bet Dagan both increased during that period. In other places in Israel, there is evidence that pan evaporation decreased slightly over the same period (Alpert and Ben-Zvi, 2004). Evaluation of changes in evaporation and its components in the past two decades awaits further research.

Rainfall

In Israel, rainfall decreases from North to South, and from West to East. Thus, mean annual rainfall exceeds 1000 mm/year in some parts of the Northern Highlands, but is only ~40 mm/year in the central eastern rift valley near the Dead Sea and the southern parts of the Arava valley (also in the African to Syrian Rift zone). This rainfall variation is associated with the differences in climate noted above.

Israel's water requirements are larger than its annually renewable water resources or average rainfall recharge capacity. So, in the long term, Israel has had to increment it's available water by desalinating sea-water and by extensive recycling of wastewater, the latter used for irrigation. These measures make the state less dependent on rainfall, which is much below potential evaporation. Even so, any change in rainfall can have a large impact on Israel's water supplies. Thus, there have been many analyses of rainfall, its distribution and intensity. In all studies, most trends are minor and many reported trends are not statistically significant. IMS (2011) analyzed rainfall trends and rainfall distribution for their network of Israeli stations from 1950–2010. They concluded that there have been no significant trends in total rainfall or in rainfall distribution in Israel. Similar conclusions were reported by Stanhill and Rosa (2010).

A more detailed analysis of rainfall data has found the following trends:

- Fewer rain events with higher rainfall and rainfall intensity per event (Ben Gai *et al.*, 1998).
- Regional trends, and in particular decreases in rainfall in the watershed of Lake Kinneret (Givati and Rosenfeld, 2004), which have been attributed to changes in anthropogenic aerosol loading in the region. These findings were subsequently contested by Alpert *et al.* (2008).
- A shift in rainfall to later in the season (Yosef *et al.*, 2009).
- Increased occurrences of long gaps with no rainfall during the regularly wet season (Saaroni *et al.*, 2012).

All of the above trends can have significant impacts on agriculture.

Future Climate Projections

Large scale simulations show that Mediterranean regions will become warmer and drier by the end of the 21st century and the local hydrology and agriculture are likely to be significantly influenced by these changes (IPCC, 2007). There have been efforts to simulate future climate for Israel in particular. However, Israel is a small country with several climate zones and a varied geography, and grid cell sizes for most Global Climate Model (or General Circulation Models, GCMs) simulations are large (200 km or more), so until recently, only regional (rather than local) projections have been made. Current and near future efforts in this field are focused on GCM's with a much reduced grid cell size for Israel.

Kitoh *et al.* (2008) reported the first results of a run of the Japan Meteorological Agency (JMA) model for the Middle East with a 20 km horizontal grid cell size. The model was run for current conditions based on measurements during the period of 1980–1999, and then for the period 2080–2099. The model predicts the disappearance of the so-called "Fertile Crescent" due to significant reductions in precipitation and stream flow. These predictions include the central and northern coastal regions of Israel. Krichak *et al.* (2011) reported simulations of future climate until 2060 in the Eastern Mediterranean, made in Trieste with the RegCM3 model using the IPCC A1B scenario. Their results show a warming of approximately 0.5°C per decade for near surface (2 m) temperature in most of Israel, with a decrease in annual rainfall of up to 30 mm/decade. The larger decreases in rainfall are for the wetter regions of Northern Israel, and the trend decreases in proportion to average annual rainfall. This decrease of rainfall is about 5% of the prior annual totals. Similarly, Jin *et al.* (2011) ran the 20 km resolution JMA model for 1979–2007 as compared to 2075–2099. They predict reductions in rainfall amounts of a similar magnitude as well as reduced rainfall duration (days per year), increased rainfall intensity and increased evaporation rates.

Climate Risks to Semi-Arid Soil Conservation

Soils are a limited resource and a majority of dryland arable soils are endangered. The main risks to soils are erosion by natural and man-made factors and degradation by inappropriate land use (Hillel and Rosenzweig, 2011). Soil conservation is a major objective today as it will be in the future. Here we discuss several soil properties that will be at risk as climate changes. Experimental and theoretical progress is necessary in order to prepare for these changes and predict their potential severity if we are to maintain soils for the agriculture needed in the future.

Soil hydraulic properties, surface runoff and soil erosion

Surface runoff and water-driven soil erosion within and from arable lands can lead to widespread land, water and environmental deterioration, in terms of water loss, soil degradation, and increased environmental hazards (Ben-Hur, 2004, 2008). Runoff water from a cultivated field is not available to the crop and it increases soil erosion and fertilizer and pesticide depletion, which could lead to decreased fertility and crop production as well as increased environmental pollution. The most common irrigation systems in Israel are drip and moving sprinkler. No runoff and soil loss is expected in drip irrigation systems because of the low impact energy of the water drops. However, due to the low structural stability of some arable soils in Israel and the high impact of the water drops in rainfall and moving sprinkler irrigation systems, runoff and soil loss are expected in these cases (Fig. 3, Ben-Hur and Assouline, 2002; Aggasi *et al.*, 1989; Agassi and Ben-Hur, 1991; Ben-Hur, 1991). The effects of runoff on spatial distribution of soil water content and peanut (*Arachis hypogaea* L.) yield under a moving sprinkler irrigation system in a field with silt loam soil and 3% slope in the western Negev of Israel were studied by Ben-Hur *et al.*, (1995), who found that local runoff increased the soil water content and the peanut pod and canopy yields in the downhill direction. Preventing surface runoff increased average pod yield by 880 kg/ha.

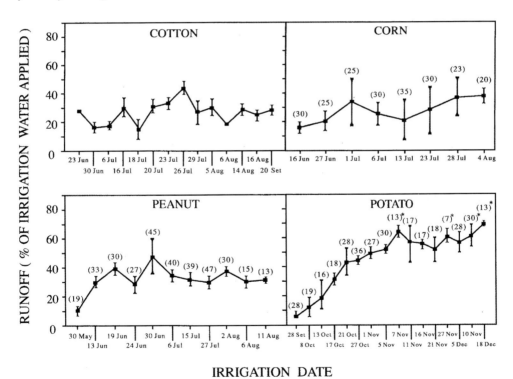

Fig. 3. Runoff percentages of representative irrigation events and rainstorms, from four different crops, as measured by means of field runoff plots <4.5 m^2 in size. Verticals bars and * represent SD and natural rainstorm, respectively, and the numbers above the symbols indicate the irrigation or rainstorm amount in mm (from Ben-Hur, 1994).

Surface runoff occurs when the intensity of rainfall or irrigation exceeds the infiltration rate (IR) and the surface storage capacity of the soil. In arid and semiarid regions like Israel, the main factor that controls the soil IR under water-drop impact conditions (e.g. rainfall or sprinkler irrigation) in cultivated fields with annual crops, (where the soil surface is bare at the beginning of the growing season, and in bare areas between tresses in orchards), is seal formation at the soil surface (Morin *et al.*, 1981; Ben-Hur, 2008). This seal is relatively thin and is characterized by greater density, higher strength, finer pores, and lower saturated hydraulic conductivity than the underlying soil (McIntyre, 1958; Chen *et al.*, 1980; Wakindiki and Ben-Hur, 2002; Assouline, 2004).

Agassi *et al.* (1981), Morin *et al.* (1981), and Lado *et al.* (2004) noted that the formation of a structural seal is a result of three complementary mechanisms: (i) physical disintegration of soil aggregates, caused by the raindrops' impact; (ii) aggregate slaking as a result of fast wetting of the soil; and (iii) the physicochemical dispersion of soil clay particles, which migrate into the soil with the infiltrating water and clog the pores immediately beneath the surface, forming a "washed-in" zone. The importance of the last mechanism depends on the electrical conductivity (EC) of the soil solution and the exchangeable sodium percentage (ESP) of the surface soil. As the EC decreases and ESP increases, the clay dispersion is enhanced and the reduction in IR caused by seal formation becomes more pronounced (Agassi *et al.*, 1981; Kazman *et al.*, 1983; Ben-Hur *et al.*, 1998). Moreover, an increase of ESP decreases the stability of the soil structure, and this, in turn, could exacerbate soil disperion, detachment, and loss (Agassi *et al.*, 1994).

Soil erosion by water involves two main processes: (i) detachment of soil materials from the soil surface by raindrop impact and surface runoff shear; and (ii) transport of the resulting sediment by raindrop splashing and overland flow (Baver *et al.*, 1972). Erosion models have separated the erosion process into two main components: rill and interrill erosion (Foster, 1982). Runoff from the soil surface may be funneled into small erodible channels known as rills. In rill erosion, the soil loss is mainly due to detachment of soil particles and their transport by the erosive forces of the flowing water. In contrast, in interrill erosion, soil detachment is caused essentially by raindrop impact, and soil transport is due to raindrop splashing and runoff flow (Watson and Laflen, 1986).

The effects of the expected climate changes on surface runoff and soil loss can be classified as two main types: (i) Short term effects — changes in water conditions in the soil and in soil hydraulic properties that occur during and shortly after rainstorms, and (ii) indirect and long term effects — lasting changes in soil physical and chemical properties, in water resources, and in vegetation types and status in cultivated fields.

Short term effects of decreasing rainfall with increased rainfall intensity

For short term effects, changes in rainfall intensity and soil water conditions are the main considerations. Surface runoff and soil loss are related to rainfall intensity (Kinnell and Cummings, 1993; Bradford and Foster, 1996; Ben-Hur and Agassi, 1997; Huang, 1998; Fox and Bryan, 1999; Kinnell, 2000; Romkens *et al.*, 2002), which, according to Jin *et al.* (2011), is expected to increase significantly in Israel in the coming decades. Increased rainfall intensity should increase runoff and soil loss rates proportionately, but when a seal is formed quickly at the soil surface, no simple relationship was found between soil loss and rainfall intensities ranging from 10–103 mm/h (Uson and Ramos, 2001).

The effects of rainfall intensity on final IR (IR value at the end of the rainstorm), runoff, and soil loss can be studied from the results presented in Fig. 4. In this study, disturbed samples of a sandy soil from the Coastal Plain of Israel were packaged in $0.3 \times 0.5 \times 0.02\,m^2$ trays and exposed to 70 mm of simulated rainfall at two intensities, 24 and 60 mm/h, and five slope gradients, 5%, 9%, 15%, 20%, and 25%. Final IR values increased with slope for both rainfall intensities (Fig. 4a), in agreement with Shanan and Schick (1980), Poesen (1984, 1986, 1987), Warrington *et al.* (1989), Ben-Hur *et al.* (1992), Bradford and Huang (1992), Janeaue *et al.* (2003) and Ben-Hur and Wakindiki (2004). The increase in final IR with the slope (Fig. 4a) can be a result of: (i) a decrease in the normal component of the impact force of the raindrops as the slope gradient increases (Poesen, 1984, 1986; Baumhardt *et al.*, 1990; Janeaue *et al.*, 2003); (ii) an increase of erosion of the seal at the soil surface, which leads to development of a thinner and less compacted seal layer at steeper slopes (Ben-Hur *et al.* 1992: Ben-Hur and Wakindiki, 2004; Assouline and Ben-Hur, 2006); (iii) increased erosion of fine particles and their decreased penetration into the washed-in zone in the seal, which, in turn, reduces clogging of the pores.

Final IR values were higher at higher rainfall intensity for all studied slope gradients, and differences became more pronounced as the slope gradient increased (Fig. 4a). Assouline and Ben-Hur (2006) reported that the increase of soil erosion with increasing rainfall intensity (from 24–60 mm/h) and slope gradient led to development of a thinner, less compacted, and more permeable seal layer at the soil surface (Fig. 4a). However, the differences in total runoff were small, (Fig. 4b) suggesting that the effect of the predicted increases in rainfall intensity due to climate change in Israel, where soils are sensitive to seal formation, are likely to be minor. Total soil loss increased exponentially with slope gradient for both rainfall intensities, and were higher at the higher rainfall intensity (Fig. 4c).

As noted, climate simulations suggest that the Mediterranean region will become warmer and drier by the end of 21st century (IPCC, 2007). In addition, Krichak *et al.*

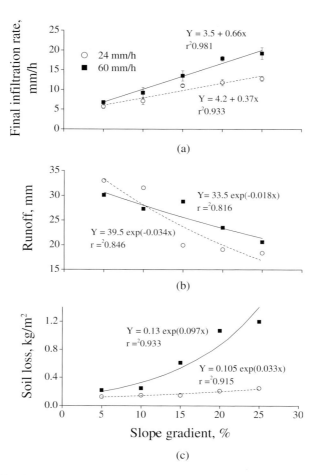

Fig. 4.   Final infiltration rate and total runoff and soil loss after 70 mm of rainfall as functions of slope gradient. Verticals bars represent two SD (after Assouline and Ben-Hur, 2006).

(2011) predict a warming of ~0.5°C per decade for near surface temperature in most of Israel, and Jin *et al.* (2011) predict a reduction in the number of rainy days per year and an increase in evaporation. These climatic changes can affect soil wetting under rainfall and irrigation, and should increase the occurences of low antecedent moisture contents in the soil before being exposed to consecutive rainstorms or sprinkler irrigation events. In this case, the wetting rates of the soil surface at the beginning of consecutive rainstorms or sprinkler irrigation events would be fast, and this could increase aggregate slaking. In addition, the duration of upper soil layer wetness (approaching field capacity) between consecutive rainstorms or sprinkler irrigation events would be shortened. The resulting differences in wetting rate and soil aging may change seal formation and IR, as well as runoff and soil loss values in Israeli soils (e.g., Levy *et al.*, 1997; Ben-Hur *et al.*, 1998; Mamedov *et al.*, 2001; Shainberg *et al.*, 2003; Lado *et al.*, 2004; Ben-Hur and Lado, 2008).

Slaking processes have been found to be important in aggregate disintegration and seal formation (Lado *et al.*, 2004). Slaking occurs when the soil aggregate is not strong enough to withstand the stresses produced by differential swelling, pressure of entrapped air, rapid release of heat during wetting, and mechanical actions of moving water (Emerson, 1977; Collis-George and Green, 1979; Kay and Angers, 1999). These stresses are termed slaking forces. Slaking is controlled by the wetting rate of the soil: the faster the wetting, the stronger the slaking forces and the more the aggregate breakdown.

Aggregate stability of six Israeli soils with different clay contents ranging from 8 to 62% and low ESP (<2) were determined under two wetting treatments: (i) fast-wetting as proposed by Le Bissonnais (1996), and (ii) slow wetting of 1 mm/h by mist. The slaking value (SV) of these soils was calculated using Eq. [X1]

$$SV = \frac{MWD_s}{MWD_f} \quad \text{[X1] (Ben-Hur and Lado, 2008)}$$

where, MWD_s and MWD_f are the mean weight diameter values of the soils after slow and fast wetting treatments, respectively. SV increases linearly with clay content (Ben-Hur and Lado, 2008). Thus increased soil clay content amplified the differential swelling, the effect of entrapped air, and the rapid release of heat during the fast wetting, which led to stronger slaking forces.

The effects of wetting rate on runoff and soil loss were studied by Lado *et al.* (2004) in three Israeli soils with low ESP (<2.3) and clay contents of 23, 41, and 62% under simulated rainstorms of 80 mm at 42 mm/h rain intensity and a kinetic energy of 18.1 J mm^{-1} m^2. Two wetting rates were studied: (i) slow, where air-dry soil was first pre-wetted at a rate of 1 mm/h to 50% field capacity with a mist of ~ 0 kinetic energy, and then subjected to a rainstorm with high impact energy; (ii) fast, where air-dry soil was exposed to a high impact energy rainstorm without being pre-wetted. No effect of wetting rates on total runoff was found for the soils with 23 and 41% clay content. In contrast, the total runoff from the soil with 62% clay and the total soil loss from the three soils were significantly higher under fast than slow wetting; the total runoff in the soil with 62% clay content under slow and fast wetting rates was 29.5 and 57.1 mm, respectively, and the total soil loss rates in the soils with 23, 41, and 62% clay contents were 360, 359, and 80 g m^{-2}, respectively, under slow wetting rate; and 900, 1,200, and 590 g m^{-2}, respectively, under fast wetting rate (Lado *et al.*, 2004). Lado *et al.* (2004) suggested that the differences in the runoff and soil loss rates in the various soils between the two wetting rates were a result of the differences in slaking forces that developed in the different wetting rates and soils. The interactive effects of soil clay content, kinetic energy of raindrops, soil ESP, and pre-wetting rate on slaking forces, infiltration rate and runoff and soil loss were discussed in detail by Ben-Hur and Lado (2008).

Another factor associated with of soil wetting that could affect soil stability is soil aging (the duration that soil remains wet and near its field capacity). Soil aging is partly attributed to the activity of microorganisms that produce chemicals that enhance soil stability during the aging process (Martin et al., 1995). In addition, cohesion forces associated with capillary water and solid-phase bonds have been hypothesized to account for soil structure strength in moist conditions (Kemper and Rosenau, 1984; Kemper et al., 1987).

Shainberg et al. (1996) studied the effect of aging on a rill erodibility factor for three Israeli soils — clay, loam, and loamy sand — using a small hydraulic flume. They found that at water contents above air-dried conditions, soil erodibility decreased with aging. This aging effect was more effective in the clay soil than in the two other soils. With no aging, soil erodibilty was 3.26×10^{-3} seconds per meter (s/m) in the clay soil and 0.52×10^{-3} s/m in the loamy sand. When aging increased from 4 to 24 hours, soil erodibility of clay decreased to 0.24×10^{-3} s/m, but that of loamy sand was not affected. They related the decrease of the soil erodibilty to the development of biological and chemical cohesion forces between soil particles during aging, which increased aggregate stability and decreased soil erodibility.

Ben-Hur et al. (1998) found that in unsealed soil (covered with vegetation) with 65% clay and high ESP (11.8%), aging had an effect on final IR after a 58 mm rainstorm. Final IR increased from 17 mm/h with no aging to 38 and 42 mm/h after four and 28 days of aging, respectively. They suggested that under near-saturation conditions, a physico-chemical mechanism supplements the microbiological mechanism to bind soil particles together in aggregates, thus countering crusting. Montmorillonite platelets (the dominant clay type in Israeli soils) are long and flexible, and can interact at several cohesive junction points, like the edges or planar surfaces (van Olphen, 1977; Keren et al., 1988). Conditions of high water content, and extended aging periods, are favorable for clay-to-clay contact, which increases the resistance of the soil aggregates to breakdown forces.

The interactive effects of soil wetting rate and aging duration on soil loss are demonstrated in Fig. 5. In this case, air-dry loam and clayey soils were first pre-wetted by a 5 mm rainstorm with a kinetic energy of 12.4 kJ/m^3 applied at three rates (1, 6, and 30 mm/h) using a drip-type rainfall simulator. The soils were left to age for either 15 min or 18 h and were then subjected to a further 60 mm rain event at an intensity of 33 mm/h. For both soils, inter-rill erosion increased linearly with pre-wetting rate, whereas aging duration reduced inter-rill erosion (Fig. 5). Differences between soil losses for the two aging durations were larger for the loam than for the clay soil, and increased as the wetting rate increased (Fig. 5).

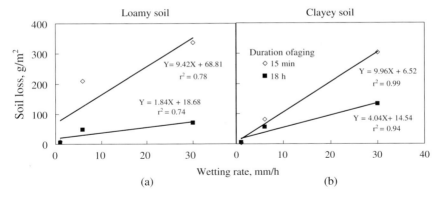

Fig. 5. Soil loss from (a) a loamy and (b) a clayey soils after two aging treatments as a function of wetting rate of the soil (from Ben-Hur and Lado, 2008; data were taken from Levy *et al.* 1997).

The above results suggest that the predicted climate changes, which will enhance fast wetting of the soils and decrease soil aging, will tend to increase runoff and rates of erosion in cultivated lands during the rainy season, particularly in soils with high clay contents (>20%).

Indirect and long term effects of irrigation with poor quality water

The fresh water resources of Israel are limited, and are expected to diminish as a result of climate changes, while the population is expected to grow, thus leading to an increasing demand for water. This trend has already led to enhanced use of marginal waters, e.g. saline water and treated domestic sewage effluents for irrigation, in order to maintain agriculture, meet the increasing demands for food, and combat desertification (Bresler *et al.*, 1982; Shainberg and Letey, 1984; Feigin *et al.*, 1991; Ben-Hur, 2004).

Large saline aquifers are located in the Negev, while treated sewage water is available in much of the country. Water quality is determined mainly by the concentration and composition of solutes. Electrical conductivity (EC) and the sodium adsorption ratio (SAR) of saline water in Israel can reach $5\,dS\,m^{-1}$, and $25\,(mmol\,L^{-1})^{-0.5}$, respectively (Bresler *et al.*, 1982; Shainberg and Letey, 1984), and that of secondary treated effluents ranges from EC 1.5–$2.3\,dS\,m^{-1}$ and SAR from 3.5–$7.3\,(mmol\,L^{-1})^{-0.5}$ (Ben-Hur, 2004). In contrast, the EC and SAR of Israeli fresh water are commonly $\sim0.9\,dS\,m^{-1}$ and $\sim2.5\,(mmol\,L^{-1})^{-0.5}$.

Long term irrigation with saline water and treated effluents can change the chemical and hydraulic properties and the structure of irrigated soil. We focus on soil salinity and sodicity. Long term (longer than seven years) irrigation with secondary-treated sewage effluent of a field with non-calcareous, sandy soil in the Coastal Plain and a field with calcareous, clayey soil in Jezreel Valley, increased SAR values of saturated paste extracts to a depth of at least 4.0 m in a non-calcareous sandy soil and to 0.7 m in a calcareous clayey soil (Lado *et al.*, 2012). The maximum SAR values in the sandy and clayey soils irrigated with treated sewage water were ~5.5 and 7.5, respectively.

Cumulative runoff and soil loss during the rainy season, which were determined in two sites in the western Negev with loess soil after long-term summer irrigation with local saline water, are presented in Fig. 6. These measurements were made in field runoff plots, each 5 m^2 in area, where the soil was cultivated and its surface was bare (without vegetation). Long term irrigation with saline water increased soil ESP to 8.4% and 22.2%. The greater ESP increased the cumulative runoff to 13.3% and 76.1% of the annual rainfall, respectively, and the soil loss rate to 0.8 and 2.5 g m^{-2} mm^{-1}, respectively (Fig. 6). Soil structure stability decreased, and the dispersion of clay particles increased during the rainstorm along with their movement into the washed-in zone in the seal. This enhanced the formation of a seal with low permeability, higher runoff (Agassi *et al.*, 1981) and exacerbated soil erodibility (Ben-Hur and Agassi, 1997).

The effects of irrigation with treated effluents on runoff from field plots, each 5 m^2 in area, during a rainy season in two sites located in the Coastal Plain of Israel are shown in Fig. 7. Soils at the sites were a non-calcareous sandy soil and a calcareous sandy clay loam. The fields were cultivated with field crops and irrigated

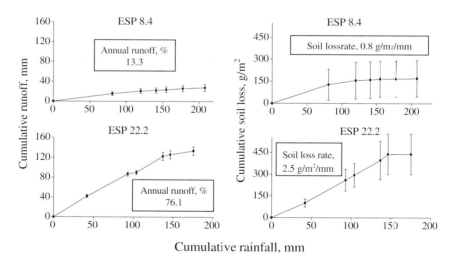

Fig. 6.   Cumulative runoff and soil loss as functions of cumulative rainfall for sandy clay loam (loess) soils with two exchangeable sodium percentage (ESP) values in two sites in the western Negev, Israel as measured by means of field runoff plots 5 m^2 in size. Verticals bars represent two SE (Ben-Hur, unpublished data).

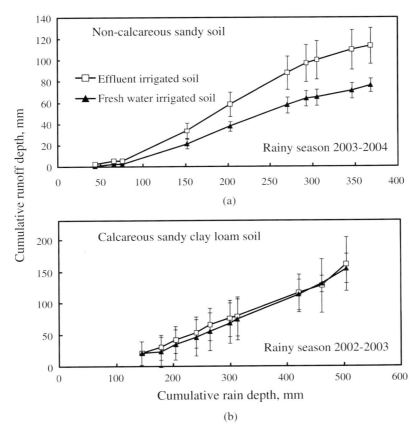

Fig. 7. Cumulative runoff depth during a rainy season in (a) non-calcareous sandy and (b) calcareous sandy clay loam soils previously irrigated with freshwater or secondary effluent. Bars indicate two SD (Ben-Hur, Lado, Gasser, Mingelgrin, and Assouline, unpublished data).

with freshwater or treated sewage effluents each summer: the field with calcareous sandy clay loam for six years, and that with non-calcareous sandy soil for two years. During the runoff measurements the soil surface in the runoff plots was bare. In the non-calcareous sandy soil, cumulative runoff was significantly higher in the effluent than in the freshwater-irrigated soils (Fig. 7a). Moreover, irrigation with effluents increased the ESP of the 0–2 cm surface layer from 1.5% to 2.5%, and this high value remained fairly constant during the entire rainy season. In contrast, in the calcareous sandy clay loam soil, no significant differences were found between the cumulative runoff from the freshwater- and the treated sewage water-irrigated soils (Fig. 7b). There, six years of irrigation with effluents increased the ESP of the 0–2 cm soil layer by the end of the irrigation season, but ESP decreased during the rainy season to values similar to those of freshwater-irrigated soil. The increased

ESP in the non-calcareous sandy soil induced seal formation and runoff during the rainy season (Lado and Ben-Hur, 2009). However, in calcareous sandy clay loam the dissolution of $CaCO_3$ during the rainstorms released Ca, which replaced the Na in the exchangeable complex, thus reducing the ESP of the upper soil layer. Consequently, seal formation and runoff during the rainy season were similar in the freshwater irrigated and effluent-irrigated soils.

Carbon Dioxide, Methane, and Nitrous Oxide Emissions and Sequestration

Recent inventories on the emissions in Israel of the three main so-called greenhouse gases (carbon dioxide, methane and nitrous oxide) were published in the Ministry of Environmental Protection of the State of Israel's Second National Communication on Climate Change. Total emissions (expressed in CO_2 equivalent units) of the three main GHGs in Israel for 2007 reached nearly 77 million tons. Most of this emissions (87%) was contributed by CO_2 (~67 million tons), with methane (CH_4) contributing 9% (6.7 million tons) and nitrous oxides (N_2O) 4% (3 million tons).

Carbon dioxide (CO_2): The direct contribution of the agricultural sector (including fishery) to CO_2 emission in Israel is negligible (0.16%). The contribution of the Agricultural sector to CO_2 emission from fuel combustion was also very small (0.2%). Hence, it is evident that total CO_2 emissions are relatively insignificant.

Methane (CH_4): Methane emission in Agriculture is associated mainly with livestock. Methane is produced via two processes: enteric fermentation and manure management. In 2007, annual methane emissions amounted to 44 thousand tons, merely 1.2% of the total CO_2 equivalents (CO_2 eq) of methane emission in Israel. About 77% of the methane emission is contributed by enteric fermentation which mostly comes from cattle. Manure management contributes 23% of the methane emissions, mainly due to cattle and poultry manure, with the contribution of the latter slightly exceeding that of the former.

Nitrous oxide (N_2O): Total N_2O emission from agriculture and its related activities was estimated in 2007 at 1.6 million ton of CO_2 eq, which is slightly more than 50% of the total N_2O emissions in Israel. There are four main contributors to N_2O emission in agriculture: (i) direct emission from cultivated soils (0.83 million), where addition of synthetic fertilizers contributes 40% of that amount; (ii) soil emission from waste produced by animal grazing (0.14 million tons); (iii) animal waste management and storage (0.27 million tons); and (iv) indirect emission induced by agricultural activities such as volatization and subsequent atmospheric deposition of NOx and NH_3 (originating from the application of fertilizers and manure) and nitrogen leaching and removal by runoff (which amounts to 0.37 million tons).

The above overview suggests that the contribution of Agriculture and the Agricultural sector to GHG emission in Israel is insignificant with respect to total CO_2 and CH_4 emissions. In the case of N_2O, the contribution of agriculture is substantial ($\sim$55%). However, the contribution of N_2O (in CO_2 eq) to the total emission of GHG in Israel is only 4%.

Comparison of the emission levels in 2007 to those in 1996 can serve as an indication of the long term trends in Israel's GHGs emission. In general, emissions of CO_2 and N_2O increased by 29% and 55%, respectively. Conversely, emission of CH_4 decreased by 24%. The contribution of agriculture (including forestry and fishery) to the emission of the three GHGs exhibited somewhat different trends. Emission of CO_2 decreased drastically by 82%, mostly due to the increase in forest area which led to increased CO_2 sequestration and removal from the atmosphere. Emission of CH_4 showed a modest 4.5% decrease. Emission of N_2O increased by 35%, probably due to increased uses of fertilizers and agro-chemicals during this period of time.

Forests

Vegetation sequesters CO_2 from the atmosphere, and hence forests are considered a sink rather than a source for atmospheric CO_2. An inventory in 2007 of the area occupied by forests in Israel showed that 99.4 thousand Ha are natural woodlands and 83.2 thousand Ha are planted forests. The removal of CO_2 by the forested area amounted to 502 thousand tons per year. Thus the contribution of the forested area to CO_2 removal is unfortunately only about 0.75% of Israel's total emission of CO_2.

Expected Impacts of Climate Change on Agriculture and Forestry

Predicted average temperature increases of two to three degrees will probably have minor impacts on plant physiology and growth, with no implications for farmers other than increased irrigation in many cases. For winter off-season vegetable crops grown in warm winter climate regions, increased temperatures may allow earlier planting, and a longer season, thus increasing profitability (Fleischer *et al.*, 2008). Some crops have a cold requirement for proper fruit growth. This includes many deciduous crops, e.g. apple and pome fruit varieties. Contemporary temperature increases have already led farmers in Northern Israel to move their apple orchards to higher (and cooler) altitudes. In the future, with more warming, this trend will probably continue. Extremely high temperatures occurring in heat waves can do significant damage to crops, and an increased incidence of heat waves is expected (Jin *et al.*, 2011). Experimental work has demonstrated that reduced radiation load, e.g. by deployment of agricultural screens above the crops, can prevent this damage. Screens are cheap enough in current economic conditions so that some orchards are already covered. Besides protection from sun-scald and thermal damage, screens are used in orchards to protect from hail storms and strong winds, and to reduce crop water use (Tanny and Cohen, in press).

Non-irrigated field crops include winter wheat for grain and silage, hay, legumes for seeds, and safflower for oil. These cover a large fraction of the land area but produce a minor part of the total agricultural output value. In these cases, crop yields are proportional to rainfall below a certain threshold. Approximately 100,000 Ha of wheat is grown in Israel on land that is not irrigated. Much of the extensive winter wheat is grown in the south central region and southward to near Beer Sheva, where average rainfall is ~300 mm. There, if rainfall is average, grain yields are obtained, while if rainfall fails, no grain grows and only hay is harvested. If the predictions of decreases in rainfall come true, then it seems inevitable that the area of production in this southern region will shift northward with the rainfall, using whatever land is available. Overall, in that scenario, non-irrigated crop production could be significantly reduced.

Similar predictions can be made for forestry. The timberline for the most arid-tolerant tree species in southern Israel, *Pinus halepensis*, is about 300 mm/year (Maseyk *et al.*, 2008; Klein *et al.*, 2011), although at this rainfall level, trees must be planted as saplings since seedlings are unable to survive the summer dry season. A consecutive series of below-average rainfall in recent years has led to approximately 20% mortality in the most southern (and arid) forest at Yatir. Trees in these water-limited areas essentially divide up the available water according to tree density, which is carefully managed. This is because trees can grow extensive root systems, while soil characteristics also play a major role in water allocations between trees. It appears that further reductions in tree density may also allow survival of existing trees in some places even when rainfall is reduced by 10–20% (Moreno-Gutiérrez *et al.*, 201).

Carbon sequestration in Israel may decline in the scenarios described above. But for food-security as well as land-security in the future, it is important to maintain an agricultural sector. What will determine this is farm profitability and not carbon uptake and gross primary productivity. The latter are already compromised since Israel 'imports fossil fuel and grain, low in monetary but high in energy content, which is paid for by exporting fruit, flowers and vegetables, crops high in monetary but low in energy content' (quoted from Stanhill, 1974). Farm profits depend on many factors, including farmer constraints, as well as local and global market economics. Small perturbations in any of these might lead to large changes in profitability and are hard to predict. Predicting the influence of minor changes in temperature, evaporative demand and water availability on this sector based on agricultural science alone is probably so inaccurate as to be foolhardy. More realistic predictions for agriculture consider farm structure and economic constraints. Fleischer *et al.* (2008 and 2011) developed a model of annual farm income for agriculture in Israel using the Ricardian technique (Mendelsohn *et al.*, 1994) based on a detailed study of 86 out of 863 rural villages and face-to-face surveys of a representative sample of farmers. The baseline meteorological data in their work was taken from 32–38 stations during the period between 1961 and 1990. The approach was to develop production functions based on linear or polynomial regressions to the many factors investigated, e.g. soil type, farm size, farmer age, water quotas and precipitation. Predictions were then made for potential climate scenarios of 2020, 2060 and 2100 based on the Parallel Climate Model (Washington *et al.*, 2000). They suggest that increased temperatures may increase profitability of farming. This is because Israeli farmers substitute capital for climate by using irrigation, crop covers (e.g. screens) and their warmer micro-climate to export to Europe. Their fruits reach the market first and thus they turn hot climates into an advantage that yields profits. An additional finding was the interaction of climate impact with water quotas (rights). Water quotas are a guarantee to farmers and allow them to invest accordingly. Thus, the benefits attainable from warmer climates were more significant when water quotas were added to the simulation.

Continuing work on the same database, Fleischer *et al.* (2011) found that farmers generally combine agricultural technologies, i.e. crop, irrigation technology and cover type. They found that in Israel six general combination pairs are used, e.g. vegetables or flowers, greenhouse and drip irrigation. After evaluating the relationship between existing climate gradients and pair selection by the farmers, they considered how moderate increases in temperature and decreases in rainfall would influence farms, and predict that as climate warms, many farmers will shift to fruits with irrigation and net covering. As long as water quotas are maintained, changes in precipitation will have only marginal impacts.

Conclusions

The coming decades will almost certainly be characterized by increases in world population and prices of food and energy. Food shortages are likely to worsen. In the semi-arid Middle East, including Israel, climate is expected to become warmer while rainfall diminishes. Increasing potential evapo-transpiration and reduced natural water flows will challenge agricultural production. Israeli agriculture and agricultural research, based on the adverse conditions of limited water resources and poor soils, have met these challenges by building a solid foundation for both agriculture and agricultural research, and have introduced largescale desalination of seawater to partly decouple water requirements from natural water supplies. These changes have come at the price of increasing energy use, but have also made Israeli agriculture a paradigm for how climate change challenges and risks can be met. In this chapter, we have focused on the national program for waste-water reuse for irrigation to show that the response time for mitigation measures in response to a step change is close to a decade. A detailed understanding of soil processes is necessary in order to anticipate and act to prevent soil deterioration as climate changes. Future changes will involve many more aspects of agriculture, including post-harvest treatments at higher temperatures, dealing with new pathogens and pests that crop up, and genomics for better and more efficient crops. Thus, this chapter has discussed only a sampling of what the future may have in store for us. We should be able to meet the portending challenges with proper planning, built on the solid foundations and further progress of climate-soil-water-crop science.

References

Agassi, M., Shainberg, I., Warrington, D. and Ben-Hur, M. 1989. Runoff and erosion control in potato fields. *Soil Sci.* 148: 149–154.

Agassi, M. and M. Ben-Hur, 1991. Effect of slope length, aspect and phosphogypsum on runoff and erosion from steep slopes. *Aust. J. Soil Res.* 29:197–207.

Agassi, M., D. Bloem and M. Ben-Hur, 1994. Effect of drop energy, and soil and water Chemistry on infiltration, and erosion. *Water Resour. Res.* 30:1187–1193.

Agassi, M., I. Shainberg and J. Morin, 1981. Effect of electrolyte concentration and soil sodicity on the infiltration rate and crust formation. *Soil Sci. Soc. Am. J.* 45:848–851.

Alpert, P., Halfon,N., Levin, Z., 2008. Does air pollution really suppress precipitation in Israel? J. Appl. Meteorol. Climatol. 47, 933–943.

Alpert, P., T. Ben-Gai, A. Baharad, Y. Benjamini, D. Yekutieli, M. Colacino, L. Diodato, C. Ramis, V. Homar, R. Romero, S. Michaelides and A. Manes, 2002. The paradoxical increase of Mediterranean extreme daily rainfall in spite of decrease in total values. *Geophys. Res. Lett.*, 29(11):31-1–31-4.

Assouline, S. 2004. Rainfall-induced soil surface sealing: a critical review of observations, conceptual models and solutions. *Vadose Zone J.* 3:570–591.

Assouline, S. and M. Ben-Hur, 2006. Effects of rainfall intensity and slope gradient on the dynamic of interrill erosion during soil surface sealing. *Catena.* 66:211–220.

Baumhardt, R.I., M.J.M. Romkens, F.D. Whisler and J.Y. Parlange, 1990. Modeling infiltration into a sealing soil. *Water Resour. Res.* 26:2497–2505.

Bar Tal, A. et al. (+16 collaborators), 2011. Evaluation of soil degradation due to long term treated waste-water irrigation and its influence on orchard crops. Research proposal and current integrative research project, Israeli Ministry of Agriculture Chief Scientist fund.

Baver, L.D., W.H. Gardner and W.R. Gardner, 1972. *Soil Physics.* John Wiley, New York.

Ben-Gai, T., A. Bitan, A. Manes, P. Alpert and S. Rubin, 1999. Temporal and spatial trends of temperature patterns in Israel. *Theoretical and Applied Climatology*, 64:163–177.

Ben-Gai, T., A. Bitan, A. Manes, P. Alpert and A. Israeli, 1998. Aircraft measurements of surface albedo in relation to climate changes in southern Israel. *Theor. and Appl. Climatol.* 61:207–215.

Ben-Gai, T., A. Bitan, A. Manes, P. Alpert and S. Rubin, 1998. Spatial and temporal changes in annual rainfall frequency distribution patterns in Israel. *Theor. and Appl. Climatol.* 61:177–190.

Ben-Hur, M. 1991. The effect of dispersants, stabilizer and slope length on runoff and water-harvesting farming. *Aust. J. Soil Res.* 29:553–563.

Ben-Hur, M. 1994. Runoff, erosions, and polymer application in moving-sprinkler irrigation. *Soil. Sci.* 158:283–290.

Ben-Hur, M. 2004. Sewage water treatments and reuse in Israel. pp. 167–180. *In* F. Zereini and W. Jaeschke (eds.), *Water in the Middle East and in North Africa: Resources, Protection, and Management* Springer-Verlage, New York.

Ben-Hur, M. 2008. Seal formation effects on soil infiltration and runoff in arid and semiarid regions under rainfall and sprinkler irrigation conditions. pp. 429–452. *In* F. Zereini and W. Jaeschke (eds.), *Climatic Changes and Water Resources in the Middle East and in North Africa* Springer-Verlang, Berlin Heidelberg.

Ben-Hur, M. and M. Agassi, 1997. Predicting interrill erodibility factor from measured infiltration rate. *Water Resour. Res.* 33:2409–2415.

Ben-Hur, M. and S. Assouline, 2002. Tillage effects on water and salt distribution in a Vertisol during effluent irrigation and rainfall. *Agron. J.* 94:1295–1304.

Ben-Hur, M., Z. Plaut, G.J. Levy, M. Agassi and I. Shainberg, 1995. Surface runoff, uniformity of water distribution and yield of peanut irrigated with a moving sprinkler system. *Agron. J.* 87:609–613.

Ben-Hur, M., M. Agassi, R. Keren and J. Zhang, 1998. Compaction, aging, and raindrop-impact effects on hydraulic properties of saline and sodic Vertisols. *Soil Sci. Soc. Am. J.* 62:1377–1383.

Ben-Hur, M. and M. Lado, 2008. Effect of soil wetting conditions on seal formation, runoff, and soil loss in arid and semiarid soils — a review. *Aust. J. Soil Res.* 46:191–202.

Ben-Hur, M., R. Stern, A.J. van der Merwe and I. Shainberg, 1992. Slope and gypsum effects on infiltration and erodibility of dispersive and nondispersive soil. *Soil Sci. Soc. Am. J.* 56:1571–1576.

Ben-Hur, M. and I.I.C. Wakindiki, 2004. Soil mineralogy and slope effects on infiltration, interrill erosion, and slope factor. *Water Resour. Res.* 40:W03303

Bradford, J.M. and G.R. Foster, 1996. Interrill soil erosion and slope steepness factors, *Soil Sci. Soc. Am. J.* 60:909–915.

Bradford, J.M. and C. Huang, 1992. Mechanisms of crust formation: physical components. pp. 55–72 *In* M.E. Summar and B.A. Stewart (eds.) *Soil Crusting: Physical and Chemical Processes.* Lewis, Boca Raton.

Bresler, B., B.L. McNeal and D.L. Carter, 1982. Saline and sodic soils, *Adv. Ser. Agric. Sci.* 10. Springer-Verlag, New York.

CBS. 2008. Israel's central bureau of statistics, Annual report. Available at www.cbs.gov.il.

CBS. 2011. Israel's central bureau of statistics, Annual report. Available at www.cbs.gov.il.

Chen, Y., J. Tarchitzky, J. Brower, J. Morin and A. Banin, 1980. Scanning electron microscope observations on soil crusts and their formation. *Soil Sci.* 130:49–55.

Cohen, S., A. Inatez and G. Stanhill, 2002. Evaporative Climate Changes at Bet Dagan, Israel, 1964–1998. *Agriculture and Forest Meteorology.* 111:83–91.

Collis-George, N. and R.S.B. Green, 1979. The effect of aggregate size on the infiltration behaviour of a slaking soil and its relevance to ponded irrigation. *Australian Journal Soil Research.* 17:65–73.

Easterling, W. E. 2010. Guidelines for adapting agriculture to climate change, in Hillel, D. and C. Rosenzweig (eds), *Handbook of Climate Change and Agroecosystems: Impacts, Adaptation, and Mitigation*, ICP Series in Climate Change Impacts, Adaptation, and Mitigation – Vol. 1, Agronomy Society of America, Imperial College Press., 450 pp.

Emerson, W.W. 1977. Physical properties and structure. pp. 78–104. *In* J.S. Russell & E.L. Greacen (eds.) *Soil factors in crop production in a semi-arid environment.* University of Queensland Press, Queensland.

Feigin, A., I. Ravina and J. Shalhevet. 1991. Irrigation with treated sewage effluent. Management for environmental protection. *Adv. Ser. Agric. Sci.* 17. Springer-Verlag, Berlin.

Fleischer, A., R. Mendelsohn and A. Dinar, 2011. Bundling Agricultural Technologies to Adapt to Climate Change, *Technological Forecasting and Social Change.* 78:982–990.

Fleischer, A., E. Lichtamn and R. Mendelsohn, 2008. Climate Change, Irrigation, and Israeli Agriculture: Will Warming Be Harmful? *Ecological Economics,* 65(3):508–515.

Foster, G.R. 1982. Modeling the erosion process, pp. 297–380. *In* C.T. Haan, (ed) *Hydrologic Modeling of Small Watersheds.* ASAE Monogr. 5. ASAE, St. Joseph, MI.

Fox, D.M. and R.B. Bryan, 1999. The relationship of soil loss by interrill erosion to slope gradient. *Catena,* 38:211–222.

Huang, C.H. 1998. Sediment regimes under different slope and surface hydrologic conditions. *Soil Sci. Soc. Am. J.* 62:423–430.

Fuch, M. 2009. Impact of research on water use for irrigation in Israel. Irrigation Science 25:443–445

Givati A. and D. Rosenfeld, 2004. Quantifying precipitation suppression due to air pollution. *Journal of Applied Meteorology* 43: 1038–56.

Goldreich, Y. 2003. The climate of Israel, observation, research and application. Kluwer publishers, New York.

Hillel, D. and C. Rosenzweig, 2011. The role of soils in climate change. pp. 1–20. Handbook of Climate Change and Agroecosystems — Impacts, Adaptation and Mitigation. Imperial College Press, London.

Huang, C.H., 1998. Sediment regimes under different slope and surface hydrologic conditions. *Soil Sci. Am. J.* 62, 423–430.

IMS 2011. Climate changes in Israel — Findings of the Meteorological Service. Pamphlet published by the Israel Meteorological Service online, Dec. 2011. Available at http://www.ims.gov.il/NR/rdonlyres/6B4C2F09-1BAF-41E8-9081-9F4D0D3F04A0/0/ClimateChangeIsrael.pdf

IPCC, 2007. Intergovernmental Panel on Climate Change (IPCC), Climate Change 2007 — Impacts, Adaptation and Vulnerability. Cambridge University Press, Cambridge, 2007.

Janeau, J.L., J.P. Briequet, O. Planchon and C. Valentin, 2003. Soil crusting and infiltration on steep slopes in northern Thailand. *Eur. J. Soil Sci.* 54:543–553.

Ji, F., A. Kitoh and P. Alper, 2011: Climatological relationships among the moisture budget components and rainfall amounts over the Mediterranean based on a super-high resolution climate mode. J. Geophys. Res., 116, D0910, doi:1.1029/2010JD014021.

Kalma, J.D. and G. Stanhill, 1972. The climate of an orange orchard: physical characteristics and microclimate relationships. *Agric. Meteorol.* 10:185–201.

Kazman, Z., I. Shainberg and M. Gal, 1983. Effect of low levels of exchangeable Na and applied phosphogypsum on the infiltration rate of various soils. *Soil Science.* 35:184–192.

Kay, B.P. and D.A. Angers, 1999. Soil structure. pp. 229–269. *In* M.E. Sumner (ed.), Handbook of Soil Science. CRC Press, New York.

Kemper, W.D. and R.C. Rosenau, 1984. Soil cohesion as affected by time and water content. *Soil Science Society of American Journal.* 48:1001–1006.

Kemper, W.D., R.C. Rosenau and A.R. Dexter. 1987. Cohesion development in disrupted soils as affected by clay and organic matter content and temperature. *Soil Science Society of American Journal.* 51:860–867.

Keren, R., I. Shainberg and E. Klein, 1988. Settling and flocculation value of sodium montmorillonite particles in aqueous media. *Soil Science Society of American Journal.* 52:76–80.

Kinnell, P.I.A. and D. Cummings, 1993. Soil/slope gradient interactions in erosion by rain-impacted flow. *Trans. ASAE*, 36:381–387.

Kinnell, P.I.A. 2000. The effect of slope length on sediment concentration associated with side-slope erosion. *Soil Sci. Soc. Am. J.* 64:1004–1008.

Kitoh, A., A. Yatagai and P. Alpert. 2008. First super-high-resolution model projection that the ancient Fertile Crescent will disappear in this century. *Hydrological Research Letters*. 2:1–4. DOI. 10.3178 HRL.2.1.

Klein, T., S. Cohen and D. Yakir, 2011. Hydraulic adjustments underlying drought resistance of *Pinus halepensis*. *Tree Physiology*. 31:637–648

Kliot, N., S. Paz, and O. Keidar (2008). Frame and framing analysis for the study of preparedness for climate change in Israel. (in Hebrew, 226 pages). Final report for the Chief Scientist of the Ministry of Environmental Protection, project no. 6-103/3000003599.

Krichak, S.O., J.S. Breitgand, R. Samuels and P. Alpert, 2011. A double-resolution transient RCM climate change simulation experiment for near-coastal eastern zone of the Eastern Mediterranean region. *Theor Appl Climatol*. 103:167–195. DOI. 10.1007/s00704-010-0279-6.

Lado, M. and M. Ben-Hur, 2009. Treated domestic sewage irrigation effects on hydraulic properties in arid and semiarid zones: A review. *Soil & Tillage Research*. 106:152–163.

Lado, M., M. Ben-Hur and I. Shainberg, 2004. Soil wetting and texture effects on aggregate stability, seal formation, and erosion. *Soil Sci. Soc. Am. J.* 68:1992–1999.

Lado, M., A. Bar-Tal, A. Azenkot, S. Assouline, I. Ravina, Y. Erner, P. Fine, S. Dasberg and M. Ben-Hur, 2012. Changes in chemical properties of semiarid soils under long-term secondary treated wastewater irrigation. *Soil Sci. Soc. Am. J.* (submitted, 2nd revision).

le Bissonnais, Y. 1996. Aggregate stability and assessment of soil crusting and erodibility: 1. Theory and methodology. *Eur. J. Soil Sci.* 47: 425–437.

Levy, G.J., J. Levin and I. Shainberg, 1997. Prewetting rate and aging effects on seal formation and interrill soil erosion. *Soil Science*. 162:131–139.

Mamedov, A.I., G.J. Levy, I. Shainberg and J. Letey, 2001. Wetting rate and soil texture effect on infiltration rate and runoff. *Australian Journal of Soil Research*. 36:1293–1305.

Martin, J.P., W.P. Martin, J.B. Page, W.A. Raney and J.D. De Met, 1995. Soil aggregation. *Advances in Agronomy*. 7:1–37.

Maseyk, K.S., T. Lin, E. Rotenberg, J.M. Grünzweig, A. Schwartz and D. Yakir. 2008. Physiology-phenology interactions in a productive semi-arid pine forest. *New Phytol.* 178:603–616.

McDowell, N.G., D.J. Beerling, D.D. Breshears, R.A. Fisher, K.F. Raffa and M. Stitt, 2011. The interdependence of mechanisms of underlying climate-driven vegetation mortality. *Trends in Ecology and Evolution*. 26(10):523–532

McIntyre, D.S. 1958. Permeability measurements of soil crust formed by raindrop impact. *Soil Sci.* 85:158–189.

McVicar, T.R. et al. (+13 authors), 2012. Global review and synthesis of trends in observed terrestrial near-surface wind speeds: Implications for evaporation. Journal of Hydrology 416–7:182–205.

Mendelsohn, R., W. Nordhaus and D. Shaw, 1994. The impact of global warming on agriculture: a Ricardian analysis. *The American Economic Review*. 84:753–771.

Moreno-Gutiérrez, C., G. Battipaglia, P. Cherubini, M. Saurer, E. Nicolás, S. Contreras and J. Ignacio Querejeta, 2011. Stand structure modulates the long-term vulnerability of *Pinus halepensis* to climatic drought in a semiarid Mediterranean ecosystem. *Plant Cell and Env.* doi: 10.1111/j.1365-3040.2011.02469.x.

Morin, J., Y. Benyamini and A. Michaeli, 1981. The effect of raindrop impact on the dynamics of soil surface crusting and water movement in the profile. *J. Hydrol.* 52:321–335.

Poesen, J., 1984. The influence of slope angle on infiltration rate and Hortonian overland flow volume. *Z. Geomorph. N.F. Suppl.Bd.* 40:117–131.

Poesen, J., 1987. The role of slope angle in surface seal formation. pp. 437–448. *In* V. Gardner (Ed.) *Proc., 1ˢᵗ International Conference on Geomorphology: Geomorphology, Resource Environment and Developing World.* John Wiley & Sons, New York.

Poesen, J. 1986. The role of slope angle in surface seal formation. in Proc. Int. Conf. Geomorphol. pp. 437–448. *In* V. Gardner. *1st: Geomorphology, Resource Environment and the Developing World* John Wiley and Sons, New York.

Roderick, M.L. and G. D. Farquhar. 2002. The cause of decreased pan evaporation over the past 50 years. *Science* 298:1410–1411.

Romkens, M.J.M., K. Helming and S.N. Prasad, 2002. Soil erosion under different rainfall intensities, surface roughness and soil water regimes. *Catena* 46:103–123.

Saaroni, H. 2012. Trends of change in temperature in Israel, 1975–2010. Lecture presented at a seminar on climate change, Jan. 2012. Slides (in Hebrew) at http://met-society.org.il/climate12.htm

Shainberg, I., D. Goldstein and G. J. Levy, 1996. Rill erosion dependence on soil water content, aging and temperature. *Soil Science Society of American Journal.* 60:916–922.

Shainberg, I. and J. Letey, 1984. Response of soils to sodic and saline conditions. *Hilgardia.* 52:1–57.

Shainberg, I., A.I. Mamedov and G.J. Levy, 2003. Role of wetting rate and rain energy in seal formation and interrill erosion. *Soil Science.* 168:54–62.

Shanan, L., Schick, A. 1980. A hydrological model for the Negev Desert highlands. Effects on infiltration, runoff, and ancient agriculture. Hydrological Sciences Bull. 25, 269–282.

Stanhill, G. 1974. Energy and agriculture: a national case study. *Agro-Ecosystems.* 1:205–217.

Stanhill, G., G.J. Hofstede and J.D. Kalma, 1966. Radiation balance of natural and agricultural vegetation surfaces. *Quart. J. Roy. Met. Soc.* 92:128–140.

Stanhill, G. 1966. Some results of helicopter measurements of the albedo of different land surfaces. 1870 *Solar Energy.* 13:59–66.

Stanhill, G., M. Israeli and D. Rosenzweig, 1973. The solar radiation balance of scrub forest and pasture on the Carmel Mountain: A comparative study. *Ecology.* 54:819–828.

Stanhill, G. and M. Fuchs, 1968. The climate of the cotton crop. Physical characteristics and microclimate relationships. *Agric. Meteorol.* 5:183–202.

Stanhill, G., Ianitz, A., 1997. Long term trends in, and the spatial variation of, global irradiance in Israel. Tellus 49 B, 112-122.

Stanhill, G. and S. Moreshet. 1992. Global radiation climate changes: the world network. *Climatic Change* 21:57-75.

Stanhill G, Cohen S (2001) Global dimming: a review of the evidence for a widespread and significant reduction in global radiation with discussion of its probable causes and possible agricultural consequences. *Agricultural and Forest Meteorology* 107, 255-278.

Stanhill, G. and R. Rosa, 2010. On the inter-annual variation of rainfall in Israel since 1931: No change, much deviation, but with a useful signal. Israel Journal of Earth Sciences 58(2):113-120

Tanny, J., A. Naor, S. Cohen, E. Raveh and A. Grava. 2009. Optimization of apple irrigation under shading screens. Final report, Ministry of Agriculture Chief Scientist Fund project no. 304-0326 (in Hebrew).

Tanny, J., and Cohen, S. (in press). Microclimate and crop water use under screen constructions. In: Agriculture in arid and semiarid zones: Soil, water, and environment aspects. edited by Dr. Meni Ben-Hur, Research Signpost

Uson, A. and M.C. Ramos, 2001. An improved rainfall erosivity index obtained from experimental interrill soil losses in soils with a Mediterranean climate. *Catena.* 43:293–305.

van Olphen. 1977. *An introduction to clay colloid chemistry.* (2nd ed). Interscience Publications, New York.

Wakindiki, I.I.C. and M. Ben-Hur, 2002. Soil mineralogy and texture effects on crust micromorphology, infiltration and erosion. *Soil Sci. Soc. Am. J.* 66:897–905.

Warrington, D., I. Shainberg, M. Agassi and J. Morin, 1989. Slope, and phosphogypsum's effect on runoff and erosion. *Soil Sci. Soc. Am. J.* 53:1201–1205.

Washington, W., *et al.* 2000. Parallel Climate Model (PCM): Control and Transient Scenarios. *Climate Dynamics.* 16:755–774.

Watson, D.A. and J.M. Laflen, 1986. Soil strength, slope, and rainfall intensity effects on interrill erosion, *Trans. ASAE*, 29:98–102.

Yermiyahu, U., A. Tal, A. Ben-Gal, A. Bar-Tal, J. Tarchisky and O. Lahav, 2007. Rethinking Desalinated Water Quality and Agriculture. *Science.* 318:920–921.

Yosef, Y., Saaroni, H., and P. Alpert. 200.Trends in Daily Rainfall Intensity Over Israel 1950/1-2003/4. *The Open Atmospheric Science Journal, 3*, 196–203.

Chapter 10

China Perspectives on Climate Change and Agriculture: Impacts, Adaptation and Mitigation Potential

Liang Tang, Yan Zhu*, Weixing Cao, and Yu Zhang

National Engineering and Technology Center for Information Agriculture
Jiangsu Key Laboratory for Information Agriculture
Nanjing Agricultural University, Nanjing 210095, China
*yanzhu@njau.edu.cn

Introduction

Global climate change and its potential impacts on agriculture have become a worldwide concern during last the few decades, as they increase the fluctuations of food production and have an important effect on food security, both globally and locally (Barry and Cai, 1996). China has experienced explosive economic growth in recent decades, but has only 7% of the world's arable land available to feed 22% of the world's population (Tong *et al.*, 2003). Agriculture in China is already under pressure both from huge and increasing demands for food, and from problems of limited land and diminishing water resources (Smit and Cai, 1996). Improved knowledge of how the past climate changes have impacted crop production in China can provide insights toward understanding the likely impacts of future changes, and is the key for development of adaptation strategies to protect vulnerable agro-ecosystems and to ensure Chinese food security.

In China, climate changes have been reported by many studies in recent decades (Ding *et al.*, 2007a; Qian and Zhu, 2001). Annual average air temperature has increased by 0.5–0.8°C during the past 100 years, which was slightly larger than the average global temperature rise (Ding *et al.*, 2006). From 1955 to 2000, annual mean maximum temperature and minimum temperature in China had increased by an average of 0.127°C /decade and 0.323°C /decade, respectively (Liu *et al.*, 2004). Annual precipitation trends in China for the period 1960–2006 (the data come from climate records at 355 meteorological stations) present strong differences between northeastern (decrease), northwestern (increase) and southeastern China (increase) (Piao *et al.*, 2010).

Crop simulation models are powerful tools for evaluating the potential impacts of climate change on agro-ecosystems and assessing the long term changes in crop potential productivity under future climate scenarios. Crop simulation models usually need site-specific input data such as weather, soil properties, variety traits and management practices (Whistler *et al.*, 1986; Penning de Vries *et al.*, 1989). Applicability of these models can be extended from field to regional scales by linking models with Geographic Information System (GIS). GIS provides an ideal tool for acquiring, managing and mapping spatial datasets when up-scaling crop simulation models. Many studies on crop growth models linked with GIS were carried out to predict regional productivity and assist decision making on farm and regional levels (Priya and Shibasaki, 2001; Rao *et al.*, 2000; Tan and Shibasaki, 2003).

Crop potential productivity is useful for assessing the scope of future crop yield increases and guiding crop production (Evans and Fischer, 1999). Spatial patterns of climate change impacts on crop productivity at regional scales are important for understanding the vulnerability and adaptation of agricultural production to climate change, which have been presented in part of or the whole of China in many studies using different methods (Tao *et al.*, 2008; Chavas *et al.*, 2009; Xiong *et al.*, 2009; Tao and Zhang, 2010).

In this chapter, we concentrate on wheat and rice, which are the major food staples in China. Our objectives are (1) to define regional patterns of climate change during the crops' growth period in past decades; (2) to consider regional patterns of rice and wheat productivity changes in the past decades and future scenarios; and (3) to propose feasible agricultural solutions for mitigating and adapting to the impacts of climate change on crops in China.

Climate Change in Recent Decades

Study area and methods

The study region of rice involved 5 rice production areas within the main rice growing region of China (Ministry of Agriculture, P.R. China, 2009), among which the areas with annual accumulated temperatures exceeding 10°C below 2000°C·d (marked in red in Fig. 1a) were eliminated. The study region of wheat involved 10 main wheat production provinces, including Heibei, Henan, Shandong, Jiangsu, Anhui, Shanxi, Shaanxi, Ganshu, Hubei, and Sichuan (Fig. 1b). The daily meteorological data, selected from 700 weather stations during the years 1961–1970 and 1996–2005 in main rice growing regions; and from 1960–1970 and 2000–2009 in the main wheat

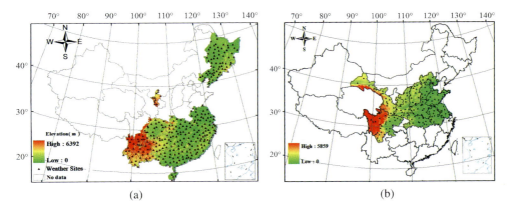

Fig. 1. Distribution map of elevation and weather stations in main rice (a) and wheat (b) growing area of China.

growing regions of China. The data from these areas were processed to generate the gridded daily meteorological surface data by using ANUSPLIN and ARCGIS (Jiang *et al.*, 2010). The spatial and temporal distribution patterns of the climate resources for the rice and wheat production, including total sunshine hours, thermal time (above 10°C), total precipitation during rice and wheat growing periods between the two decades (1960s and 2000s) were calculated and analyzed based on the grid cell (0.05° resolution) (Jiang *et al.*, 2011). Based on the analysis of weather data, the crop potential photosynthetic, photo-thermal and climatic productivity were calculated using an ecological model (Jiang *et al.*, 2012).

Climate change between 1960s and 2000s

The spatial distributions of total sunshine hours indicate a trend that rose gradually from north to south in the single-cropping rice growing region at two historical stages (Fig. 2 Aa, Ba). Total sunshine hours is higher in the north and south of the double-cropping rice-growing region, totaling more than 1300 hours. As compared with the 1960s, the average total sunshine hours during the rice growth period in 2000s decreased by 11.93% in the main rice growing regions of China. Most areas exhibited a decreasing trend (blue area in Fig. 2 Ca), which would reduce the biomass production.

    The spatial distributions of thermal time (above 10°C) had trends that rose gradually from north to south in single-cropping and double-cropping rice growing regions at two historical stages (Fig. 2 Ab, Bb). Northeast and Southwest China had lower total thermal time, ranging from 1000 to 2000°C · d. As compared with the 1960s, the average total thermal time during the rice growth period in 2000s in most areas had increasing trends (red area in Fig. 2 Cb), which increased by an average 9.40%

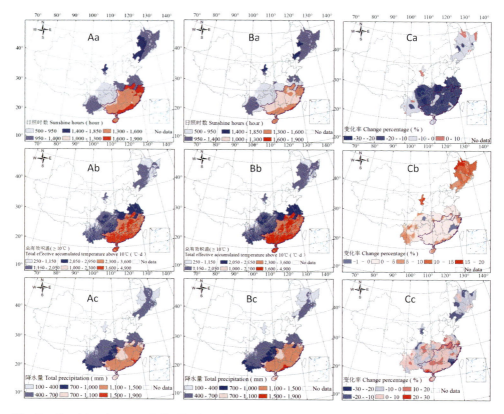

Fig. 2.   Spatial and temporal distribution of total sunshine hours (a), thermal time (b), precipitation (c), during the rice growing periods in the main rice growing regions of China (A. 1960s, B. 2000s, C. percentage change from 1960s to 2000s). The region outlined with purple line is the double-crop growing region; the unmarked region is the single-crop rice-growing region.

in the main rice growing regions of China. The area with thermal time lower than 1000°C · d diminished. In addition, some areas were significantly increased, such as in northeast China, where the increases were 15~20%, similar to the results of previous studies (Piao, *et al.*, 2010; Ding, *et al.*, 2007).

The spatial distributions of total precipitation had the trends that rose gradually from north to south in single-cropping and double-cropping rice growing regions at two historical stages (Fig. 2 Ac, Bc). Total precipitation during the rice growth period were below 550 mm in most single-cropping rice-growing regions and more than 950 mm in most double-cropping rice growing regions. As compared with the 1960s, the average total precipitation during the rice growth period in 2000s increased by 1.59% in main rice growing regions of China. However, most of north China had a decreasing trend, the decrease being no more than 10% (Fig. 2 Cc).

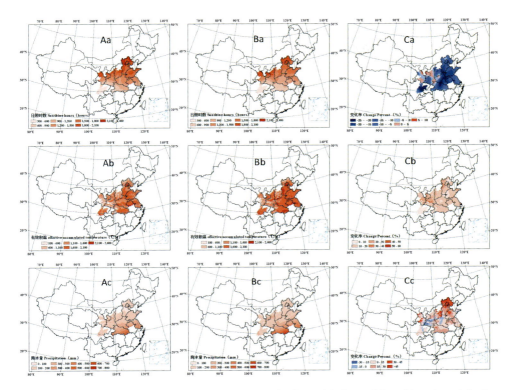

Fig. 3.   Spatial and temporal distribution of total sunshine hours (a), thermal time (b), and precipitation (c) during the wheat growing periods in the main wheat growing regions of China (A. 1960s, B. 2000s, C. percentage change from 1960s to 2000s).

The spatial distributions of total sunshine hours, thermal time and precipitation in wheat growing region at 2 historical stages were showed in Fig. 3. Total sunshine hours decreased from north to south (Fig. 3 Aa, Ba), thermal time and precipitation had the trends that went up gradually from north to south (Fig. 3 Ab, Bb, Ac, Bc). As compared with 1960s, the average total sunshine hours, thermal time and precipitation during wheat growth period in 2000s decreased by 9.38%, increased by 16.69% and 17.16%, respectively, in the main wheat growing regions of China (Fig. 3 Ca, Cb, Cc).

Crop potential productivity between the 1960s and 2000s

The spatial distribution characteristics of rice potential photosynthetic productivity in both single-cropping and double-cropping rice-growing regions had the trends that rose gradually from north to south in different historical stages (Fig. 4 Aa, Ba). As compared with the 1960s, the potential photosynthetic productivities of 2000s

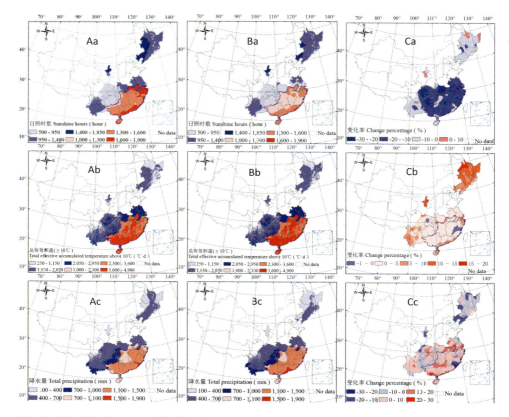

Fig. 4.   Spatial and temporal distribution of rice potential photosynthetic, photo-thermal, and climatic productivity in main rice growing regions of China (A. 1960s, B. 2000s, C. percentage change between the 1960s and 2000s). The region outlined with purple line is the double-cropping growing region, the unmarked region is the single-cropping rice growing regions.

decreased by 5.40% in main rice growing region of China, while the decreasing rates in Northeast and Southwest China were lower than those in Central and South China (Fig. 4 Ca).

In different historical stages, the spatial patterns of rice potential photo-thermal productivity in single-cropping rice growing regions had the trends that increased gradually from north to south and from west to east. However, the patterns in double-cropping rice growing region had the trend that increased gradually from north to south (Fig. Ab, Bb). From the 1960s to 2000s, the potential photo-thermal productivities decreases by 2.56% in the main rice growing region of China, and the decreases in the potential photo-thermal productivity over the single-cropping and double-cropping rice growing regions were 1.17% and 5.46%, respectively (Fig. 4 Cb).

The spatial distributions of the potential climatic productivity in both single- and double-cropping rice growing regions had the trend that rose gradually from north to south in two historical stages (Fig. 4 Ac, Bc). Compared with the 1960s, the potential climatic productivities of the 2000s decreased by 7.44% in the main rice growing regions of China, and the decreases in the potential climatic productivities over the single-cropping and double-cropping rice growing regions were 8.13% and 6.04%, respectively (Fig. 4 Cc).

The spatial distributions of wheat potential photosynthetic, photo-thermal and climatic productivity in the wheat growing region in two historical stages are demonstrated in Fig. 5. Rice potential photosynthetic productivity decreased gradually from north to south (Fig. 5Aa, Ba). Photo-thermal and climatic productivity had the trends that rose gradually from north to south (Fig. 5Ab, Bb, Ac, Bc). As compared with the 1960s, the average rice potential photosynthetic, photo-thermal and climatic productivity during the wheat growth period in the 2000s decreased by 6.95%, 6.99% and 1.67%, respectively in main wheat growing regions of China (Fig. 5Ca, Cb, Cc).

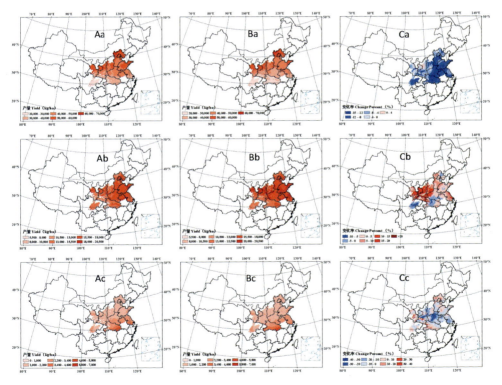

Fig. 5. Spatial and temporal distribution of rice potential photosynthetic, photo-thermal, and climatic productivity in the main rice growing regions of China (A. 1960s, B. 2000s, C. percentage change between the 1960s and 2000s).

Crop Productivity Under Future Scenarios

Crop yield changes under future climate scenarios were estimated by the use of wheat and rice growth models. The models are WheatGrow (Cao and Moss, 1997; Cao, *et al*., 2002) and RiceGrow (Tang, *et al*., 2009), which are specified with data based on GIS technology. The operation of RiceGrow normally needs site-specific data for weather, soil properties, cultivar parameters, and cultural practices. Based on the spatial variability of environmental variables (soil and climate data) and management levels, this study area was divided into a number of well-proportioned basic simulation grids at a grid resolution of $0.1° \times 0.1°$ through interpolating and stacking basic data by GIS technology (Fig. 6). Wheat and rice cropping regions were divided into several specifications of sub-regions. Moreover, different management data and cultivar data were estimated in each sub-region. Weather data from 700 meteorological stations in China from 1980–2005 were interpolated by using the IDW method. Soil characteristics of China in the form of a map in GIS (scale; 1:1 million) were used, including properties like soil structure, bulk density, organic matter, total nitrogen, pH, etc.

Rice productivity

The effects of climate change on rice yields were assessed under 3 transient climate change scenarios in 2020, 2030 and 2050 for southern China (not including Hannan and Taiwan). The CO_2 concentrations assumed in 2020, 2030, 2050 and baseline

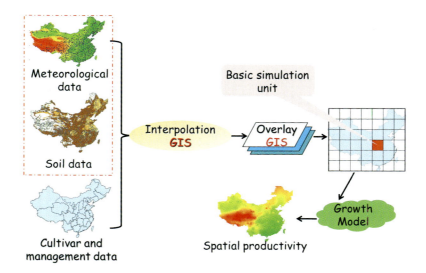

Fig. 6.   Flow chart of predicting crop productivity.

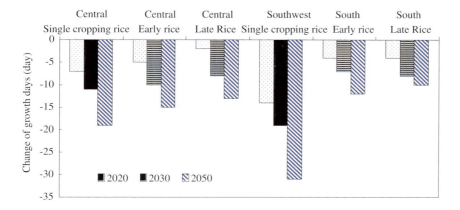

Fig. 7.   Changes of rice growth duration under three climate change scenarios.

scenarios were 450 ppm, 500 ppm, 610 ppm and 370 ppm, respectively, and the temperatures increased by 1.7°C, 2.1°C, and 2.8°C, respectively. The study area was southern China, including three main rice planting regions (Central, Southwest and South China). Figure 7 showed the changes of rice growth duration under three climate change scenarios compared with the baseline scenario. The phenology of single and double rice both decreased under three climate change scenarios due to climate warming; the decreases in days ranged from 2–14 days, 8–19 days, and 10–31 days, respectively. The phenology of single rice decreased by more days than double rice because single rice phenology is inherently longer. Besides, single rice in Southwest China was more sensitive to climate warming due to high altitude and low temperatures.

The spatial distributions of yield changes in rice for climate change scenario in 2050 are shown in Fig. 8. Whether considering direct effects of CO_2 fertilization or not, the yields in most of Southern China decreased in comparison with the baseline scenario. The yields of high altitude mountain areas in Sichuan, Yunnan, and Guizhou Provinces increased compared with the baseline scenario. The areas of increased yields were enlarged after considering the direct effect of CO_2 fertilization. Moreover, the regions north of Jiangsu and south to Yangtze River showed yield increases.

Wheat productivity

Using two IPCC (Intergovernmental Panel on Climate Change) SRES (Special Report on Emissions Scenarios) emission scenarios as drivers (A2 and B2), which was assumed that mean temperature increased by 3.9°C and 2.0°C, precipitation increased by 12.9% and 10.2%, and CO_2 concentrations increased to 721 ppm and

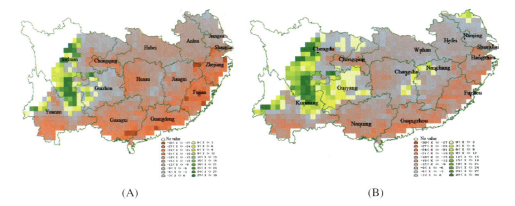

(A)   (B)

Fig. 8.   Spatial distributions of changes in rice yields for climate change scenario in 2050 (A: without direct effect of CO_2 elevation; B: with direct effect of CO_2 elevation).

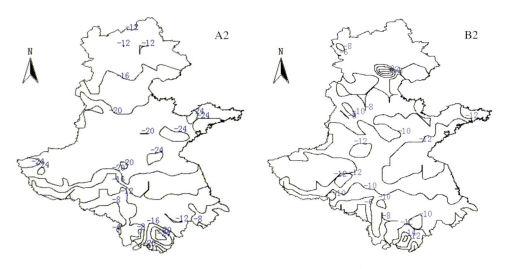

Fig. 9.   Spatial distribution of changes in rained wheat growth durations in Huang-huai-hai plain under A2 and B2 emission scenarios in 2080s.

561 ppm, respectively in the 2080s. Baseline scenario was set at the average weather data from 1999–2003. The study pertains to the main wheat production provinces in Huang-huai-hai plain which is one of the most important agricultural regions in China. The phenology of rain-fed wheat had an accelerating trend due to temperature enhancement of growth in Huang-huai-hai plain under A2 and B2 emission scenarios in the 2080s (Fig. 9). The growth days under the A2 and B2 emission scenarios were shortened 3–31 days (average 17 days), and 5–18 days (average 10 days), respectively. Henan and Shandong province were most affected by temperature rise.

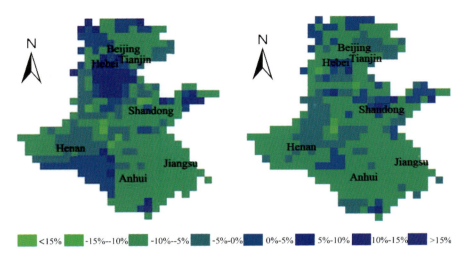

Fig. 10. Spatial distribution of yield changes of rained wheat in Huang-huai-hai plain under A2 and B2 scenarios in 2080s (without direct physiological effects of CO_2 elevation).

Figure 10 showed the spatial distribution of predicted yield changes of rain-fed wheat in Huang-huai-hai region of China under the A2 and B2 scenarios in the 2080s. The rainfed wheat yields under the A2 and B2 scenarios (without direct effects of CO_2 elevation) in the 2080s are predicted to decrease in most areas in the Huang-Huai-Hai region in China (Fig. 10), by 7.34% and 6.05% relative to the baseline value, respectively. The yield are predicted to increase especially significantly under the A2 scenarios in some areas of Henan and Hebei Provinces which have less precipitation under the baseline scenario. The two future climate scenarios are predicted to have increased precipitation, which should cause rain-fed wheat yields to increase. The rain-fed wheat yields under A2 and B2 scenarios in the 2080s were increased in most areas of Huang-Huai-Hai region in China (Fig. 11) with direct effects of CO_2 enhancement.

Agricultural Responses Affecting Mitigation and Adaptation

Adjustment of cropping systems

Studies have shown that changing cropping systems is one of the important solutions to increase crop yield and rational utilization of agricultural climatic resources (Wang, 1997; Deng *et al.*, 2008). The effect of climate change on the cropping system in China is likely to be so great that the boundary of multiple cropping will shift northward and westward, and its area will expand (Li *et al.*, 2010; Yang *et al.*, 2010). For example, according to the temperature requirement (above 10°C) for different

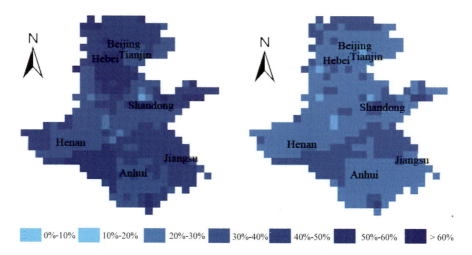

| | 0%-10% | | 10%-20% | | 20%-30% | | 30%-40% | | 40%-50% | | 50%-60% | | > 60% |

Fig. 11.   Spatial distribution of yield changes of rained wheat in Huang-huai-hai plain under A2 and B2 emission scenarios in 2080s (with direct physiological effects of CO_2 elevation).

rice types (China National Rice Research Institution, 1988), single rice varieties require thermal time 2000~4500°C, whereas double rice varieties require thermal time 4500~7000°C. Climate warming gives more thermal resources to the rice cropping region, and enlarges the boundary of rice cropping area in China. From this point of view, climate warming is beneficial to crop yields, especially in the area that has only enough thermal resources to grow one-season crop but cannot grow two-seasons crop, such as the area north of Yangtze River in Central China, which may be able to grow double rice in the future scenarios. Thus, an efficient solution for China to increase crop yield is to increase the cropping index according to the present cropping system in the coming fifty years.

Adjustment of sowing date

Under the background of global warming, the phenology of crop growth has evidently shifted toward earlier maturation (Tao *et al.*, 2006; Menzel *et al.*, 2006). In addition, the crops were also influenced by daily extreme temperatures, especially during the flowering period, which could cause the crop sterility and other temperature stress. If a present crop variety is to be planted in future climate scenarios on present sowing dates, the yield would be inevitably decreased. Figure 12 showed simulated rice yields under full irrigation and rainfed conditions from 1981–2009 by setting a typical sowing date at Zhenjiang (32°11′N, 119°28′E) and Wuchang (44°54′N, 127°09′E) in China. A declining trend in both potential and rainfed yields would result, and this declining trend is likely to be significant for both potential

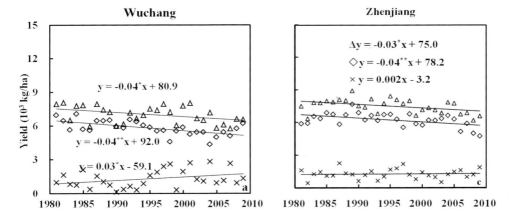

Fig. 12. Simulated rice yields under full irrigation and rainfed conditions at the two sites (1981–2009). Straight lines show the linear trends. **Significant at $P < 0.01$; *Significant at $P < 0.05$. $\triangle$ indicates potential yield, $\lozenge$ indicates rainfed yield, $\times$ indicated the difference between the two.

and rainfed yields at Wuchang and Zhenjiang in China. This phenomenon implies that past climate change had a negative impact on both potential and rainfed yields at these two sites. Another study has shown that optimal sowing dates (the sowing date that gave the highest yield in a given year was regarded as the optimal sowing date for that year) in the last three decades (1981–2010) were variable at Xuzhou (30.1°N, 119.9°E), Yizheng (132.3°N, 19.2°E) and Hangzhou (30.1°N, 119.9°E) in China (Fig. 13). Generally, radiation, heat and water resources are changed and redistributed due to global climate change. Therefore, adjusting the sowing date to adapt to this redistribution may be an effective solution to mitigate and adapt to the impacts of climate change.

Breeding new crop varieties

Introduction and cultivation of new crop varieties adapted to climate change are the future trends in agriculture development. Two objectives of breeding are: (1) to breed stress-resistant varieties; and (2) to breed new varieties having suitable phenology and able to adapt to new climate change scenarios. Over the past several decades, China has already experienced some devastating climate extremes (Ding *et al.*, 2006; Editorial board, 2007). Many environmental stresses such as extreme high and low temperatures, floods, and droughts have occurred more frequently due to climate warming. Thus, it is necessary to introduce and cultivate new stress-resistant cultivars to mitigate the negative impacts. Another study we conducted showed that if we changed the genetic parameter that represents the temperature causing 50%

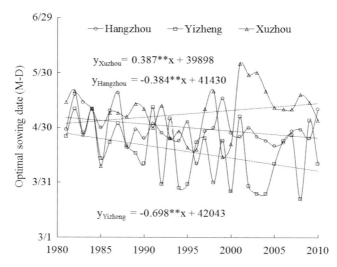

Fig. 13.   Optimal sowing date with highest yield from 1981–2010 in Hangzhou, Yizheng and Xuzhou.

spikelet sterility in RiceGrow model from 35°C to 37°C, the yield under 2020, 2030 and 2050 scenarios in southern China would increase by 18.28%, 23.65% and 12.4% respectively. In addition, since climate warming has a significant influence on crop phenology, it will inevitably change cropping systems in different sites in the future. A good solution is to adjust the crop phenology so as to adapt to new cropping systems and thus to utilize the redistribution of resources in the future.

References

Barry, S. and Y.l. Cai. 1996. Climate change and agriculture in China. *Global Environmental Change* 6(3):205–214.

Cao, W.X., T.M. Liu, W.H. Luo, S.H. Wang, J. Pan and W.S. Guo. 2002. Simulating organ growth in wheat based on the organ-weight fraction concept. *Plant Prod. Sci.* 5:248–256.

Cao, W.X. and D.N. Moss.1997. Modeling phasic development in wheat: A conceptual integration of physiological components. J. *Agric. Sci.* 129:63–172.

Chavas, D.R., R.C. Izaurralde, A.M. Thomson and X.J. Gao. 2009. Long-term climate change impacts on agricultural productivity in eastern China. *Agricultural and Forest Meteorology* 149:1118–1128.

China National Rice Research Institution. 1988. Rice Cropping Regionalization in China, Hangzhou: Zhejiang science and technology press.

Deng, Z.Y., Q. Zhang and J.Y. Pu. 2008. The impact of climate warming on crop planting and production in northwestern China. *Acta Ecologica Sinica* 28(8):3760–3768.

Ding,Y.H, G.Y. Ren and G.Y. Shi, *et al.* 2006. National assessment report of climate change (I): Climate change in China and its future trend. *Advances in Climate Change Research* 2:3–8.

Ding,Y.H., G.Y. Ren, G. Y. Shi, P. Gong, X.H. Zheng, P.M. Zhai, D.E. Zhang, Z.C. Zhao, S.Y. Wang, H.J. Wang, Y. Luo, D.L. Chen, X.J. Gao and X.S. Dai. 2007. China's National Assessment

Report on Climate Change (I): Climate change in China and the future trend. *Advances in Climate Change Research* 1673–1719.

Ding, Y.H., G.Y. Ren, Z.C. Zhao, Y. Xu, Y. Luo, Q.P. Li and J. Zhang. 2007. Detection, causes and projection of climate change over China: An overview of recent progress. *Adv. Atmos. Sci.* 24:954–971.

Editorial board. 2007. China's National Assessment of Report on Climate Change. Beijing: Science Press.

Evans, L.T. and R.A. Fischer. 1999. Yield potential: Its definition, measurement, and significance. *Crop Sci.* 39:1544–1551.

Jiang, X.J., L.Tang, X.J. Liu, F. Huang, W.X. Cao and Y. Zhu. 2011.Spatial and temporal characteristics of rice production climatic resources in main growing regions of China. *Transactions of the CSAE* 27(7):238–245.

Jiang, X.J., L. Tang, X.J. Liu, W.X. Cao and Y. Zhu. 2012. Predicting the Impact of Climate Change on Rice Production: Spatial and Temporal Characteristics of Rice Potential Productivity in the Main Growing Regions of China. *Agricultural Sciences in China* (in press).

Jiang, X.J., X.J. Liu, F. Huang, H.Y. JIANG, W.X. Cao and Y. Zhu. 2010. Comparison of spatial interpolation methods for daily meteorological elements. *Chin. J. Appl. Ecol.* 21(3):624–630.

Liu, B.H., M. Xu, M. Henderson, Y. Qi and Y. Li. 2004. Taking China's temperature: Daily range, warming trends, and regional variations, 1955–2000. *Journal of climate* 17:4453–4462.

Li, K.N., X.G. Yang, Z.J. Liu, W.F. Wang and F. Chen. 2010. Analysis of the potential influence of global climate change on cropping systems in China III. The change characteristics of climatic resources in northern China and its potential influence on cropping systems. *Scientia Agriicultura Sinica* 43(10):2088–2097.

Menzel, A., T. Sparks, N. Estrella and D.B. Roy, 2006. Geographic and temporal variability in phenology. Glob. *Ecol. Biogeogr.* 15:498–504.

Ministry of Agriculature of the P.R. China, 2009. The technical specification model http://www.moa.gov.cn/ztzl/gl/jsms/200904/P020090410579290158332.doc

Penning de Vries, F.W.T., D.M. Jansen, H.F.M. ten Berge and A. Bakema. 1989. Simulation of Ecophysiological Processes of Growth in Several Annual Crops. The Netherlands, Pudoc, Wageningen.

Piao, S.L.P. Cias, Y. Huang, Z.H. Shen, S.S. Peng, J.S. Li, L.P. Zhou, H.Y. Liu, Y.C. Ma, Y.H. Ding, P. Friedlingstein, C.Z. Liu, K. Tan, Y.Q. Yu, T.Y. Zhang and J.Y. Fang. 2010. The impacts of climate change on water resources and agriculture in China. *Nature*. 467:43–51.

Priya, S. and R. Shibasaki. 2001. National spatial crop yield simulation using GIS-based crop production model. *Ecological Modelling* 135:113–129.

Qian, W.H. and Y.F. Zhu. 2001. Climate Change in China from 1880 to 1998 and its Impact on the Environmental Condition. *Climatic Change* 50:419–444.

Rao, M.N., D.A. Waits and M.L. Neilsen, 2000. A GIS-based modeling approach for implementation of sustainable farm management practices. *Environ. Model. Software* 15:745–753.

Smit, B. and Y.L. Cai. 1996. Climate Change and Agriculture in China. *Global Environmental Change* 6(3):205–214.

Tan, G.X. and R. Shibasaki. 2003. Global estimation of crop productivity and the impacts of global warming by GIS and EPIC integration. *Ecological Modelling* 168:357–370.

Tang L., Y. Zhu, D. Hannaway, Y.L. Meng, L.L. Liu, L. Chen and W.X. Cao. 2009. RiceGrow: A rice growth and productivity model. *NJAS -Wageningen Journal of Life Sciences* 57:83–92.

Tao, F.L., M. Yokozawa, Y.L. Xu, Y. Hayashi and Z. Zhang. 2006. Climate changes and trends in phenology and yields of field crops in China, 1981–2000. *Agricultural and Forest Meteorology* 138:82–92.

Tao, F.l., M. Yokozawa, J. Liu and Z. Zhang. 2008. Climate-crop yield relationships at province scale in China and the impacts of recent climate trend. *Climate Research* 38:83–94.

Tao, F.L. and Z. Zhang. 2010. Adaptation of maize production to climate change in North China Plain: Quantify the relative contributions of adaptation options. *Europ. J. Agronomy.* 33:103–116.

Tong, C., C.A.S. Hall and H. Wang. 2003. Land use change in rice, wheat and maize production in China (1961–1998). *Agric. Ecosyst. Environ*. 95:523–536.

Wang, F.T. 1997. Impact of climate change on cropping system and its implication for agriculture in China. *Acta Meteorological Sinica*. 11(4):407–415.

Whistler, F.D., B. Acock, J. Lemmon, M. Mckinnion and V.R. Reddy. 1986. Crop simulation models in agronomic systems. *Advances in Agronomy* 40:141–208.

Xiong, W., D. Conway, E.D. Lin and I. Holman. 2009. Potential impacts of climate change and climate variability on China's rice yield and production. *Climate Research* 40:23–35.

Yang, X G, Z.J. Liu and F. Chen. 2010. The possible effects of climate warming on northern limits of cropping systems and crop yields in China. *Scientia Agriicultura Sinica* 43(2):329–336.

Chapter 11

South Asia Perspectives on Climate Change and Agriculture: Adaptation Options

Pramod Aggarwal*, Himanshu Pathak[†], Soora N. Kumar[†], and Pradeep Sharma**

*CGIAR Research Program on Climate Change,
Agriculture and Food Security,
International Water Management Institute,
New Delhi-110012, India
**HP Krishi Vishwavidyalaya, Palampur, India
[†]Indian Agricultural Research Institute,
New Delhi-110012, India
*p.k.aggarwal@cgiar.org

Introduction

South Asia includes 2.4% of the world's land surface area but is home to almost 23% of the world's population. Agriculture employs more than 50% of the labor force in South Asia and contributes 14–25% of the region's GDP. The region has achieved tremendous progress in the last four decades in food production and availability, yet 25% of the world's hungry and 40% of the world's malnourished children and women live here. Achieving the Millennium Development Goals (MDG) has remained a daunting task for most of the countries of the region.

South Asia's population continues to grow rapidly, creating an ever greater demand for food. It is estimated that by 2030, food grain requirements in South Asia will be almost 35% greater than the current demand (Paroda and Kumar, 2000). The additional quantities of food will have to be produced, from the same or even shrinking land resources due to increasing competition for land from the non-agricultural sectors. The tasks of alleviating poverty and attaining food security at the household and sub-national/regional level are thus major challenges.

Climate change is likely to compound this situation further. IPCC has projected that the global mean surface air temperature increase by the end of this century is likely to be in the range of 1.8–4.0°C (IPCC, 2007a). For South Asia, the projections are for 0.5–1.2°C rise in mean annual temperature by 2020, and 1.56–5.44°C by

2080, depending upon the scenarios of future developments (IPCC, 2007b). It is very likely that heat waves and extremes, as well as heavy precipitation events, will become more frequent. The sea level rise by the end of this century is likely to be 0.18–0.59 meters. The absolute amount of precipitation is likely to increase in South Asia in the future in all months except in the period of December–February when it is likely to decrease. This precipitation increase may be associated with heavier precipitation events and fewer rainy days leading to increased frequency of floods and droughts in the region.

Millions of people in South Asia are vulnerable to climate change because of depleting glaciers, increasing coastal erosion, frequent floods and droughts associated with global warming. High population density, dependent on agricultural sectors for livelihood that in turn are prey to monsoons, and extreme poverty (large sections of the population living under $1–2 a day).

The IPCC (2007b) showed that densely populated megadeltas of the region, such as in India and Bangladesh, and islands and coastal areas such as in Sri Lanka, Maldives, India and Bangladesh are more vulnerable to global climate change. Enhanced glacier melting in the Himalayas is projected to reduce availability of water resources in the agriculturally important Indo-Gangetic plains of Pakistan, India, Nepal and Bangladesh.

Producing enough food for the increasing demand against the background of diminishing resources under a changing climate scenario, while also minimizing further environmental degradation, will be a challenging task. Addressing climate change is central for the region's future food security and the attainment of Millennium Development Goals, especially for poverty alleviation. The region will need to implement strategies linked with its development plans, to enhance its adaptive capacity, mitigate emissions of greenhouse gases (GHG) and sequester more carbon. This paper reviews the current understanding of climate change impacts on South Asian agriculture and the potential options available for increasing climatic resilience and mitigation while maintaining/increasing food production.

Impacts of Climate Change on Agriculture in South Asia

Relatively few studies have been done to understand the potential impacts of climate change on South Asia's agricultural systems. Table 1 summarizes the key impacts for this region.

Adaptation Strategies for Climate Change

Adaptation strategies need to take a number of factors into consideration, including globalization, population and income growth, and the resulting changes in food

Table 1. A summary of actual or predicted impacts of climate change on South Asian agriculture.

Sector	Impact
Crop	• Increase in ambient CO_2 to 550 ppm may lead initially to a 10–15% increase in the yields of wheat, rice, chickpea and mustard. • An estimated 10–40% crop production could be lost of by 2100 AD despite the beneficial effects of higher CO_2 on crop growth, unless agriculture adapts (Rosenzweig and Parry, 1994, Fischer *et al.*, 2002, Parry *et al.*, 2004, IPCC 2007b, Cline, 2007; Stern, 2007). India could lose 4–5 million tons of wheat with every rise of 1°C temperature (Aggarwal, 2008). Rice production is projected to reduce by 2–10% in different scenarios of climate change (Pathak *et al.*, 2003; Knox *et al.*, 2011). Indeed, rice yields have shown a declining trend during the last two decades related to the gradual change in weather conditions (Aggarwal *et al.*, 2000; Pathak *et al.*, 2003). • Irrigated maize, sugarcane, and sorghum yields are likely to decrease by 7–25%, depending upon scenario and time (Knox *et al.*, 2011). Rainfed crops are likely to benefit in some regions due to the projected increase in rainfall. However, Nelson *et al.* (2010) report a significant decrease in the yields of all crops. • In Northwestern parts of Himalayas, a significant reduction in average productivity of apples has been observed in recent years due to inadequate chilling, crucial for good apple yields. It has resulted in a shift of the apple belt to higher elevations (Ranbir *et al.*, 2010). • The projected increase in drought and flood events could result in greater instability in food production and threaten the livelihood security of farmers. This is well-illustrated by the fact that in the more recent drought events of 2002, 15 million hectares of the rainy-season crop area and more than 10% of production was lost. Similar losses were noticed in the more recent drought of 2009 in India. Increased production variability could perhaps be the most significant effect of global climate change on Indian agriculture and food security. • Cold waves cause significant losses to crops such as mustard, mango, guava, papaya, brinjal, tomato, and potato in northern India, Nepal, Afghanistan and Pakistan. There are indications that such cold waves and frost events could decrease in the future due to global warming and hence yield losses in these crops associated with frost damage are likely to diminish. • The nutritional quality of cereals and pulses may also be moderately affected by the increase in temperatures which, in turn, will have consequences for nutritional security. • Nelson *et al.* (2009), for example, have shown that prices of most cereals will rise significantly due to climatic changes leading to a fall in consumption and hence decreased calorie availability and increased child malnutrition.
Water	• A significant increase in runoff is projected in the wet season that, however, may not be very beneficial unless storage infrastructure could be greatly expanded (Moors *et al.*, 2011). This extra water in the wet season, on the other hand, may lead to increase in frequency and duration of floods (Gosain *et al.*, 2006).

(Continued)

Table 1.   (*Continued*)

Sector	Impact
	• Remote sensing has shown that glaciers in Himalayas are retreading (Scherler *et al.*, 2011). The increased melting and recession of glaciers associated with global climate change could further change the runoff pattern (Kulkarni *et al.*, 2002). The increased glacier melting in the Himalayas could affect the availability of irrigation in the Indo-Gangetic plains, which, in turn, would have consequences on food security. • Climate change is likely to increase the demand for groundwater to manage the increasing periods of limited water availability. Lower groundwater tables and the resulting increase in the energy required to pump water will make irrigation more expensive.
Soil	• Organic matter content, which is already quite low in most soils of South Asia, would become still lower. Quality of soil organic matter may be affected. • The residues of crops under elevated CO_2 concentration will have higher C:N ratios, and this may reduce their rate of decomposition and nutrient release. • Increase of soil temperature will increase N mineralization rate, but its availability may diminish due to increased gaseous losses through processes such as volatilization and denitrification. • Change in rainfall volume and frequency, and of wind, may alter the severity, frequency and extent of soil erosion. • Rise in sea level may lead to salt-water ingression in the coastal lands, making them less suitable for conventional agriculture.
Livestock	• Global warming would increase water, shelter, and energy requirement of livestock for meeting projected milk demands. • Climate change is likely to aggravate the heat stress in dairy animals, adversely affecting their reproductive performance. • Heat stress may also directly impact forage quality, ingestion of food and feed, declines in physical activity, and ultimately reduce dairy milk yield.
Fishery	• Increasing sea- and river-water temperature is likely to affect fish breeding, migration, and harvests. Migration of oil sardines to northern latitudes and along the east coast of India in last few decades due to warming of the sea surface by 0.5–1.0°C is a typical example (Vivekanadan *et al.*, 2009). • Coral bleaching is likely to increase due to higher sea surface temperature. Heat stress in 1998 and 2002 caused considerable bleaching of corals in the Indian Ocean (Wilkinson *et al.*, 1999; Kumarguru *et al.*, 2002). Such events are likely to become more frequent in the future due to global climate change.

Source: Largely adapted from Aggarwal *et al.* (2009). Although most of these results are based on studies done in India, it is expected that the responses in other South Asian countries will be similar.

preferences and demand, as well as the socioeconomic and environmental consequences of alternative adaptation options. These strategies should be climate smart, taking care of not only adaptation but also resilience, mitigation and sustainable intensification (FAO, 2010). Some of these strategies are discussed below.

Augmenting production

The climatic factors in South Asia allow reasonably high yield potentials of most crops. For rice and wheat these are estimated to be 10 and 8 t/ha, respectively (Aggarwal *et al.*, 2000). National Demonstration plots also show considerably higher yield potential of most crops than what is being achieved. At the same time, national average yields of most crops remain low in South Asia. For example, the national average yields of rice and wheat crops in India are less than 3 t/ha today, indicating large yield gaps in the country. In the eastern parts of South Asia, gaps are especially large, and this region can be a future source of food security for the sub-continent. For meeting increased food demands of the future and to overcome the constraints poised by climatic changes, there is a need to raise yield potential for regions such as Punjab in India and Pakistan, where yield gaps are small, and increase stability and bridge yield gaps in other regions. Fragile seed sector, poor technology dissemination mechanisms, and lack of adequate capital and information are the principal reasons for yield gaps. Reducing even some of the yield gap could strengthen the food security in the region and reduce the vulnerability of the populations to climate change.

Developing climate-ready crops

Future crop breeding efforts would need to address multiple stresses — droughts, floods, heat waves and pest loads — imposed by changing climate. There will be a need to stack several adaptive traits in a suitable agronomic background. This would require substantial breeding efforts, which will depend on the collection, conservation and distribution of appropriate crop genetic material among plant breeders and other researchers. There is a need for a better understanding of wild relatives, landraces, and their distributions, sensitivity of wild and cultivated species, and genetic material currently in the gene banks, to climatic variables; creating trait-based collection strategies, and establishing pre-breeding as a public good for providing a suitable response to the multiple challenges posed by global climate change. Exploitation of genetic engineering and molecular breeding for 'gene pyramiding' becomes essential to pool all desirable traits in one plant to get the 'ideal plant type' which may also be an 'adverse climate tolerant' genotype (Varshney *et al.*, 2011). Using these approaches, a 'sub1' gene has been introduced in the rice plant that has provided a great opportunity to develop submergence tolerance without apparent ill-effects on productivity, or grain quality (Bailey-Serres *et al.*, 2010).

Changes in land-use management

Historically, land use changes have been an important source of adaptation to climatic stresses. For example, to cope with a period of drought, rainfed farmers

diversify from monocropping to intercropping, increase their dependence on livestock, or migrate to other regions. Strategic land use changes are however guided
not only by the need to cope with climatic stresses but also to adapt to opportunities provided by technology, institutions and markets. Cultivation of almost 13
million hectares of rice-wheat system in the Indo-Gangetic plain of South Asia is
a typical example (Timsina and Connor, 2001). During last few decades, temperatures in hilly areas of Himachal Pradesh have risen resulting in declining yields
of apple, which need a critical chilling temperature. Consequently apple belt has
shifted to higher elevation regions and the area so vacated has now been occupied
by vegetables. Such a land use change has not only helped farmers to adapt but
also increase their incomes (Ranbir *et al.*, 2010). The rapid rise in shrimp farming at the cost of rice in Bangladesh and coastal India in the last two decades is
an example of market driven land use changes (Islam, 2006). In recent times, the
South Asia region has been witnessing rapid growth in gross income and as a result
of the increase in demand for vegetables, fruits, meat and milk. In future, climatic
changes together with such changes in demand are likely to result in major changes in
land use.

For ecosystem resilience, farmers will need crop varieties with greater tolerance
to stresses such as drought and heat, excess soil moisture, as well as photo- and
thermal-insensitive varieties. Adapted seeds may no longer be found in existing
local seed systems and, therefore, interventions must involve long-distance seed
exchange and be facilitated by genebanks. Modern GIS-based tools such as climate
analogues, can be used to identify materials that have been successful around the
world in regions with similar climatic profiles. Nevertheless, there is still a big need
for scientific tools to assist in regional land use planning that can help farmers and
societies to adapt to climate change, meet food demands and environmental goals;
and also lead to higher income.

Adjusting the cropping sequence, including changing the timing of sowing,
planting, spraying, and harvesting, to take advantage of the changing duration of
growing seasons and associated heat and moisture levels is another option for
increasing adaptation to climate change. Altering the time at which fields are
sown or planted can also help farmers regulate the length of the growing season to better suit the changed environment. But in the Indo-Gangetic Plains, this
option will be limited by low winter temperatures. Adjustment of planting dates
to minimize the effect of temperature increase-induced spikelet sterility can be
used to reduce yield instability, by avoiding having the flowering period coincide
with the hottest period. Reliable seasonal weather forecasts could also be seen as
one supportive measure to optimize planting and irrigation patterns (Gadgil *et al.*,
2002).

For ecosystem resilience, farmers will need crop varieties with greater tolerance
to stresses such as drought and heat, excess soil moisture, as well as photo- and
thermal-insensitive varieties. Adapted seeds may no longer be found in existing
local seed systems and, therefore, interventions must involve long-distance seed
exchange and be facilitated by genebanks. Modern GIS-based tools such as climate
analogues, can be used to identify materials that have been successful around the
world in regions with similar climatic profiles. Nevertheless, there is still a big need
for scientific tools to assist in regional land use planning that can help farmers and
societies to adapt to climate change, meet food demands and environmental goals;
and also lead to higher income.

Resource conservation and use efficiency

There is an increasing competition for resource use (such as land, water, capital and labor) in agriculture in South Asia with industry and urban sectors. This pressure on agriculture will increase further in future due to continuously increasing populations and income. Climate change will further exacerbate this competition. Efforts need to be made now to conserve natural resources or increase their use efficiencies.

Agriculture in South Asia uses more than 70% available water. This fraction needs to be reduced in future. The efficiencies of water-use, therefore, needs to be improved at the farm, community and regional scales. It has been demonstrated that proper leveling of farms could improve water application efficiencies by over 20%. Greater realism in water and energy pricing could also promote efficient use of natural resources (Shah, 2009). Use of modern irrigation methods such as micro-irrigation could also be promoted to enhance water-use efficiency. Small changes in climatic parameters can often be better managed by altering dates of planting, spacing and input management. Early planting of wheat, facilitated after rice harvest by surface seeding or zero-tillage may offset most of the losses associated with increased temperatures in South Asia. This reduces costs of production and the use of natural resources such as fuel, and improves the efficiency of water and fertilizers. Other resource-conserving technologies such as systems of rice intensification, bed planting, and alternate wetting and drying also need to be carefully assessed. On a regional scale, Managed Aquifer Recharge has large potential to enhance natural recharge rates of groundwater (Shah, 2009). The use of efficient water utilization methods such as micro-irrigation coupled with groundwater use may lead to reduction in depletion of groundwater. Conjunctive management of surface and groundwater in Punjab offers large opportunities for improving water productivity as well as for saving energy (Shah, 2009). Such resource-conserving technologies in rice-wheat system also have pronounced effects on mitigation of greenhouse gas emission and adaptation to climate change (Aggarwal and Pathak, 2009).

The sector in South Asia contributes to about 20% of total GHG emissions. The emissions from agriculture are primarily due to enteric fermentation in ruminant animals, rice paddies, and nitrous oxides from application of manures and fertilizers to agricultural soils (Fig. 1). In future, emissions from soils are likely to increase significantly because of increased intensification of input use required to meet increased food demand. Increasing resource-use efficiency also has huge co-benefits and implications for GHG mitigation (Aggarwal, 2008). Some of these are listed below:

- Altering water management in paddy, particularly promoting mid-season aeration by short-term drainage and improving organic matter management by promoting

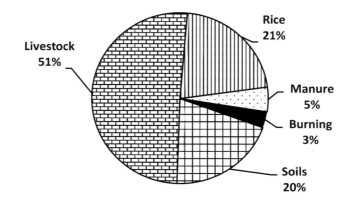

Fig. 1.   Contribution of various agricultural sectors to GHG emissions in South Asia in 2005.
Source: Smith *et al.* (2007).

aerobic degradation have the potential for reducing methane emissions. However, drainage of rice fields may lead to increased emission of nitrous oxides.

- Soil/crop management practices that increase fertilizer N-use efficiency can significantly reduce N_2O emissions. Appropriate application of nitrate-N in upland conditions and ammoniacal-N in wetland soils can help decrease N_2O emissions. Another option to mitigate N_2O production is to use nitrification inhibitors, such as nitrapyrin and dicyandiamide (DCD), along with N fertilizers to curtail the process of nitrification. Plant-derived organics such as neem oil, neem cake and karanja seed extract which can also act as nitrification inhibitors.

- Resource-conservation technologies restrict release of soil carbon, thus mitigating increase of CO_2 in the atmosphere (Grace *et al.*, 2003). According to an estimate, zero tillage saves about 30 l/ha of diesel compared to conventional tillage, thereby leading to a reduction of CO_2 production by 80 kg/ha/year. If such savings could be translated even partially to larger arable lands, substantial reduction in CO_2 emission to the atmosphere could be achieved. Lal (2004) estimates that such practices can lead to carbon sequestration of 25–50 Tg C/year in soils of South Asia for several decades, including 11–22 Tg C/year on croplands.

- Improving the efficiency of energy use in agriculture by using better designs of machinery, increasing fuel efficiency in agricultural machinery, commercialization of wind/solar power potential, and use of laser levelers also contribute to mitigation.

- Changing land use by increasing the area under biofuels or agro-forestry could also mitigate GHG emissions (Smith *et al.*, 2007a). This, however, may have trade-offs with the goal of increasing food production. South Asian countries that are scarce of land need to examine the consequences of diverting land to biofuels on their food security (Aggarwal *et al.*, 2007).

Table 2.   Potential and constraints of greenhouse gas mitigation options.

Option	Mitigation potential	Constraints
Methane from rice field		
Intermittent drying	25–30%	Assured irrigation
Direct-seeded rice	30–40%	Machine, herbicide
System of rice intensification	20–25%	Labor, assured irrigation
Nitrous oxide from soil		
Site-specific N management	15–20%	Awareness, fertilizer policy
Nitrification inhibitor	10–40%	Cost, incentive

Source: Pathak *et al.* (2010)

The GHG mitigation potential of the most promising resource-conserving technologies and their constraints are summarized in Table 2. Some technologies such as intermittent drying and site-specific N management can be easily adopted by farmers without extra investment, whereas other technologies may need economic incentives and policy support.

Harnessing the local technical knowledge of farmers

Farmers in South Asia, often poor and marginal, have been experimenting with climatic variability for centuries. There is a wealth of knowledge of a range of measures that can help in developing technologies to overcome climate vulnerabilities. Sharing such experiences accumulated over centuries could be useful at the household, community as well as regional levels in many parts of the world. Several regions have earlier minimized their exposure to climatic stresses by resorting to mixed cropping, changing varieties and planting times, by diversifying sources of income for farmers, and maintaining buffer stocks of food for managing periods of scarcity. These management strategies would also help in the future climate change scenarios but may not be enough in view of the increasing intensity of climatic risks and pressure on land to produce more food and also with much higher efficiency. In some parts of the world such as in Bangladesh and eastern parts of India, recurring cyclones and floods have almost become a rule. In such instances, traditional risk minimization approaches might no longer apply. Instead, a focus on adaptation action is needed and would require considerable investment to increase resilience of the communities. It must also be pointed out that although local knowledge can often help individual subsistence farmer to survive periods of climatic stress, but in today's world, farmers need to produce food not only for themselves but also for 50% of South Asia's population living in urban areas. It remains to be evaluated if local knowledge can assist the farmers in ensuring food security for all.

Improved risk management though early warning system and crop insurance

The increasing probability of floods and droughts and other uncertainties in climate may seriously increase the vulnerability of Eastern India and of resource-poor farmers to global climate change. Policies that encourage crop insurance can provide protection to the farmers in the event their farm production is reduced due to natural calamities. In view of these climatic changes and the uncertainties in future agricultural technologies and trade scenarios, it will be very useful to have an early warning system of environmental changes and their spatial and temporal magnitude. Such a system could help in determining the potential food-insecure areas and communities given the type of risk. Modern tools of information technology could greatly facilitate this. India has made remarkable progress in the areas of crop insurance, index-based insurance and mobile telephone based dissemination of weather forecasts and related agro-advisories. Such systems need to be developed and implemented in other countries of South Asia.

Enabling institutional and policy support

Both current and future climatic risks can be managed better if there is appropriate policy and institutional support together with technological interventions. Several technological options recommended by the scientists are generally not implemented by the governments because of very limited knowledge available today on the costs and benefits associated with these and other competing social and economic activities. Integrating perspectives on climatic risks in current national policies and programs in different sectors such as disaster management, water resources management, land use, biodiversity conservation, and agricultural development would lead to increased adaptive capacity to current as well as future climatic variability. Incentives should be provided to the farmers and industry for increasing the efficiency of water, fertilizer, and energy use and sequestration of carbon (FAO, 2008). Rational pricing of surface and groundwater, for example, can arrest its excessive and injudicious use. Availability of assured prices and infrastructure could create a situation of better utilization of groundwater. Policies should to be evolved that would encourage farmers to enrich organic matter in the soil and thus improve soil health such as financial compensation/incentive for green manuring.

Although the share of agriculture in gross domestic product has declined in many countries, a large fraction of the population is dependent on it for livelihood. For example, in India, the share of agriculture in the national gross domestic product has declined to 15% but more than 50% population continues to remain dependent on this. Such trends have resulted in fragmentation and decline in size of land holdings leading to inefficiency in agriculture and rise in unemployment, underemployment, and low volume of marketable surplus and therefore increased vulnerability to global change. Institutional arrangements, such as cooperatives and contract farming that can bring small and marginal farmers together for increasing production and marketing efficiencies are needed.

Conclusion

Global climate change is likely to affect food and livelihood security of millions of poor farmers and landless laborers in South Asia and other developing countries. The climatic changes have the potential to limit the future capacity of South Asia to remain agriculturally self-sufficient. Urgent steps are needed to increase the region's adaptive capacity toward current as well as future climatic risks. The fight against hunger and climate change in South Asia needs to be complementary and synergized. There are ways by which adverse impacts may be mitigated and agriculture can be adapted to the changed climate. Development and implementation of adaptation strategies necessitate empowerment of farmers besides developing competencies in acquiring knowledge and skills. This will require increased adaptation research, capacity building, development activities, changes in policies, and support of global adaptation and mitigation funds and other resources.

To facilitate this, CCAFS (CGIAR Research program on Climate Change, Agriculture and Food Security) is currently testing a scalable model of participatory climate smart agriculture (climate smart villages/farms) in partnership with major CGIAR institutes, civil society and NARS across several sites in the Indo-Gangetic plains (Fig. 2). Elements of value-added weather forecasts; index based insurance; food, feed, forage, seed and water banks; adapted varieties; RCTs; and a range of low GHG emitting options are currently being evaluated. Climate analogues are being used to provide knowledge of adaptation options and to facilitate farmer exchange for their learning pathways. Such climate smart villages need to be established across the world in regions with significant climatic risks.

There is an urgent need for action since it takes time for adaptive practices to become effective. The costs of adaptation and mitigation are not clearly known today but these are expected to be high (Stern, 2007; Cline *et al.*, 2007; Nelson *et al.*, 2009). A universally preferred solution is to start with such adaptation strategies that are needed anyway for sustainable development. Strategies that maximize synergies between adaptation, mitigation, food production and sustainable development would be most appropriate. For example, increasing the efficiency of fertilizer and water-use will lead to higher profits for the farmers as well as sequester more carbon, which will improve soil health and lead to higher production. Such practices would become more attractive if farmers could be granted incentives for environmental services that

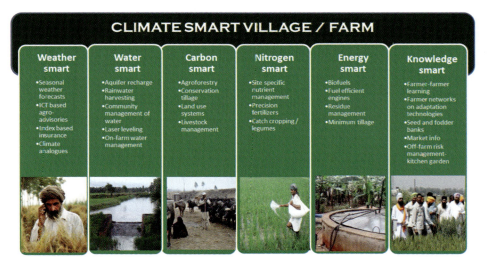

Fig. 2.   Linking knowledge to action — climate smart villages/farms: Key agricultural activities for managing risks, increasing adaption and mitigation to climate change.

agriculture provides. It is time that society considers such incentives to farmers in the interest of the global environment, food security, and poverty alleviation.

References

Aggarwal, P.K. 2008. Global climate change and Indian agriculture: Impacts, adaptation and mitigation. *Indian J. Sci.* 78:911–919.

Aggarwal, P.K., and S.K. Bandyopadhyay, H. Pathak, N. Kalra, S. Chander and S. Sujith Kumar. 2000. Analysis of the yield trends of rice-wheat system in north-western India. *Outlook on Agriculture* 29(4):259–268.

Aggarwal, P.K. and H. Pathak. 2009. Conservation cultivation to combat climate change. *Indian Farming* 59(3):5–10.

Aggarwal, P.K. (Ed.). 2009. *Global Climate Change and Indian Agriculture. Case Studies from the ICAR Network Project.* Indian Council of Agricultural Research: New Delhi, India.

Aggarwal, P.K., H.C. Joshi, S. Kumar, N. Gupta and S. Kumar. 2007. Fuel ethanol production from Indian agriculture: Opportunities and constraints. *Outlook on Agriculture* 36:167–174.

Aggarwal, P.K., K.K. Talukdar and R.K. Mall. 2000. Potential yields of rice-wheat system in the Indo-Gangetic plains of India. *Rice-Wheat Consortium Paper Series 10.* New Delhi, India. RWCIGP, CIMMYT. pp. 16.

Bailey-Serres, J., T. Fukao, P. Ronald, A. Ismail, S. Heuer and D. Mackill. 2010. Submergence tolerant rice: SUB1's journey from landrace to modern cultivar. *Rice* 3: 138–147.

Cline, W. 2007. *Global Warming and Agriculture: Impact Estimates by Country.* Peterson Institute for International Economics.

FAO. 2008. *High-Level Conference on World Food Security: Climate Related Transboundary Pests and Diseases.* FAO, Rome. Available at http://www.fao.org/foodclimate/expert/em3/outputs-em3/en/

FAO. 2010. *"Climate-Smart" Agriculture: Policies, Practices and Financing for Food Security, Adaptation and Mitigation.* 41pp. Rome.

Fischer, G., M. Shah and H. van Velthuizen. 2002. *Climate Change and Agricultural Vulnerability, International Institute for Applied Systems Analysis.* Laxenburg, Austria.

Gadgil, S., P.R.S. Rao and K.N. Rao. 2002. Use of climate information for farm-level decision making: rainfed groundnut in southern India. *Agric. Syst.* 74:431–457.

Gosain, A.K., S. Rao and D. Basuray. 2006. Climate change impact assessment on hydrology of Indian River basins. *Current Science* 90(3):346–353.

Grace, P.R., M.C. Jain, L. Harrington and G. Philip Robertson. 2003. The long-term sustainability of tropical and sub-tropical rice and wheat systems: An environmental perspective. In: *Improving the Productivity and Sustainability of Rice-Wheat Systems: Issues and Impacts.* ASA Special publication 65. Madison, USA.

IPCC. 2007a. Climate Change 2007: The Physical Science Basis. Summary for Policymakers. Inter-Governmental Panel on Climate Change.

IPCC. 2007b. Climate Change 2007: Climate Change Impacts, Adaptation and Vulnerability. Summary for Policymakers. Inter Governmental Panel on Climate Change.

Islam, R.M. 2006. Managing diverse land uses in coastal Bangladesh: Institutional approaches. pp. 237–248. In C.T. Hoanh, T.P. Tuong, J.W. Gowing and B. Hardy (eds.) *CAB International 2006. Environment and Livelihoods in Tropical Coastal Zones.*

Knox, J.W., T.M. Hess, A. Daccache and M. Perez Ortola. 2011. What are the projected impacts of climate change on food crop productivity in Africa and South Asia? DFID Systematic Review, Final Report. Cranfield University. pp. 77.

Kulkarni, A.V. and I.M. Bahuguna. 2002. Glacial retreat in the Baspa basin, Himalaya, monitored with 17 satellite stereo data (Correspondence). *Journal of Glaciology* 48:171–172.

Kumaraguru, A.K., K. Jayakumar and C.M. Ramakritinan. 2002. Coral bleaching 2002 in the Palk Bay, southeast coast of India. *Current Science* 85:1787–1793.

Lal, R. 2004. Soil carbon sequestration impacts on global climate change and food security. *Science* 304:1623–1627.

Moors, E.J., A. Groot, H. Biemans, C.T. van Scheltinga, C. Siderius, M. Stoffel, C. Huggel, A. Wiltshire, C. Mathison, J. Ridley, D. Jacob, S. Pankaj Kumar, Bhadwal, A.K. Gosain and D.N. Collins. 2011. Adaptation to changing water resources in the Ganges basin, northern India. *Environmental Science & Policy* 14:758–769.

Nelson, G.C., M.W. Rosegrant, J. Koo, R. Robertson, T. Sulser, T. Zhu, C. Ringler, S. Msangi, A. Palazzo, M. Batka, M. Magalhaes, R. Valmonte-Santos, M. Ewing and D. Lee. 2009. *Climate Change: Impact on Agriculture and Costs of Adaptation.* Washington, DC, IFPRI.

Paroda, R.S. and P. Kumar. 2000. Food production and demand in South Asia. *Agric. Econ. Res. Rev.* 13:1–24.

Parry, M.L., C. Rosenzweig, Iglesias, A.M. Livermore and G. Fischer. 2004. Effects of climate change on global food production under SRES emissions and socio-economic scenarios. *Global Environmental Change* 14:53–67.

Pathak, H., A. Bhatia, N. Jain and P.K. Aggarwal. 2010. Greenhouse gas emission and mitigation in Indian agriculture — A review. pp. i–iv and 1–34. *In ING Bulletins on Regional Assessment of Reactive Nitrogen.* SCON-ING, New Delhi.

Pathak, H., J.K. Ladha, P.K. Aggarwal, S. Peng, S. Das, Y. Singh, B. Singh, S.K. Kamra, B. Mishra, A.S.R.A.S. Sastri, H.P. Aggarwal, D.K. Das and R.K. Gupta. 2003. Trends of climatic potential and on-farm yields of rice and wheat in the Indo-Gangetic Plains. *Field Crops Research* 80:223–234.

Ranbir Singh Rana, R., M. Bhagat, V. Kalia, and H. Lal. 2010. Impact of climate change on shift of apple belt in Himachal Pradesh. pp. 131–137. In Sushma Panigrahy, Shibendu Shankar Ray and Jai Singh Parihar (eds.), *ISPRS WG VIII/6 Archives XXXVIII-8/W3 Workshop Proceedings: Impact of Climate Change on Agriculture*. Space Application Centre Ahmadabad.

Rosenzweig, C. and M.L. Parry. 1994. Potential impact of climate change on world food supply. *Nature* 367:133–138.

Scherler, D., B. Bookhagen, M.R. Strecker. 2011. Spatially variable response of Himalayan glaciers to climate change affected by debris cover. *Nature Geoscience* 4.

Shah T. 2009. Climate change and groundwater: India's opportunities for mitigation and adaptation. *Environ. Res. Lett.* 4:13pp.

Smith P., D. Martino, Z. Cai, D. Gwary, H. Janzen, P. Kumar, B. McCarl, S. Ogle, F. O'Mara, C. Rice, B. Scholes and O. Sirotenko. 2007a. Agriculture. In (B. Metz, O.R. Davidson, P.R. Bosch, R. Dave, L.A. Meyer (eds.), *Climate Change 2007: Mitigation. Contribution of Working Group III to the Fourth Assessment Report of the Intergovernmental Panel on Climate Change*, Cambridge University Press, Cambridge, United Kingdom and New York, NY, USA.

Stern, N. 2007. *Stern Review on the Economics of Climate Change*. Cambridge University Press. Cambridge.

Timsina, J. and D.J. Connor. 2001. Productivity and management of rice-wheat cropping systems: Issues and challenges. *Field Crops Res.* 69:92–132.

Varshney, R.K. K.C. Bansal, P.K. Aggarwal, S.K. Datta and P. Craufurd. 2011. Agricultural biotechnology for crop improvement in a variable climate: hope or hype? *Trends in Plant Science.* 16:363–371.

Vivekanandan, E., M. Rajagopalan and N.G.K Pillai. 2009. Recent trends in sea surface temperature and its impact on oil sardine. pp. 89–92. *In* Aggarwal, P.K. (ed.), *Global Climate Change and Indian Agriculture. Indian Council of Agricultural Research.* New Delhi.

Wilkinson, C.R., O. Linden, H. Cesar, G. Hodgson, J. Rubens and A.E. Strong. 1999. Ecological and socioeconomic impacts of 1998 coral mortality in the Indian Ocean: An ENSO impact and a warning of future change. *Ambio.* 28:188–196.

Section III

Programs and Projects

Chapter 12

Research Needs for Agriculture Under Elevated Carbon Dioxide

Mary Beth Kirkham

Department of Agronomy,
2004 Throckmorton Plant Sciences Center,
Kansas State University, Manhattan,
KS 66506-5501, USA

mbk@ksu.com

Introduction

Water and carbon dioxide (CO_2) are the two most important compounds affecting plant growth. In introductory botany textbooks, we have seen the familiar equation for photosynthesis, which shows CO_2 joining with water (H_2O), in the presence of light and chlorophyll, to form glucose, a sugar ($C_6H_{12}O_6$), and oxygen (O_2), as follows:

$$6CO_2 + 6H_2O \xrightarrow[\text{chlorophyll}]{\text{light}} C_6H_{12}O_6 + 6O_2 \tag{1}$$

This chapter reviews research from experiments that have been done with elevated CO_2 and suggest future research needs. It is important to recognize that elevated CO_2 has direct effects on the water in soils and plants, whether or not it may affect the temperature on earth. It has been shown by experimentation that elevated CO_2 is already having effects on the water in soils and plants. Carbon dioxide can be elevated both in the soil and the atmosphere. Therefore, we need to distinguish the effect of elevated CO_2 in the soil from elevated CO_2 in the atmosphere.

The first accurate measurements of CO_2 in the atmosphere were made by Charles Keeling starting in 1957 at Mauna Loa in Hawaii and in Antarctica at the South Pole (Kirkham, 2011, p. xiii and p. 26). His measurements showed that, in 1958, the concentration of CO_2 in the air was 316 ppm. The CO_2 concentration in the air in 2010, the last year for which complete data are available, was 389 ppm (Dlugokencky, 2010, 2011). Therefore, the concentration has increased 73 ppm in 52 years for an average increase of 1.4 ppm per year.

For controlled agricultural production, growers have long elevated CO_2 in greenhouses to stimulate growth. One of the first scientists to show increased growth was the Frenchman Demoussey (1904), who grew several species in normal air and in atmospheres containing 1500 ppm CO_2 and saw increase in plant weight as a result of the additional CO_2 (Hicklenton, 1988, p. 2). In the USA in 1918, the beneficial effects on CO_2 on growth rate and yield of plants in greenhouses were established after several years of work at the Vermont Agricultural Experiment Station by Cummings and Jones (1918), who demonstrated the feasibility of CO_2 enrichment in greenhouses. Despite these pioneering studies, commercial adoption of the technique was slow both in Europe and in North America. It was not until the late 1950's that a number of new studies reawakened interest in CO_2 enrichment and resulted in worldwide commercial application (Hicklenton, 1998, p. 2). Today, the emphasis of research has shifted to development of improved methods of control, and using the optimal concentrations of CO_2 for different crops (Nelson, 2012, p. 320). Researchers have found that levels of 2000 ppm evoke growth responses in some crops (Nelson, 2012, p. 320). Apparently, this goes back to the earlier adaptation of plants to higher CO_2 levels in primordial times (Nelson, 2011, p. 320). The CO_2 concentration in the atmosphere in the Cretaceous, 120 million years ago when angiosperms appeared, was 3000 ppm (Kirkham, 2011, p. 252). The high concentration of CO_2 in the Cretaceous shows that plants grow well in much higher concentrations of CO_2 than now exist. In a greenhouse, if environmental conditions are optimal (e.g., light, temperature, and water), then elevating the CO_2 concentration will increase growth. For example, Gaastra (1962) showed that, under optimal conditions, the photosynthetic rate of cucumber (*Cucumis sativus* L.) was increased three times (from 90 mm^3 CO_2 cm^{-2} h^{-1} to 280 mm^3 CO_2 cm^{-2} h^{-1}) when the CO_2 level of the atmosphere was increased from 300 ppm to 1300 ppm. The commonly injected levels of CO_2 into greenhouses around the world today are 900 to 1200 ppm (Nelson, 2012, p. 321). These concentrations are two to three times more than the current CO_2 concentration in the open air. As Hicklenton (1988, p. 2) points out, "global CO_2 concentrations remain far below those necessary to sustain maximum rates of growth in green plants."

Elevated CO_2 in the Soil

The four soil physical factors that affect root growth are mechanical impedance (related to bulk density), soil water, soil aeration, and soil temperature (Shaw, 1952, p. vii). There is little information concerning elevated CO_2 in the soil and its effects on these four soil physical factors. Tackett and Pearson (1964) in Auburn, Alabama, varied the level of CO_2 and bulk density in columns of soil to study their effects on root growth of cotton (*Gossypium hirsutum* L.) seedlings at constant oxygen concentration (21%). They compacted a sandy loam soil to bulk densities of 1.3, 1.5, 1.6, and 1.7 g/cm^3. They supplied CO_2 to the soil at four levels (0.03%; 6.1%; 11.9%, and 24.0% by volume). When results of all densities were averaged, CO_2 did not affect the rate of root elongation until the concentration of CO_2 reached a level of about 12%. The absence of extreme toxicity at high levels of CO_2 (up to 24% in the soil) indicated that CO_2 would rarely limit root growth in this sandy loam. Mechanical impedance to root penetration would supersede aeration effects. The best root growth at 24% CO_2 in the soil occurred at the lowest bulk density (1.3 g/cm^3). Because elevated CO_2 in the soil and bulk density interact to control root growth, tillage may need to be increased. We need experiments to resolve this issue.

There is little information concerning the interaction of elevated CO_2 in soil with the other three factors that control root growth: water, aeration, and temperature. Oxygen in the soil has been much studied, but not CO_2. We know the minimum amount of oxygen in the soil essential for roots. Gases in the soil move by diffusion (Kirkham, 1994). If the soil pores are filled with water and have no air, then roots lack oxygen. Roots of most crops need at least 10% by volume of air space in the soil to survive. This is because diffusion of gases in the soil stops when the air-filled pore space is less than 10% (Kirkham, 2005, p. 129). Knowing that there is 21% oxygen in the air (Kirkham, 2011, p. 30), we can calculate that the minimum amount of oxygen necessary in the soil for roots is 2.1% (21% $\times$ 0.10 = 2.1%). We do not know the maximum concentration of CO_2 in soil that still allows plant growth.

The paucity of information concerning the interaction of the soil physical factors that control root growth and elevated levels of CO_2 in the soil is remarkable. Perhaps the reason for the lack of research is that studies focus on elevated CO_2 in the atmosphere, not in the soil. We need to answer the following questions: How great a tolerance do crop plants have to soil physical conditions under elevated CO_2? What is the interaction of elevated temperature in the soil and elevated CO_2 in the soil on root growth? What is the interaction of soil matric potential and elevated CO_2 in the soil? If the soil is dry, is a seed more susceptible to damage under elevated CO_2 in the soil compared to a seed in a well-watered soil? What is the maximum concentration of CO_2 in the soil that limits the growth of roots? A wide variety of crops needs to be studied, and they need to be grown in soils of different textures. How is seed germination affected by elevated CO_2? What are the best soil physical conditions for seed germination under elevated CO_2? What volume of soil about a seed is critical for diffusion of CO_2 away from the root and oxygen toward it? Will row spacing need to be increased under elevated CO_2 in the soil? What is the heterogeneity of CO_2 concentration in a soil?

We need instruments that can measure not only CO_2 in the soil averaged over large areas, but also ones that can measure CO_2 in small areas. Can an instrument like the dual-probe heat-pulse sensor, which measures water content over a few millimeters (Song *et al.*, 1998) be developed to measure CO_2? The instruments must be commercially available, so anyone can measure the CO_2. The instrumentation needs to be simple, so the measurements can be made without extensive training and servicing. Can the agricultural engineers develop an instrument that monitors CO_2 as the tillage equipment rolls over the soil surface? Such equipment is being developed to monitor water content. A heavy-duty time domain reflectometry soil moisture probe has been made for use in intensive field sampling (Long *et al.*, 2002). Contours of soil moisture have been obtained with the probe. If probes could be attached to cultivators or other such equipment, water and CO_2 could be monitored simultaneously. The problem is slowing a tractor to a low enough speed so that there is time enough for a measurement to be taken at a certain location. The instruments that measure CO_2 could be hooked up to a computer to monitor spatial variability of CO_2 across a field. If there are localized spots of high CO_2 (i.e., too high to support root growth), could oxygen be injected into the soil, or the soil tilled, to reduce the high CO_2? Large amounts of farmyard manure applied to soil increase the production of CO_2 in the soil (Kirkham, 2011, p. 43), so much so that the CO_2 may reduce root growth. Can the non-limiting water content range (Letey, 1985) be modified to include measurements of CO_2? The non-limiting water range, which includes three of the four soil physical factors that control root growth (water, aeration, and temperature), is being widely used to predict crop yields (Asgarzadeh *et al.*, 2010). The oxygen diffusion rate is measured to determine the aeration status of the soil for use in calculating the non-limiting water range. Can measurements of CO_2 be incorporated into these calculations?

Elevated CO_2 in the Atmosphere above the Soil

Because apparently no measurements of the soil water content under augmented atmospheric CO_2 had been made, we in the Evapotranspiration Laboratory at Kansas State University measured over a 2-year period (1989–1990) the soil water content in a tall grass prairie exposed to ambient (337 ppm) and twice ambient (658 ppm) levels of CO_2 (Kirkham *et al.*, 1993; Kirkham, 2011, p. 83). The dominant species on the grassland is big bluestem (*Andropogon gerardii* Vitman), a warm-season grass. About 70% of the botanical composition of the grassland is big bluestem. The plants were grown in closed-top chambers (1.5 m in diameter and 1.8 m tall). Half of the plots (chambers) were kept at field capacity (high soil moisture level), and half of the plots were kept at half field capacity (low soil moisture level). The soil was a silty clay loam. We measured soil water content weekly during the two growing seasons with a neutron probe between the 100 and 2000 mm depth. Soil water contents in the 0–50 and 50–100 mm depths were determined gravimetrically.

The data were put on a volumetric basis by multiplying by the bulk density (Kirkham, 2011, p. 83). There was more water in the soil under the elevated CO_2 than in the soil under the ambient CO_2. For example in the 1989 season, for the elevated and ambient concentrations of CO_2, the average amounts of water in the soil were 722 and 705 mm, respectively, for the high water level. For the low water level, these values were 669 and 638 mm, respectively. Therefore, with twice-ambient CO_2, the soil contained an average of 17 and 31 mm more water at field-capacity and half-field capacity water regimes, respectively. The higher soil water content in the plots with elevated CO_2 was probably due to the fact that elevated CO_2 closes stomata and reduces transpiration.

Other data have confirmed that soil water content is greater under elevated CO_2 than under ambient CO_2. Some studies show no effect of elevated CO_2 on soil water content (Kirkham, 2011, p. 91–95). Where the soil is extremely dry (deserts), no effect of elevated CO_2 is seen (Kirkham, 2011, p. 91–92). This is probably because under the dry conditions of a desert the stomata remain closed, so that large amounts of water are not transpired. In dry periods, desert plants may not be transpiring at all, and the water loss from vegetated land might be similar to that from barren land. Studies are needed in which soil water content is measured in bare soil under ambient and under elevated atmospheric CO_2 concentrations.

In arid and semi-arid regions where crops are irrigated, we need studies on different crops and different soils under different weather conditions to know when to irrigate in a CO_2 augmented world.

Elevated CO_2 and Stomata

Elevated CO_2 concentrations stimulate stomatal closure, while lowered CO_2 concentrations have the opposite effect (Kirkham, 2011, p. 155). The mechanism of the CO_2 response is unknown (Assmann, 1993). The regulation of stomatal aperture is under active investigation (e.g., see Chen *et al.*, 2010). Elucidating the mechanism of stomatal closure under elevated CO_2, along with understanding how water rises in tall trees (defying the law of gravity), are two major unsolved problems in plant physiology. We need further studies to understand why stomata close and open in response to elevated and lowered concentrations of CO_2. We also need to know if and how elevated CO_2 may affect the rise of water in plants. Is water in the xylem of plants more susceptible to cavitation in an atmosphere of elevated CO_2 compared to an atmosphere with lower CO_2? Does it take longer for xylem to become refilled in the spring in an atmosphere with elevated CO_2 compared to an atmosphere with lower CO_2?

The data show that stomatal density may decrease, increase, or stay the same under elevated atmospheric CO_2. Results apparently depend on several factors, including species, CO_2 concentration, and time of exposure to elevated CO_2 concentration (Kirkham, 2011, p. 195). Several authors have demonstrated, both experimentally (growing plants at different elevations) and from fossil and herbarium material, an inverse relationship between stomatal density and increasing ambient CO_2 levels. These investigations suggest that reductions in stomatal density are a physiological response by plants that increases the water use efficiency and may, therefore, influence their growth and survival under moisture-limited conditions (Kirkham, 2011, p. 195). We need more studies elucidating the interaction between water supply (wet and dry periods) and CO_2 affecting stomatal density during long time periods. The studies need to consider crop plants that have not been genetically altered, so one can determine how increasing CO_2 concentrations in the atmosphere are affecting stomatal density without human manipulation of the genetic material (Kirkham, 2011, p. 194).

In these studies, it is essential that herbarium specimens be studied. Researchers need herbariums to study the historical record to see how stomatal density has changed over time. Many herbariums are being lost as monies go to molecular biology rather than taxonomy (Kirkham, 2011, p. 180). But herbariums need to be maintained. Plants in herbariums are irreplaceable records of past history, and the information in them is lost forever when the specimens are thrown away.

Evapotranspiration

Because elevated carbon dioxide induces closure of stomata, evapotranspiration is usually reduced under elevated carbon dioxide. The extent to which evapotranspiration is reduced depends on leaf area. Leaf area is often increased by elevated CO_2, because growth is increased (see Eq. 1). No savings of water can be expected in canopies where elevated CO_2 stimulates an increase in leaf area relatively more than it closes stomata (Kirkham, 2011, p. 221). Even though some models predict that evapotranspiration will be increased (e.g., see Jung *et al.*, 2010, for data modeled between 1982 and 1997), experimental data show that evapotranspiration is usually decreased because of reduced transpiration due to stomatal closure. On average, experiments show that doubled CO_2 reduces stomatal conductance by about 40%, transpiration by about 20%, and evapotranspiration by about 10–14% (Kirkham, 2011, p. 221).

Because water lost from plants is decreased under elevated atmospheric CO_2, canopy temperature is increased. For example, when we grew our rangeland plants under ambient and doubled CO_2, elevated CO_2 increased average seasonal canopy temperature in 1989 by 1.7°C and 0.8°C for the field capacity and half-field capacity treatments, respectively. Models suggest that the decreased cooling from reduced transpiration will add to global warming (McDowell, 2010). Researchers in the

Department of Global Ecology at the Carnegie Institution for Science in Stanford, California, developed a model in which they doubled the concentration of atmospheric CO_2 and recorded the magnitude and geographic pattern of warming from different factors (McDowell, 2010; Cao *et al.*, 2010). They found that, averaged over the entire globe, the reduced evapotranspiration of plants under elevated CO_2 may account for 16% of the warming of the land surface.

However, how far away from a field does the warming of the air due to reduced transpirational cooling persist? We need to measure the temperature of the air in small increments away from fields with elevated CO_2 and in fields with ambient CO_2. Because even under calm conditions, the air is still moving, mixing of air is probably going to dissipate any warming of the air close to the field. But we need quantification. Allen (1973), in his pioneering studies with Lemon, his Ph.D. advisor at Cornell, found it impossible to elevate CO_2 in a field using pipes on soil that emitted CO_2. Turbulent diffusion processes mixed the air immediately. Warming of the air due to reduced transpirational cooling may be measurable only right above a field because of turbulence. How far away is temperature increased in the air above fields with elevated CO_2? Data need to be obtained for different crops, soils, and environmental conditions.

Water use Efficiency

Water use efficiency is defined as the biomass gained divided by the water used. The biomass may be either vegetative yield or grain yield, and the type of yield must be stated. Because under elevated CO_2 transpiration is reduced and biomass is increased, water use efficiency is increased. Water requirement, the reciprocal of water use efficiency, is decreased. In 1984, we determined the water requirement of winter wheat (*Triticum aestivum* L.) and sorghum [*Sorghum bicolor* (L.) Moench] under different CO_2 concentrations (Kirkham, 2011, p. 231–234). The ambient CO_2 concentration in 1984 was 330 ppm. Using data from these 1984 experiments, we can calculate what the water requirements of wheat and sorghum are 24 years later, based on the CO_2 concentration in the atmosphere in 2008 (385 ppm) (Kirkham, 2011, p. 234). Wheat and sorghum are now using 41 and 43 mL, respectively, less water to produce a gram of grain than they were in 1984. The water requirements to produce a gram of wheat grain and sorghum grain have been reduced by 6% and 4%, respectively, due to the 55 ppm increase in CO_2 since 1984. These may seem like small percentages. But spread over the large acreage of wheat and sorghum grown in Kansas, for example, the more efficient use of water now compared to the mid-1980s should have a large impact. Improving water use efficiency has long eluded agronomists. The increasing level of atmospheric CO_2 concentration appears to offer an opportunity to increase water use efficiency of plants where other methods have failed (Kirkham, 2011, p. 245). We need more information stating exactly how much water use efficiency is increased for each increment of CO_2 increase for different crops.

Final Comment

My book (Kirkham, 2011) is only a beginning in understanding how elevated CO_2 affects soil and plant water relations. My hope is that many more studies can be done with different plants and soils under elevated CO_2. We need a handbook giving the results of such studies. A handbook dealing with climate change and agroecosystems has been written by the leading researchers in the world (Hillel and Rosenzweig, 2011). However, we need handbooks that give tables, such as the *Handbook of Chemistry and Physics* (Haynes, 2011). We need tables of data showing exactly how every crop responds to elevated CO_2. For example, many studies show that wheat yield is maximal at 890 ppm CO_2 (Amthor, 2001; Kirkham, 2011, p. 310). Upper threshold levels of CO_2, above which growth is reduced, are reported to be 2200 ppm for tomato (*Lycopersicon esculentum* Mill.) and 1500 for cucumber (Nelson, 2012, p. 320). What are these values for other crops? We have handbooks of biology (e.g., Altman and Dittmer, 1966, 1971). My father, Don Kirkham, contributed data to the 1966 handbook to show how hydraulic conductivity changes with soil type and depth (Kirkham and Powers, 1966, p. 462–463). My co-major thesis advisor at the University of Wisconsin, Wilford R. Gardner, contributed data from my Ph.D. research that showed osmotic adjustment of barley (*Hordeum vulgare* L.) under varying concentrations of salinity (NaCl) (Altman and Dittmer, 1971, p. 728). How will soil hydraulic conductivity and osmotic adjustment change in a CO_2-enriched world? A handbook can perhaps be written telling exactly how every plant and soil changes in response to elevated CO_2. The information in the handbook must be based on experimental data.

Summary

We need more experiments to answer the following questions: How do the four soil physical factors that affect root growth (water, aeration, temperature, and mechanical impedance) interact with elevated CO_2 to affect crop growth? Because elevated CO_2 apparently increases the soil water content, what are irrigation recommendations for different crops under elevated CO_2? Why does elevated CO_2 close stomata? How does elevated CO_2 change stomatal density? How far away from a canopy is the increase in canopy temperature due to elevated CO_2 measurable? What is the increase in water use efficiency for different crops under elevated CO_2? What is the optimal concentration of CO_2 in the air for the growth of different crops?

Acknowledgments

I thank K.A. Williams, Professor, Department of Horticulture, Forestry, and Recreational Resources, Kansas State University, for pointing out the book by P.V. Nelson and the use of elevated CO_2 in greenhouse management. I thank Cynthia Rosenzweig and Daniel Hillel for inviting me to give a talk at their symposium, "Agriculture's Contributions to Climate Change Solutions: Mitigation and Adaptation at Global and Regional Scales," held in San Antonio, TX, October 18, 2011. This is Contribution No. 12-217-B from the Kansas Agric. Exp. Sta., Manhattan, Kansas.

References

Allen, L.H. Jr. 1973. Crop Micrometeorology: A. Wide-Row Light Penetration. B. Carbon Dioxide Enrichment and Diffusion. Ph.D. Thesis. Cornell University, Ithaca, New York. *Dissertation Abstracts International* 33(12). Order No. 73–14, 716.

Altman, P.L. and D.S. Dittmer (eds). 1966. *Environmental Biology*. Federation of American Societies for Experimental Biology, Bethesda, Maryland.

Altman, P.L. and D.S. Dittmer (eds). 1971. *Respiration and Circulation*. Federation of American Societies for Experimental Biology, Bethesda, Maryland.

Amthor, J.S. 2001. Effects of atmospheric CO_2 concentration on wheat yield: Review of results from experiments using various approaches to control CO_2 concentration. *Field Crops Res.* 73:1–34.

Asgarzadeh, H., M.R. Mosaddeghi, A.A. Mahboubi, A. Nosrati and A.R. Dexter. 2010. Soil water availability for plants as quantified by conventional available water, least limiting water range and integral water capacity. *Plant Soil*. 335:229–244.

Assman, S.M. 1993. Signal transduction in guard cells. *Annu. Rev. Cell Biol.* 9:345–375.

Cao, L., G. Bala, K. Caldeira, R. Nemani and G. Ban-Weiss. 2010. Importance of carbon dioxide physiological forcing to future climate change. *Proc. Nat. Acad. Science* 107:9513–9518.

Chen, Y.-H., L. Hu, M. Punta, R. Bruni, B. Hillerich, B. Kloss, B. Rost, J. Love, S.A. Siegelbaum and W.A. Hendrickson. 2010. Homologue structure of the SLAC1 anion channel for closing stomata in leaves. *Nature*. 467:1074–1080.

Cummings, M.B. and C.H. Jones. 1918. The aerial fertilization of plants with carbon dioxid [sic]. University of Vermont and State Agricultural College, Burlington, Vermont. *Vermont Agr. Expt. Sta. Bull.* 211. 56 p. + 8 plates on 4 pages.

Demoussy, E. 1904. Sur la vegetation dans les atmospheres riches en acide carbonique. *Comptes Rendus Acad. Sci. Paris*. 139:883–885 (cited by Hicklenton, 1988).

Dlugokencky, E.J. 2010. Carbon dioxide (CO_2) [in "State of the Climate in 2009"]. *Bull. Amer. Meteorol. Soc.* 91(7):541–542.

Dlugokencky, E.J. 2011. Carbon dioxide [in "State of the Climate in 2010"]. *Bull. Amer. Meteorol. Soc.* 92(6):559–560.

Gaastra, P. 1962. Photosynthesis of leaves and field crop [sic]. *Netherlands J. Agr. Sci.* 10(5, Special issue):311–324.

Haynes, W.M. (ed.). 2011. *CRC Handbook of Chemistry and Physics*. 92nd Edition. CRC Press, Taylor and Francis Group, Boca Raton, Florida.

Hicklenton, P.R. 1988. *CO_2 enrichment in the greenhouse: Principles and Practice*. Timber Press. Portland, Oregon.

Hillel, D. and C. Rosenzweig. 2011. *Handbook of Climate Change and Agroecosystems: Impacts, Adaptation and Mitigation*. Imperial College Press, London.

Jung, M., M. Reichstein, P. Ciais, S.I. Seneviratne, J. Sheffield, M.L. Goulden, G. Bonan, A. Cescatti, J. Chen, R. de Jeu, A.J. Dolman, W. Eugster, D. Gerten, D. Gianelle, N. Gobron, J. Heinke, J. Kimball, B.E. Law, L. Montagnani, Q. Mu, B. Mueller, K. Oleson, D. Papale, A.D. Richardson, O. Roupsard, S. Running, E. Tomelleri, N. Viovy, U. Weber, C. Williams, E. Wood, S. Zaehle and K. Zhang. 2010. Recent decline in the global land evapotranspiration trend due to limited moisture supply. *Nature.* 467:951–954.

Kirkham, D. and W.L. Powers. 1966. Factors influencing hydraulic conductivities of soil. Section 149, p. 462–463. In: P.L. Altman and D.S. Ditlmer (eds). *Environmental Biology.* Federation of American Societies for Experimental Biology, Bethesda, Maryland.

Kirkham, M.B., D. Nie, H. He and E.T. Kanemasu. 1993. Responses of plants to elevated levels of carbon dioxide, p. 130–161. *In Proceedings of the Symposium on Plant Growth and Environment, October 16, 1993, Suwon, Korea.* Korean Agricultural Chemical Society, Suwon, Korea.

Kirkham, M.B. 1994. Streamlines for diffusive flow in vertical and surface tillage: A model study. *Soil Sci. Soc. Amer. J.* 58:85–93.

Kirkham, M.B. 2005. *Principles of Soil and Plant Water Relations.* Elsevier, Amsterdam.

Kirkham, M.B. 2011. *Elevated Carbon Dioxide: Impacts on Soil and Plant Water Relations.* CRC Press, Taylor and Francis Group, Boca Raton, Florida.

Letey, J. 1985. Relationship between soil physical properties and crop production. *Adv. Soil Sci.* 1:277–294.

Long, D.S., J.M. Wraith and G. Kegel. 2002. A heavy-duty time domain reflectometry soil moisture probe for use in intensive field sampling. *Soil Sci. Soc. Am. J.* 66:396–401.

McDowell, T. (ed.). 2011. CO_2 Makes Plants Increase Global Warming. Carnegie Science. *The Newsletter of the Carnegie Institution.* Summer, 2010, p. 14. Carnegie Institution for Science, Washington, D.C.

Nelson, P.V. 2012. *Greenhouse Operation and Management.* Seventh Edition. Prentice Hall, Boston. 607 p.

Shaw, B.T. (ed.). 1952. *Soil Physical Conditions and Plant Growth. Agronomy. A Series of Monographs.* Volume II. Academic Press, New York.

Song, Y., J.M. Ham, M.B. Kirkham and G.J. Kluitenberg. 1998. Measuring soil water content under turfgrass using the dual-probe heat-pulse technique. *J. Amer. Soc. Hort. Sci.* 123:937–941.

Tackett, J.L. and R.W. Pearson. 1964. Effect of carbon dioxide on cotton seedling root penetration of compacted soil cores. *Soil Sci. Soc. Am. Proc.* 28:741–743.

Chapter 13

Contributions of WMO Programs in Addressing Climate Change and Agriculture

Mannava V.K. Sivakumar*, Oksana Tarasova, Slobodan Nickovic,
Deon Terblanche, and Ghassem Asrar

World Meteorological Organization (WMO)
7bis Avenue de la Paix, 1211 Geneva 2, Switzerland
*msivakumar@wmo.int

Introduction

The emerging challenges imposed by climate and climate change demand a holistic view of the earth as a complex system. The earth system is composed of interacting physical, chemical, and biological processes that move and exchange materials and energy on earth. The system provides the conditions necessary for life on the planet. For example, plants, which are part of the living system, use solar energy to convert carbon dioxide into organic carbon. Less carbon dioxide in the atmosphere helps cool the planet. Winds and ocean currents move heat from the tropics polewards, helping to warm the higher latitudes.

Earth's climate system is controlled by a complex set of interactions among the atmosphere, oceans, continents and living systems. We are increasingly discovering the sources and causes of its natural variability and changes, in our quest to explain which changes are due to nature and which to human activities. Advances in observation networks from space, at the Earth's surface, in the atmosphere and in the oceans have helped to document changes that had occurred in the past or are taking place presently. Computer-based Earth system models are providing projections of the range and relative magnitude of future changes under different socioeconomic development scenarios. A combination of the technological innovations in observations, information and telecommunication technologies, and high performance and cloud computing, have enabled us to begin to assess routinely the changes and variability of the Earth's climate and to understand the relative contributions of nature and humans. This scientific knowledge is also being translated routinely in the form of scientific assessments for policy decision makers and those who will use them in planning adaptation and/or managing the risks of climate variability and change.

Although climate is a global phenomenon, the effects of its variability and change are experienced most acutely at the regional, local and community levels around the world: According to the WMO Statement on the Status of the Global Climate in 2010, average global temperatures were estimated to be $0.53°C \pm 0.09°C$ above the 1961–1990 annual average of 14°C. This makes 2010 one of the warmest years in records dating back to 1880. The decade 2001–2010 was also the warmest on record. Recent warming has been especially strong in Africa. Temperatures for the 2001–2010 decade averaged 0.85°C above normal, 0.49°C warmer than any previous decade, and the five hottest years on record for Africa have all occurred since 2003. In some regions, there are changes in the frequency and intensity of extreme events such as high-intensity rainfall, tropical storms, high winds, extreme temperatures, and droughts. Agricultural productivity in tropical Asia is sensitive not only to temperature increases, but also to changes in the nature and characteristics of monsoons. In the semi-arid tropics of Africa, which are already having difficulty coping with environmental stress, climate change resulting in increased frequencies of drought poses the greatest risk to agriculture. In Latin America, agriculture and water resources are most affected through the impact of extreme temperatures and changes in rainfall. The need for regional and local climate information, while a major challenge, has been identified by decision makers and accepted by scientists as a high priority.

World Meteorological Organization (WMO)

The World Meteorological Organization (WMO) is a specialized agency of the United Nations. It is the UN system's authoritative voice on the state and behaviour of the Earth's atmosphere, its interaction with the oceans, the climate it produces and the resulting distribution of water resources.

WMO originated from the International Meteorological Organization (IMO), which was founded in 1873. Established in 1950, WMO became the specialized agency of the United Nations in 1951 for meteorology (weather and climate), operational hydrology, and related geophysical sciences. As weather, climate, and the water cycle transcend national boundaries, international cooperation at a global scale is essential for the development of meteorology and operational hydrology, as well as to reap the benefits from their application. WMO provides the framework for such international cooperation.

Under WMO leadership and within the framework of WMO programmes, National Meteorological and Hydrological Services (NMHSs) contribute substantially to the protection of life and property against natural disasters, to safeguarding the environment, and to enhancing the economic and social well-being of all sectors of society in areas such as food security, water resources, and transport.

WMO promotes cooperation in the establishment of networks for making meteorological, climatological, hydrological, and geophysical observations, as well as the exchange, processing, and standardization of related data. WMO assists technology transfer, training, and research. It also fosters collaboration between the NMHSs of its Members and furthers the application of meteorology to public weather services, agriculture, aviation, shipping, the environment, water issues, and the mitigation of the impacts of natural disasters.

WMO facilitates the free and unrestricted exchange of data and information, products and services in real- or near-real time on matters relating to safety and security of society, economic welfare, and the protection of the environment. It contributes to policy formulation in these areas at national and international levels.

Contiributions of WMO Programmes in Addressing Climate Change and Agriculture

The following programmes in WMO address the issues of climate change and agriculture:

a) Agricultural Meteorology Programme
b) Global Atmosphere Watch Programme
c) World Climate Research Programme
d) Global Observing Systems Programme

Agricultural meteorology programme

The importance of meteorology to agriculture, internationally, was probably recognized at least as early as 1735 when the Directors of European meteorological services first met to discuss meteorology on an international scale. The first reference to co-operation between meteorology and agriculture was in correspondence between the IMO and certain national institutes of agriculture and forestry, seeking an exchange of meteorological Information and data.

The purpose of the Agricultural Meteorology Programme (AGMP) of WMO is to support food and agricultural production and activities. The Programme assists Members in provision of meteorological and related services to the agricultural community to help develop sustainable and economically viable agricultural systems; improve production and quality; reduce losses and risks; decrease costs; increase efficiency in the use of water, labour and energy; conserve natural resources; and decrease pollution by agricultural chemicals or other agents that tend to degrade the environment.

A formal Commission for Agricultural Meteorology (CAgM) of IMO was appointed in 1913, but its meeting was delayed by the First World War. The Commission was re-constituted in 1919 and held its first meeting at Utrecht, The Netherlands, in 1923. Subsequently it held six additional meetings; the seventh and last being at Toronto in Canada, 1947. CAgM provides scientific and technical support for the implementation of the Agricultural Meteorology Programme.

The CAgM of WMO, which meets approximately once in four years, has so far held fifteen sessions since 1951. The technical work of the Commission is performed mainly by working groups appointed at the sessions and these groups undertake work between sessions. At its thirteenth session, held in Ljubljana, Slovenia, CAgM adopted a new structure of Open Programme Area Groups (OPAGs), Implementation and Coordination Teams (ICTs), and Expert Teams (ETs). Based on the activities of the various working groups, rapporteurs and expert teams over the year, a number of reports were published and distributed by WMO.

Agricultural meteorology programme activities in addressing climate change and agriculture

The close association between climate and agricultural issues was recognized by the CAgM-VII session held in Sofia, Bulgaria, in 1979 through the establishment of working groups to study the role of forests in the global balances of carbon dioxide, water, and energy; as well as meteorological aspects of agriculture in humid and sub-humid tropical areas. For the first time, a working group was established on the interaction between climate variability and agricultural activities, an effort that was followed up by in-depth studies over the next sessions. Other topics of relevance to climate issues that received new or renewed attention included remote sensing techniques in agrometeorology, and the role of CAgM in the World Climate Programme.

The gradual change in agricultural meteorology that continued in and through CAgM in the 1980's was expressed by emphasis on themes such as agrometeorological forecasts and warnings, formulation of data requirements for agricultural purposes, and introduction of effective methods for disseminating agrometeorological information, advice and warnings to address climate variability and climate change. In the 1990's, the Commission spent all its efforts in giving more thrust to transfer of knowledge and methodology in the developments in agricultural meteorology, in the dynamic realities of agricultural production, with an increasing emphasis on developing countries.

In the second WMO Long Term Plan (LTP) for the period 1988–1997, one of the six specific objectives of the AGMP was the formulation of relationships between climate/weather and agricultural production. The third LTP (1992–2001) put emphasis on the protection of the environmental resource base used for agricultural production. This additional necessity was of course induced by the realities of agricultural production, endangered by land degradation of various kinds, water waste and water pollution, air pollution, increasing climate variability, and the possibilities of climate change.

Following are the specific contributions of WMO's Agricultural Meteorology Programme in the areas of climate change and agriculture:

(a) *Assessment and management of current and future climate-related risks and impacts, including those related to extreme events affecting specific sectors*

In many parts of the world climate change and extreme climatic events (such as severe droughts, floods, storms, tropical cyclones, heat-waves, freezes and extreme winds) are among the biggest production risks and uncertainty factors impacting agricultural systems performance and management. Coping with agrometeorological risks and uncertainties requires assessing their probabilities and potential effects and then developing strategies to cope with these risks. CAgM has always been paying a lot of attention to climate related risks and impacts, including those related to extreme events in the agriculture sector. Several working groups were established by CAgM over the years and five WMO Technical Notes were published (WMO 1994; Salinger *et al.*, 1997; Salinger *et al.*, 2000; Das *et al.*, 2003; WMO 2004).

An International Workshop on Climate Prediction and Agriculture (CLIMAG), held in 1999 in WMO (Sivakumar, 2000) in collaboration with the Global Change System for Analysis, Research and Training (START), led to the establishment of a number of pilot projects demonstrating the application of seasonal to interannual climate forecasts. A second international workshop, held in 2005, discussed the advances and challenges in climate prediction and agriculture, and developed useful recommendations on science, capacity building, network development, and institutional partnership (Sivakumar and Hansen, 2007).

CAgM established an Expert Team on the Reduction of the Impact of Natural Disasters and Mitigation of Extreme Events in Agriculture, Forestry and Fisheries. A meeting of this Expert Team was held in Beijing, China, in February 2004. The proceedings of this Expert Team Meeting was published as a book entitled "Natural Disasters and Extreme Events in Agriculture" (Sivakumar *et al.*, 2005).

To address the issue of agrometeorological risk management, WMO, in collaboration with a number of co-sponsors, organized an International Workshop on Agrometeorological Risk Management: Challenges and Opportunities, in 2006 in New Delhi, India. Papers on approaches to dealing with risks highlighted preparedness planning, risk assessments, and improved early warning systems which can lessen the vulnerability of society to weather and climate risks. Enterprise diversification, contract hedging, crop insurance, weather derivatives and weather index insurance play key roles in developing agricultural risk management strategies. A special session examined the use of crop insurance strategies and schemes to reduce the vulnerability of the farming communities to risks posed by weather and climate extremes.

A number of strategies were identified to cope with risks. These include the use of seasonal forecasts in agriculture, forestry, and land management to assist alleviation of food shortages, drought, and desertification. The use of integrated agricultural management and crop simulation models with climate forecasting systems gives the highest benefits. Strategies to improve water management and increase the efficient use of water include crop diversification and better irrigation. Especially important was recognizing the value of local indigenous knowledge. A combination of locally adapted traditional farming technologies, seasonal weather forecasts, and warning methods is important for improving yields and incomes. Challenges to coping strategies were many and identified in several papers. Particularly important is the impact of different sources of climate variability and change on the frequency and magnitude of extreme events. Lack of systematic data collected from disasters impeded future preparedness, as did the need for effective communication services for the timely delivery of weather and climate information to enable effective decision making. Finally, a range of policy options to cope with such risks was presented. These included contingency planning, use of crop simulation modelling, and use of agrometeorological services (Sivakumar and Motha, 2007).

(b) *Identification of gaps, needs, opportunities, barriers and constraints to predicting climate variability, impacts, and extreme events across regions and hazards*

Climate variability affects all economic sectors, but the agricultural and forestry sectors are perhaps two of the most vulnerable and sensitive activities to such climate fluctuations. Climate change and variability, drought and other climate-related extremes have a direct influence on the quantity and quality of agricultural production and in many cases adversely affect it, especially in developing countries, where technology generation, innovation and adoption are too slow to counteract the adverse effects of varying environmental conditions. For example, inappropriate management of agroecosystems, compounded by severe climatic events such as recurrent droughts in West Africa, have tended to make the drylands increasingly vulnerable and prone to rapid degradation and hence desertification. Even in the high rainfall areas, increased probability of extreme events can cause increased nutrient losses due to excessive runoff and waterlogging. Projected climate change can influence pest and disease dynamics with subsequent crop losses. Improved adaptation of food production to climate changes, particularly in areas where climate variability is large, holds the key to improving food security for local communities and for the global population as a whole.

The range of adaptation options for managed systems such as agriculture and forestry is generally increasing because of technological advances, thus opening the way for reducing the vulnerability of these systems to climate change. However, some regions of the world, particularly developing countries, have limited access to these technologies and to appropriate information on how to implement them. Here, successful traditional technologies used over the centuries should be maintained. Incorporation of climate change concerns into resource-use and development decisions and plans for regularly scheduled investments in infrastructure will facilitate adaptation.

Agriculture and forestry are not optimally managed with respect to today's natural climate variability because of the nature of policies, practices and technologies currently in vogue. Decreasing the vulnerability of agriculture and forestry to natural climate variability through a more informed choice of policies, practices and technologies will, in many cases, reduce the long-term vulnerability of these systems to climate change. For example, the introduction of seasonal climate forecasts into management decisions can reduce the vulnerability of agriculture to floods and droughts caused by the ENSO phenomena.

WMO organized the International Workshop on Reducing Vulnerability of Agriculture and Forestry to Climate Variability and Climate Change in conjunction with the 13th Session of CAgM. The workshop reviewed the latest assessments of the science of climate variability and climate change, and their likely impacts on agriculture and forestry in different agroecological regions during the 21st century. It also surveyed and presented a range of adaptation options for agriculture and forestry and recommended appropriate adaptation strategies required to reduce vulnerability of agriculture and forestry to climate variability and climate change (Salinger *et al.*, 2005).

(c) Developing regional adaptation frameworks for climate change

Coping with climate change in the agriculture sector through appropriate adaptation strategies is an effective way to deal with the impacts of global warming. In order to help vulnerable regions develop such adaptation strategies, WMO brought together experts from NMHSs, agricultural departments, international and regional organizations and institutions; and policy makers from national planning/financial departments to present state-of-the art expositions, real-world applications, and innovative techniques for coping with climate change and to offer recommendations for planning and implementing an effective Regional Framework for Adaptation of Agriculture to Climate Change.

WMO, in collaboration with a number of UN Agencies and international and regional organizations and institutions, organized three international conferences in three vulnerable regions: (1) South Asia; (2) West Africa; and (3) West Asia and North Africa. These included the International Symposium on Adaptation to Climate Change and Food Security in South Asia in Dhaka, Bangladesh (25–29 August 2008); the International Workshop on Adaptation to Climate Change in West African Agriculture in Ouagadougou, Burkina Faso (27–30 April 2009); and the International Conference on Adaptation to Climate Change and Food Security in West Asia and North Africa in Kuwait City, Kuwait (13–16 November 2011).

These international conferences provided a central forum to develop an improved understanding and assessment of climate change impacts on agriculture and the associated vulnerability in South Asia (Rattan Lal *et al.*, 2011), West Africa, and West Asia and North Africa, and to discuss and develop informed decisions on practical adaptation strategies for the agricultural sector in different agro-ecosystems in these three regions. By the end of these Conferences, participants adopted Regional Frameworks for Adaptation of Agriculture to Climate Change. They also discussed and recommended policy and financial innovations to enable smooth implementation of the regional frameworks and their integration into sustainable development planning in different countries in the three regions.

This approach helped enhance the capacity to: identify/understand impacts, vulnerability and adaptation; select and implement adaptation actions; and enhance cooperation among the countries in South Asia, West Africa, and West Asia-North Africa to better manage climate risks; and enhance integration of climate change adaptation with sustainable agricultural development in these three regions.

(d) *Promoting understanding of impacts of, and vulnerability to, climate change.*

CAgM established an Open Area Programme Group 3 (OPAG 3) on Climate Change/Variability and Natural Disasters in Agriculture at it's thirteenth session held in Ljubljana, Slovenia, in October 2002. OPAG 3 had the responsibility to maintain an active and responsive overview of all the activities related to improving short-, medium- and long-term weather forecasting for agriculture; determining the impact of climate change/variability on climate forecasting; research on the impact of natural climate variability and the reduction of the impact of natural disasters on agriculture; and means to help reduce the contributions of agricultural production to global warming.

Three Expert Teams (ETs) under the OPAG 3 [i.e., the ET on the Impact of Climate Change/Variability on Medium- to Long-Range Predictions for Agriculture (ETCMLP); the ET on the Reduction of the Impact of Natural Disasters and Mitigation of Extreme Events in Agriculture, Rangelands, Forestry and Fisheries (ETRND); and the ET on the Contribution of Agriculture to the State of Climate (ETCAC)] held their meetings during 2003–2006 and produced useful reports (Salinger *et al.*, 2005; Sivakumar *et al.*, 2005; Desjardins *et al.*, 2007).

Experts in this programme area identified the following three potential pilot projects for implementation in different regions that would promote an understanding of the impacts of, and vulnerability to, climate change:

- Assessment of Natural Disaster Impacts on Agriculture (ANADIA)
- Climate Forecasts for User Communities in Agriculture
- Contribution of Agriculture to the State of Climate (CONASTAC)

In view of the high complexity of communicating climate-related information to non-professionals, as well as the multiplicity of end-users, OPAG 3 recommended the implementation of a number of actions that may significantly help to bridge the agricultural and climate-science communities. These include:

- Efforts to accommodate the needs of the various end-users wherever possible;
- Timely production of forecast products;
- Classification of end-users in order to provide them with the information they need;
- Regular provision of updated products to make them available to the broadest audience possible;
- Need for climate information producers to provide information on the characteristics of season types in terms of length, dates of onset and termination, frequency of dry spells, chances of flooding, storms, etc.;
- Regular dissemination of weather information, especially of hazardous weather, to help in confidence building;
- Education of end-users to distinguish between short- and longer-term predictions and how these are compiled.

(e) *Organization of roving seminars on weather and climate for farmers*

CAgM is organizing Roving Seminars on Weather, Climate and Farmers in different regions of the world to bring together farmers from several villages at a centralized location and apprise them of the important features of weather and climate in their cropping regions, especially those related to climate variability and climate change, and to educate them about the need for applications of weather and climate information in their operational activities. Feedback from the farmers at these seminars is helping refocus the efforts of agrometeorologists to provide improved and relevant information to the farming community.

From 2008–2011, the Government of Spain through its State Agency of Meteorology (AEMET) funded the organization of 149 Roving Seminars on Weather, Climate, and Farmers in fifteen countries in West Africa. Over 5,700 farmers participated in these seminars and over 3,100 simple raingauges were distributed. Based on the success of this project, a new project called METAGRI-Operational is being funded by Norway to strengthen activities such as developing crop advice based on historical climate data run through a crop model, standardizing raingauges, and formalizing dissemination methods to farmers. Also, the Rockefeller Foundation is funding a project in Ethiopia to train agricultural extension workers in weather and climate information and products.

Global Atmosphere Watch Programme

Agriculture is highly sensitive to climate variability and weather extremes, such as droughts, floods and severe storms, as well as to atmospheric composition factors, such as enhanced levels of carbon dioxide (CO_2) and surface ozone (Olesen *et al.*, 2011). Interactions between atmospheric composition and agriculture are non-linear and include numerous feedbacks. The atmospheric content of some species (including ozone, nitrogen oxides or sulphur oxides) impacts the health and productivity of crops plant, which in turn leads to change in biogenic emissions of those plants (like VOCs or methane).

Improved understanding of the complex feedback mechanisms between atmospheric composition and agriculture requires comprehensive measurement programs. The Global Atmosphere Watch (GAW) Programme was established within WMO in 1989 to coordinate international efforts on the observations and analysis of the chemical composition of the atmosphere and some of its physical properties. GAW provides data for scientific assessments and for early warnings of changes in the chemical composition and related physical characteristics of the atmosphere that may have adverse affects upon the environment. The highlighted grand challenges shaped the observational priorities within the programme, which now includes, six groups of variables (ozone, UV radiation, reactive gases, greenhouse gases, aerosol, and precipitation chemistry) and an urban air quality programme.

The GAW monitoring programme includes a coordinated global network of observing stations along with supporting facilities and expert groups. Currently GAW coordinates activities and data from 28 Global Stations (Fig. 1), more than 400 Regional and about 100 Contributing Stations (http://gaw.empa.ch/gawsis/). To ensure the quality of the observations and data availability, the programme coordinates the activity of a numbers of Central Facilities (Central Calibration

Fig. 1.   Global stations in the Global Atmosphere Watch Programme of WMO. Definition of the GAW Global stations can be found in the GAW Strategic plan: 2008–2015 (GAW, 2008) and its Addendum (GAW, 2011).

Laboratories, World and Regional Calibration Centers and World Data Centers), and organizes regular trainings and workshops.

The GAW Programme has an open data policy and the data in support of agricultural impacts assessment can be found in the GAW Data Centers. The links to the Data Centers and other Central Facilities used to ensure harmonization and quality of observations are available on the programme web page (www.wmo.int/gaw).

The GAW Programme requires observations to be performed under unpolluted conditions (away from the direct pollution sources) in order to characterise the background or undisturbed state of the atmosphere. In many regions of the world, these "unpolluted conditions", away from the, cities and towns, are often situated in agricultural areas. Hence, monitoring within the GAW Programme includes observations in agricultural areas and this helps in the understanding of the complex relations between agriculture and air composition.

The following most critical issues between atmospheric chemistry and agriculture will be discussed in more detail:

- emissions of greenhouse gases from agriculture
- impact of the land use change on emissions of the reactive gases and aerosols
- contribution of utilization of agricultural wastes (burning) and biomass burning to the air pollution levels
- ozone impact of the crops' productivity

Agriculture and greenhouse gases

Increasing atmospheric CO_2 levels, driven by emissions from human activities, can act as a fertilizer and enhance the growth of some crops such as wheat, rice and soybeans. CO_2 can be one of a number of limiting factors that, when increased, can enhance crop growth. Other limiting factors include water and nutrient availability. While it is expected that CO_2 fertilization will have a positive impact on some crops, other aspects of climate change (e.g., temperature and precipitation changes) may offset the beneficial CO_2 fertilization effect (IPCC, 2007).

Agricultural activities are a substantial source of the non-CO_2 greenhouse gases. In the global budget of methane (CH_4) and nitrous oxide (N_2O), agriculture is one of the main contributors (Fig. 2). USEPA (2006) estimated that agricultural activities were responsible for approximately 52% and 84% of global anthropogenic CH_4 or N_2O emissions, respectively, in the year 2000. Methane and nitrous oxide are the second (about 18%) and the third (about 6%) most important anthropogenic contributors to the increase of the radiative forcing, hence contributing directly to climate change.[1] On the other hand, emissions of these gases from agriculture are climate sensitive as well (i.e., with temperature increase CH_4 emissions from rice cultivations will also increase) and not-linearly depend on the atmospheric content

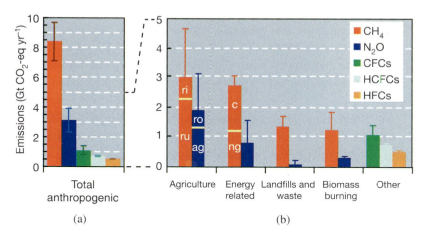

Fig. 2. Recent emissions of the non-CO_2 greenhouse gases (from Montzka *et al.*, 2011): (a) Emissions by compound; (b) Emissions by compound and sector (note scale change). For CH_4 ruminants (ru); rice production (ri); natural gas, oil and industry (ng); and coal mining (c) are the main contributors. Emissions of N_2O are from agriculture directly (ag) and from run-off (ro) to aquatic ecosystems of excess nitrogen attributable to agriculture. For details please, consult Fig. 2 in Montzka *et al.* (2011).

---

[1]These percentages are calculated as the relative contributions of the mentioned gases to the increase in global radiative forcing caused by all long-lived greenhouse gases since 1750 (http://www.esrl.noaa.gov/gmd/aggi).

Table 1.   Photosynthesis rate (μ mol CO_2 m^{-1}s^{-1}) in selected plants grown (Mean $\pm$ 1σ) in different areas.

Site	Mustard	Wheat	Pea	Mung
Reference area	13.75 ± 0.3	20.7 ± 0.31	11.51 ± 0.41	10.03 ± 0.28
Industrial and urban area	7.24 ± 0.35	13.9 ± 0.67	4.56 ± 0.64	5.26 ± 0.26
Periurban area	11.65 ± 0.34	15.2 ± 0.42	5.68 ± 0.38	8.11 ± 0.14
Urban area	10.21 ± 0.49	14.2 ± 0.50	4.96 ± 0.19	7.29 ± 0.32
Rural area	13.55 ± 0.26	18.0 ± 0.52	7.62 ± 0.09	8.34 ± 0.11

of CO_2. Van Groenigen *et al.* (2011) showed that CH_4 and N_2O emissions from plants and soil will increase with increasing levels of CO_2.

Fertilizer application to agricultural lands is one of the primary causes of nitrogen compounds emissions in the atmosphere. Nitrification is a process in agricultural soils resulting in N_2O and NO releases. In this process, the biological oxidation of ammonia into nitrite is followed by the oxidation of these nitrites into nitrates. Ammonia (NH_3) in the atmosphere results primarily from the anaerobic decomposition and volatilisation of organic matter (including animal wastes).

Plants and soil are not only a source but can also act as a sink of greenhouse gases from the atmosphere. CO_2 uptake by the biosphere depends on the state of the climate and levels of some atmospheric species (e.g., air pollution level). CO_2 uptake by photosynthetic activity critically depends on the level of pollutants (Beig *et al.*, 2007). Table 1 shows a clear decrease in CO_2 uptake from the atmosphere by some cultivated species being exposed to polluted environments.

Land-use change

Ecosystem changes associated with land-use and land-cover change are complex, involving a number of feedbacks (Lepers *et al.*, 2005). Changes of land-use, in particular due to agricultural practices, influence climate and atmospheric composition in different ways. For example, conversion of natural vegetation to agricultural land drives climate change by altering regional albedo and latent heat flux (which causes additional summer warming in key regions in the boreal and Amazon regions, and winter cooling in the Asian boreal zone), by releasing CO_2 via losses of biomass and soil carbon, and through a 'land-use amplifier effect'.

Land-use change alters biogeochemical cycles, resulting in direct emissions of CO_2 into the atmosphere due to deforestation. Cumulative emissions from historical landcover conversion for the period 1920–1992 have been estimated to be between 206 and 333 Pg CO_2 and as much as 572 Pg CO_2 for the entire industrial period 1850–2000. This is roughly one-third of total anthropogenic carbon emissions over this period (IPCC, 2007).

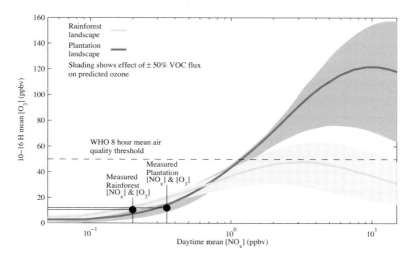

Fig. 3.   Sensitivity of daytime (10.00–16.00 hours) average ozone concentration to concentrations of NOx in the boundary layer, for the isoprene and monoterpene emission rates measured at the rainforest and oil palm landscapes, as computed by CiTTyCAT (from Hewitt *et al.*, 2009).

In addition to the direct effect on CO_2, changes in agricultural practices can lead to air quality problems. An interesting example of this feedback was reported by Hewitt *et al.* (2009). More than half the world's rainforest has been lost to agriculture since the industrial revolution. The rainforest has been replaced by the most widespread tropical crop, namely oil palm (*Elaeis guineensis*): global production of palm oil now exceeds 35 million tonnes per year. In Malaysia, for example, 13% of land area is now oil palm plantation, compared with 1% in 1974. The paper demonstrates (Fig. 3) that oil palm plantations in Malaysia directly emit more oxides of nitrogen and volatile organic compounds than rainforest. These emissions can leads to the local and regional production of ozone. Elevated ozone levels negatively affect crop productivity. The model calculations presented in Hewitt *et al.* (2009) predict that if concentrations of oxides of nitrogen in Borneo are allowed to rise (e.g. due to increase of traffic), ground-level O_3 mixing ratios will reach or exceed 100 ppb and thus constitute levels known to be harmful to human health.

Burning of agricultural wastes results in the emission of carbon monoxide (CO) and smoke particles to the atmosphere, which can transport them over long distances and cause pollution episodes even in remote areas. For example, in spring 2006, agricultural waste burning (Fig. 4) caused severe pollution in the Arctic (Stohl *et al.*, 2007). This pollution event corresponded to the highest ozone, carbon monoxide and aerosol optical depth measurements ever observed at the Zeppelin station in Ny Ålesund, Svalbard, located far north of the Arctic Circle. Clear effects were observed on snow chemistry and albedo as a result of this pollution event. As can be seen in Fig. 5, visibility has substantially decreased during the episode.

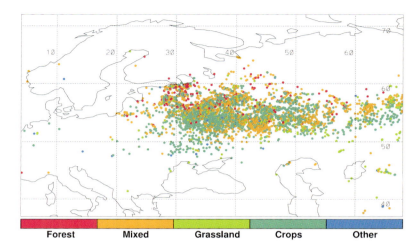

Fig. 4.   MODIS fire detections between 21 April and 5 May 2006. The color indicates the dominant land cover where the detection occurred (Stohl *et al.*, 2007).

Fig. 5.   View from the Zeppelin station during the smoke episode on 2 May 2006. (Image courtesy of Ann-Christine Engvall., from Stohl *et al.*, 2007).

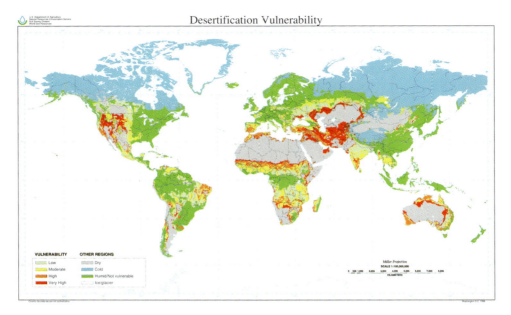

Fig. 6.   Global desertification map (Source: USGS; http://soils.usda.gov/use/worldsoils/mapindex/
desert.html).

Agriculture, desertification, and mineral dust

Desertification is a global process (Fig. 6) that takes place in dryland areas where the soil is fragile and suffers from lack of rainfall. As a result, top soil is destroyed due to wind and water erosion, followed by loss of the land's ability to sustain crops and livestock. The UN estimates that 70% of the 5.2 billion hectares of drylands used for agriculture around the world has already degraded (UNEP, 2000).

Desertified land, along with natural dust-producing soils, is the major emission source of atmospheric mineral dust. The Atmospheric transport of this dust has various effects on the local and global environment. For example, Neff *et al.* (2008) links the expansion of livestock grazing in the early twentieth century with a 500% increase in dust deposition in the western United States. Mahowald *et al.* (2010) have shown that the global atmospheric dust load has doubled in the 20th century due to anthropogenic activities.

In addition to adverse effects of desertification and consequent dust emission, long-range transport of dust-borne nutrients such as iron and phosphorus and their deposition might also have beneficial effects on the environment. Dust deposited in the ocean contributes to marine productivity, supporting fish stock sustainability through the marine food chain. Within the GESAMP[2] project, entitled *The*

---

[2]The Group of Experts on Scientific Aspects of Marine Environmental Protection.

Atmospheric Input of Chemicals to the Ocean, WMO has through its GAW and SDS-WAS[3] activities coordinated research on the atmospheric deposition of dust mineral nutrients to the ocean (GAW, 2011). It was found that an increase of the dust deposition led to an increase of ocean productivity by about 6% over the 20th Century, drawing down an additional 4 ppm (8 PgC) of carbon dioxide into the oceans (Mahowald *et al.*, 2010).

Dust, and especially phosphorus carried by it, when deposited downwind over land also plays a role as fertilizer. There is evidence that some tropical forests and savanna ecosystems are phosphorus-limited because of the age of their soil (e.g., Okin *et al.*, 2008). For example, the Amazon Basin is highly dependent on deposition of dust originating from Sahara for the maintenance of long-term productivity.

Agriculture and air pollution

Air pollution impacts on vegetation can be direct and indirect. Direct impact constitutes damage resulting directly from exposure to the pollutant (e.g. gaseous uptake of the pollutant by vegetation resulting in internal cellular damage or negative changes to biochemical or physiological processes). The negative impact of air pollution on plants can be evident in several ways. Damage to foliage may be visible in a short time and appear as necrotic lesions (dead tissue), or it can develop slowly as a yellowing or chlorosis of the leaf. The plant damage can lead to the decrease of productivity of some cultivated species. Among the atmospheric compounds associated with air pollution are ozone, nitrogen oxides and sulphur dioxide. O_3 has the highest impact on plants due to its high reactivity. O_3 penetrates leaves through the stomata, where it reacts with various compounds to yield reactive odd-oxygen species that oxidize plant tissue and result in altered gene expression, impaired photosynthesis, protein and chlorophyll degradation, and changes in metabolic activity (Booker *et al.*, 2009; Fuhrer, 2009). Ozone exposure for plants is estimated based on the AOT40 index, which indicates the number of hours of ozone-mixing in excess of 40 ppb level. Present day global relative yield losses due to ozone are estimated to range between 7% and 12% for wheat, between 6% and 16% for soybean, between 3% and 4% for rice, and between 3% and 5% for maize (range resulting from different metrics used) (Dingenen *et al.*, 2009). Under the 2030 "current legislation (CLE) scenario", the global situation is expected to deteriorate mainly for wheat (additional 2–6% loss globally) and rice (additional 1–2% loss globally). India, for which no mitigation measures have been assumed by 2030, accounts for 50% of these global decreases in crop yield. On a regional-scale, significant reductions in crop losses by CLE-2030 are predicted to occur in Europe (soybean) and China (wheat).

---

[3] Sand and Dust Storm Warning Advisory and Assessment System.

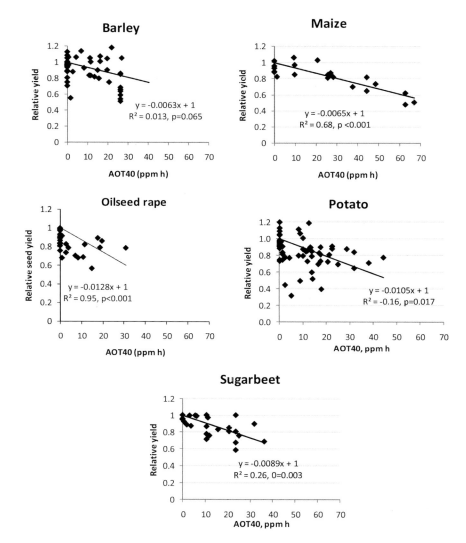

Fig. 7.   AOT40 — response functions for barley, maize, oilseed rape, potato and sugar beet (from Mills and Harmens, 2011).

The response of plants to ozone stress depends on the species (Fig. 7) and on meteorological conditions during the crop growing season, as well as on combinations with other pollutants. Hence, a comprehensive measurement programme is required at various agricultural sites to assess the complex effect of ozone on crops productivity.

SO_2 and NOx loads impact crop productivity as well. The impacts of these pollutants on agriculture is assessed via observations of atmospheric concentrations (and their comparison with *critical levels, i.e. levels in the* atmosphere above which direct adverse effects on receptors such as plants, ecosystems, or materials may occur) and deposition (critical loads). Both are regulated by WHO norms. SO_2 exposure leads to visible foliar injury, altered plant growth, and elimination of lichens and bryophytes. NOx exposure leads to altered plant growth and enhanced sensitivity to secondary stresses. SO_2 and NOx can also contribute to acidification of sensitive soils, which may be accompanied by a depletion of base cations, affecting the local vegetation over relatively long timescales. NOx and ammonia emissions can also cause long-term eutrophication of nutrient-poor terrestrial ecosystems, although the additional nitrogen deposition may also lead to short-term stimulation of growth.

World Climate Research Programme

The World Climate Research Programme (WCRP) was established in 1980, under the joint sponsorship of the International Council for Science (ICSU) and the World Meteorological Organization (WMO), and since 1993 has also been sponsored by the Intergovernmental Oceanographic Commission (IOC) of UNESCO.

The main objectives, set for the WCRP at its inception and still valid today, are to determine the predictability of climate and the effects of human activities on climate.

Since 1980, the WCRP has made enormous contributions to advancing climate science. As a result of WCRP efforts, it is now possible for climate scientists to monitor, simulate and project global climate with unprecedented accuracy so that climate information can be used for governance, in decision-making, and in support of a wide range of practical end-user applications (Asrar *et al.*, 2009).

Coordinated regional climate downscaling experiment (CORDEX)

In 2009, WCRP established the Coordinated Regional Climate Downscaling Experiment (CORDEX, http://wcrp.ipsl.jussieu.fr/RCD_Projects/CORDEX/) project to provide a more reliable scientific basis for regional assessments, similar to today's international global assessments by the IPCC.

CORDEX builds on existing regional networks of capabilities and expertise to:

- Better support decision making on impacts adaptation, and vulnerabilities associated with climate variability and change;
- Better communicate the scientific uncertainty inherent in regionalized climate projections;
- Develop a coordinated framework for evaluating and improving Regional Climate Downscaling (RCD) techniques; and
- Produce a new generation of fine-scale climate projections for identified regions worldwide.

A major challenge for CORDEX is to engage and coordinate ongoing observations, research, modeling and prediction carried out by networks of regional and local experts, especially in less-developed and developing regions of the world (Giorgi *et al.*, 2009). CORDEX will link closely with the 5th Coupled Model Intercomparison Project (CMIP5), which will deliver new global climate-change projections using improved global climate models and new greenhouse-gas emission scenarios.

CORDEX takes full advantage of the solid scientific foundation established during the past 30 years to bridge the existing gap between the climate modelling community and the end-users of climate information. The intent of CORDEX is to produce a framework valid for multiple domains across the world. Since this task requires considerable time and resources, Africa has been chosen as the first priority region. Africa is especially vulnerable to climate change, both because of the dependence of many vital sectors (e.g. agriculture, water management, health) on climate variability, and because of the relatively limited adaptive capacity of its economies.

Climate change may have significant impacts on temperature and precipitation patterns over Africa, which, in turn, can interact with other environmental stressors such as land-use change, desertification, and aerosol emissions. Africa stands to benefit greatly from the CORDEX framework, both with respect to further developing local research capabilities and introducing more detailed climate information into the African adaptation and policy arena where such information is needed for use in impact, adaptation, and vulnerability studies. Early results from CORDEX Africa initiative are becoming available (Nikulin *et al.*, 2012).

Major Scientific Challenges and Opportunities

The first challenge is to have stable, accurate, and comprehensive observations of the Earth's climate system to characterise and quantify the uncertainties concerning the evolution of past climate and to have a much better understanding of the present and future climate conditions. Unfortunately, such a comprehensive system does not exist today, although some critical elements are either in place or in development.

We must also assemble the best possible "climate data records" from all existing and available observing systems, despite their diversity and lack of optimum attributes. Therefore, WCRP together with other sister international observations and research programmes, has been promoting the adoption of internationally agreed-upon "climate monitoring principles" together with its sister programs that are focused on Earth observations (e.g. Global Climate Observing System (GCOS)). The main objective is to develop uniform, consistently evaluated and validated, and well documented climate datasets.

For example, a major effort was initiated by WCRP and GCOS in 2010 to facilitate a rigorous analysis, inter-comparison and documentation of the current and future climate data records to enhance their use especially in conjunction with Earth system models by researchers and other users. The goal is to have self-organized efforts for consistent and routine evaluation, inter-comparison, and documentation of multiple climate records of similar parameters that are developed and distributed by various organizations and nations. The intent is to minimize confusion and avoid misinterpretation of such observational records by users who are less familiar with these datasets. The ultimate objective is to promote adoption and use of standard procedures for evaluating and documenting long-term climate observational records similar to the processes now used routinely by the global climate modelling community for inter-comparison and evaluation of simulations resulting from their models.

The next challenge is long-term continuity, coverage, stability and quality of the future climate observations. The climate observing system(s) are expected to obtain essential climate variables identified by GCOS and GEOSS, and to:

- Detect subtle changes due to natural variability and to human contributions;
- Be adaptable to support changing and emerging research requirements;
- Observe trends and patterns in variables to assist in understanding the underlying processes and mechanisms;
- Use them as initial conditions for climate predictions — both experimental and operational, support activities aimed at adaptation to climate change and its mitigation; and
- Be useful in generating user-tailored climate information for managing risks and opportunities associated with climate/environmental changes of concern.

These requirements call for a reliable, well maintained and sustained basic climate observing system with a possibility to expand the scope of observations where and when it is required. Such an integrated and dedicated system does not exist today, but some major components of it have been in development during the past several decades (Manton *et al.*, 2010).

Most methods that are widely used for converting Earth observations into geophysical parameters/datasets are subject to continuous sophistication, calibration and other improvements. Thus, there is a need to apply the improved techniques to the original observations to obtain a more accurate, consistent and longer time series of climate variables. WCRP advocates for a coordinated reprocessing of long-term observational records with community-wide agreement on processing techniques and comprehensive validation, evaluation and assessment of results, and the description of the entire process to the benefit of users of these datasets. Together with the

GCOS, WCRP is promoting the principles of rigorous evaluation and documentation of climate datasets by providing, for example, an estimate of their attributes
and uncertainties (WCRP, 2011).

Earth climate system models are the only means available for projection and prediction of climate conditions. To make the climate projections and prediction useful
to decision-makers, we need to include major contributing processes to climate variability and change in them, but we must also translate the output from these models
into readily understandable and useful information for decision makers. The best
example we can offer is the improvements in numerical weather prediction models,
which led to extending useful predictions from 2–3 days to 10–12 days during the
past few decades. In addition, their systematic errors have been reduced significantly
through comprehensive use of observations. The weekly and biweekly weather predictions with associated likelihood are made available routinely to the public and a
vast array of users through national and international weather service centers and the
news media. This information is used routinely in agriculture, energy, transportation, tourism, water resources management, etc. The potential for use of climate
information for decision making also exists for these sectors and regions around the
world. The climate models, however, must be even more encompassing, accurate,
and diverse than weather prediction. Thus, a major challenge is the development,
evaluation, verification, and maintaining the diversity of most advanced climate
models to support such services. This is a major area of support and coordination by
the WCRP and other sister international research programs under the Global Framework for Climate Services (GFCS), which was established by the World Climate
Conference-3.

Specifically, promoting education and training of the next generation of
Earth and climate-system model developers to improve the existing models and
develop future ones is a major challenge facing the global climate-change research
community.

The tasks of climate research and prediction require a strategy for development
and evaluation of models. The WCRP and its partners have been instrumental in facilitating the development of such a strategy. The World Modelling Summit for Climate
Prediction (Shukla *et al.*, 2010) was convened for this purpose and recommended: (1)
a community-wide consultation on Model Evaluation and Improvement; (2) establishing a Climate Modelling Project; and (3) promoting investment in computational
capabilities to enable high resolution and more sophisticated climate model simulations to meet the needs for regional/local climate information by decision-makers.
There are several very successful and ongoing modelling efforts under WCRP aus-
pices. These efforts have been instrumental in learning how to quantify inherent
model uncertainties, continue to reduce such uncertainties, and develop quantitative criteria for estimating model skills in simulating various aspects of the climate

system. WCRP also coordinates a wide range of user-oriented model-based experiments to provide climate prediction data for multi-sectoral and multi-regional analysis. The results from all these efforts are published in the open literature for use by scientists around the world in support of science-based environmental assessments such as the IPCC Fifth Assessment. The latest efforts include several categories of experiments in response to the needs for adaptation planning and risk management, and for further improvement of the models and studies to determine the theoretical limits of climate predictability.

For example, more than twenty climate modelling groups from around the world are participating in the fifth phase of the Coupled Model Intercomparison Project (CMIP5) coordinated by the WCRP's Working Group on Coupled Modelling (WGCM). CMIP5 will identify sources of model difficulties in simulating the global carbon and water cycles, and, more generally, differences in model response to similar forcing, and advance research on climate predictability by exploring the ability of models to predict climate on decadal and longer time scales (CLIVAR Exchanges, 2011).

The WCRP coordinated climate prediction and projection experiments together with re-analyses of historical observations will produce a very large amount of data that if not managed properly will be hard to access, interpret, and use by the target scientists and decision makers. User access to such data is therefore resource-demanding. For example, a combination of re-analyses of historical observations together with CMIP5, CHFP, and CORDEX experimental archives start a new era in the availability of climate information for adaptation to climate change and variability, risk management, and other climate-related aspects of decision-making. The total volume of these datasets combined together is comparable to the observations obtained by a combination of all space-based and in situ observation networks established during the past three decades. To facilitate easy access to these datasets for research, to model inter-comparisons and analysis, and to minimize the need for multiple re-processing, WCRP is promoting common standards for numerically generated and observed datasets and the use of common projections and supporting documents that describe how these data were developed along with their error characteristics. Equally important is the availability of tools and techniques for analysis, interpretation, and display of these datasets to allow users to decide what portion of these large datasets needs to be ported to their computers and institutions. In the absence of such capabilities, if every user decides to transfer the entire archive(s) to their location, major delays and log-jams will frustrate both users and providers of these data. WCRP has been promoting and facilitating the establishment of distributed portals equipped with analysis tools for CMIP5, CORDEX and several other international projects sponsored by major funding agencies in Europe, Asia and the Americas to overcome such limitations and enhance greater use of these data

sets by scientists from around the world, especially those from developing regions/
nations.

The development of climate information services poses challenges and
opportunities for interpretation of various climate data, regardless of their source,
for use in products relevant to specific sectors (e.g. agriculture and food, water
resources, energy and transportation, etc.) and regions. It is very important for
the users to understand the levels of accuracy and uncertainty associated with
climate data and the resulting information for their applications. In this regard,
availability of climate data and information will be enhanced by the full implementation of the WMO Information System (WIS) and the ICSU new initiatives in observations and data management that will ensure the necessary level
of interoperability of the data and information systems world-wide, along with
corresponding metadata and cataloguing for ease of access and application by the
users.

The GFCS is a system that relies on unrestricted and easy exchange of data
and information to enable decision making. Various types of data, information, and
knowledge-sharing will be facilitated through these initiatives to ensure the best
available scientific knowledge on one hand, and consideration of stakeholders and
users' needs as a matter of high priority in establishing the observing networks and
research priorities, on the other hand. For this model to be effective and successful,
it must be implemented with due attention to all of its components (i.e. observations,
research and prediction, information system, users interface, etc.) in an integrated
and seamless manner.

Global Observing System Programme

The Global Observing System (GOS) is the coordinated system of methods and facilities for making meteorological and other environmental observations on a global
scale in support of all WMO Programmes. The system consists of operationally
reliable surface-based and space-based subsystems. The GOS comprises observing
facilities on land, at sea, in the air and in outer space. These facilities are owned
and operated by the Member countries of WMO, each of which undertakes to meet
certain responsibilities in the agreed global scheme, so that all countries can benefit
from the consolidated efforts.

WMO Integrated Global Observing System (WIGOS)

The WMO Integrated Global Observing System (WIGOS) is an all-encompassing
approach to the improvement and evolution of WMO global observing systems.
It will foster the orderly evolution of the present WMO global observing systems,
in particular the Global Observing System (GOS), the Global Atmosphere Watch

(GAW), and the World Hydrological Cycle Observing System (WHYCOS), into an integrated, comprehensive and coordinated system. It will satisfy, in a cost-effective and sustainable manner, the observing requirements of WMO Members, while enhancing coordination of the WMO observing system with systems operated by international partners. Together with the WMO Information System (WIS), WIGOS will provide the basis for accurate, reliable and timely weather, climate, water and related environmental observations and products by all Members and WMO Programmes, leading to improved service delivery.

New observing capabilities and impacts

The delivery of high-quality climate services requires a coordinated, comprehensive observing component that can be supplied only by WMO Members and the Organization's national and international partners. One goal of WIGOS is to meet this need by providing compatible, quality-assured, quality-controlled, and well-documented long-term observations.

Implementation of WIGOS will enable Members, in collaboration with national agencies, to meet countries' observational requirements for improving timely advisories and early warnings of extreme weather and climate events. It will also enable them to improve weather, climate, water, and related environmental monitoring and forecast services, and to adapt to and mitigate climate change, especially in developing and least-developed countries.

The experience gained from WIGOS Demonstration and Pilot Projects has paved the way for the system's implementation, which is to take place between 2012 and 2015. The process will focus on developing a framework for better governance and improved management, integration and optimization of the present observing systems. Thus, it will lay the groundwork for an operational system from 2016 onward.

WIGOS will build upon and add value to the existing surface and space-based subsystems, while providing a foundation for integration of new and emerging observational technologies. Meeting the quality requirements and expectations of users is critical to the success of WIGOS. This will require an in-depth examination of current practices used by existing WMO observing systems. The implementation strategy will outline all processes involved in the Quality Management System, including guidance on effective oversight. A top priority in the implementation of WIGOS will be to ensure a sustained and strong governance framework and improved management efficiency at all levels. WIGOS addresses an urgent need for strong collaborative mechanisms, and it will promote greater mutual commitment and expanded roles and responsibilities among all parties involved. An effective capacity-building strategy is an essential component of WIGOS implementation. Specialized training activities should be reflected in the regional, subregional and national WIGOS implementation plans, especially for NMHSs of least developed countries, landlocked developing countries, and small island developing States. Capacity-building is not limited to scientific and technological concerns, but includes human resources development, resource mobilization, and communications and outreach activities as well.

Conclusions

Climate science is interdisciplinary and a comprehensive study, aimed at understanding of land-ocean-atmosphere interactions. WMO's programmes, including the Agricultural Meteorology Programme, the Global Atmosphere Watch, the World Climate Research Programme and the Global Observing System Programme. Their activities contribute to a better understanding of climate change and agriculture. Global cooperation and a unified global perspective are crucial in addressing soil and crop management in a changing climate.

References

Asrar, G.R., J.A. Church and A.J. Busalacchi. 2009. The World Climate Research Programme: Climate Information for Decision Makers. Science and Technology: Public Service Review, Issue 3, pp. 115–120.
Beig G., S. Gunthe and D.B. Jadhav. 2007. Simultaneous measurements of ozone and its precursors on a diurnal scale at a semi urban site in India. *Journal of Atmospheric Chemistry* 57:239–253.
Booker, F.L., R. Muntifering, M. McGrath, K. Burkey, D. Decoteau, E. Fiscus, W. Manning, S. Krupa, A. Chappelka and D. Grantz. 2009. The ozone component of global change: potential effects on agricultural and horticultural plant yield, product quality and interactions with invasive species. *Journal of Integrative Plant Biology*. 51:337–351.
CLIVAR Exchanges. 2011. WCRP Coupled Model Intercomparison Project-Phase 5 (CMIP5). Vol. 16, No. 2, May 2011.
Das, H.P., T.I. Adamenko, K.A. Anaman, R. Gommes and G. Johnson. 2003. *Agrometeorology Related to Extreme Events. Technical Note No. 201 (WMO No. 943)*. Geneva: World Meteorological Organization.
Desjardins, R.L., M.V.K. Sivakumar and C. de Kimpe (eds.). 2007. The Contribution of Agriculture to the State of Climate. *Special Issue of the Agricultural and Forest Meteorology*. 142(2–4).
Dingenen R.V, F.J. Dentener, F. Raes, M.C. Krol, L. Emberson and J. Cofala. 2009. The global impact of ozone on agricultural crop yields under current and future air quality legislation. *Atmospheric Environment* 43:604–618.
Fuhrer, J. 2009. Ozone risk for crops and pastures in present and future climates. *Naturwissenschaften* 96:173–194.
GAW. 2008. *WMO/GAW Strategic Plan: 2008–2015 — A Contribution to the Implementation of the WMO Strategic Plan: 2008–2011, GAW report 172 (WMO TD No. 1384)*. Geneva: World Meteorological Organization.
GAW. 2011a. *Addendum for the Period 2012–2015 to the WMO Global Atmosphere Watch (GAW) Strategic Plan 2008–2015, GAW report 197*. Geneva: World Meteorological Organization.

GAW. 2011b. *Workshop on Modelling and Observing the Impacts of Dust Transport/Deposition on Marine Productivity, Sliema, Malta, 7–9 March 2011), GAW report 202.* (Prepared by Robert A. Duce and Prof. Peter Liss), Geneva: World Meteorological Organization.

Giorgi, F., C. Jones and G.R. Asrar. 2009. Addressing climate information needs at the regional level: the CORDEX framework. *World Meteorological Organization Bulletin.* 58(3):175–183.

Hewitt C.N., A.R. MacKenzie, P. Di Carlo, C.F. Di Marco, J.R. Dorsey, M. Evans, D. Fowler, M.W. Gallagher, J.R. Hopkins, C.E. Jones, B. Langford, J.D. Lee, A.C. Lewis, S.F. Lim, J. McQuaid, P. Misztal, S.J. Moller, P.S. Monks, E. Nemitz, D.E. Oram, S.M. Owen, G.J. Phillips, T.A.M. Pugh, J.A. Pyle, C.E. Reeves, J. Ryder, J. Siong, U. Skiba and D.J. Stewart. 2009. Nitrogen management is essential to prevent tropical oil palm plantations from causing ground-level ozone pollution, *PNAS*, 106:18447–18451.

IPCC. 2007. Climate Change 2007: Impacts, Adaptation, and Vulnerability. Contribution of Working Group II to the Third Assessment Report of the Intergovernmental Panel on Climate Change [Parry, Martin L, Canziani, Osvaldo F, Palutikof, Jean P, van der Linden, Paul J and Hanson, Clair E. (eds.)]. Cambridge University Press, Cambridge, United Kingdom.

Lepers, E., E.F. Lambin, A.C. Janetos, R. DeFries, F. Achard, N. Ramankutty and R.J. Scholes. 2005. A synthesis of information on rapid land-cover change for the period 1981-2000. *Bio. Science* 55:115–124.

Mahowald, N.M., S. Kloster, S. Engelstaedter, J. Keith Moore, S. Mukhopadhyay, Joseph R. McConnell, S. Albani, Scott C. Doney, A. Bhattacharya, M.A.J. Curran, M.G. Flanner, F.M. Hoffman, D.M. Lawrence, Keith Lindsay, P.A. Mayewski, Jason C. Neff, D. Rothenberg, E. Thomas, Peter E. Thornton and Charles S. Zender. 2010. Observed 20th century desert dust variability: impact on climate and biogeochemistry. *Atmos. Chem. Phys.* 10:10875–10893.

Manton, M.J., A. Belward, D.E. Harrison, A. Kuhn, P. Lefale, S. Rösner, A. Simmons, W. Westermeyer, J. Zillman. 2010. Observation Needs for Climate Services and Research. World Climate Conference-3, Elsevier, *Procedia Environmental Sciences* 1:184–191.

Mills, G. and H. Harmens (eds). 2011. *Ozone pollution: A hidden threat to food security (Report prepared by ICP Vegetation), September 2011.* Available at http://icpvegetation.ceh.ac.uk

Neff. J.C., A.P. Ballantyne, G.L. Farmer, N.M. Mahowald, J.L. Conroy, C.C. Landry, J.T. Overpeck, T.H. Painter, C.R. Lawrence and R.L. Reynolds. 2008. Increasing eolian dust deposition in the Western United States linked to human activity. *Nature Geoscience* 1:189–195.

Nikulin, G., C. Jones, F. Giorgi, G. Asrar, M. Büchner, R. Cerezo-Mota, O.B. Christensen, M. Déqué, J. Fernandez, A. Hänsler, E. Meijgaard, P. Samuelsson, M.B. Sylla and L. Sushama. 2012. Precipitation climatology in an ensemble of CORDEX-Africa regional climate simulations. *Jour. of Climate*, (In print).

Okin, G.S., N. Mahowald, O.A. Chadwick and P. Artaxo. 2004. Impact of desert dust on the biogeochemistry of phosphorus in terrestrial ecosystems. *Global Biogeochemical Cycles* 18, GB2005, doi:10.1029/2003GB002145.

Olesen, J.E., M. Trnka, K.C. Kersebaum, A.O. Skjelvåg, B. Seguin, P. Peltonen-Sainio, F. Rossi, J. Kozyra and F. Micale. 2011. Impacts and adaptation of European crop production systems to climate change. *European Journal of Agronomy.* 34:96–112.

Rattan, Lal., Mannava V.K. Sivakumar, S.M.A. Faiz, A.H.M. Mustafizur Rahman and Khandakar R. Islam (Eds.). 2011. *Climate Change and Food Security in South Asia.* Berlin: Springer.

Salinger, J., M.V.K. Sivakumar and R.P. Motha (eds.). 2005. *Increasing Climate Variability and Change: Reducing the Vulnerability of Agriculture and Forestry.* Berlin: Springer.

Salinger, M.J., R.L. Desjardins, H. Janzen, P.H. Karing, S. Veerasamy and G. Zipoli. 2000. *Climate Variability, Agriculture and Forestry: Towards Sustainability.* Technical Note No. 200 (*WMO No. 928*). Geneva: World Meteorological Organization.

Salinger, M.J., R.L. Desjardins, M.B. Jones, M.V.K. Sivakumar, N.D. Strommen, S. Veerasamy and W. Lianhai. 1997. *Climate Variability, Agriculture and Forestry: An Update. Technical Note No. 199 (WMO No. 841).* Geneva: World Meteorological Organization.

Shukla, J., R. hagedorn, B. Hoskins, J. Kinter, J.Marotzke, M. Miller, T, Palmer and J. Slingo. 2010. Revolution in climate prediction is both necessary and possible: A decleration at the World Modelling Summit for Climate prediction. *Bulletin American meteorol. Soc.* 90:175–178.

Sivakumar, M.V.K. (ed.). 2000. *Climate Prediction and Agriculture. Proceedings of the START/WMO International Workshop held in Geneva, Switzerland, 27–29 September 1999.* Washington DC, USA: International START Secretariat.

Sivakumar M.V.K and J.W. Hansen (eds.). 2007. *Climate Prediction and Agriculture: Advances and Challenges.* Berlin: Springer.

Sivakumar M.V.K and R.P. Motha (eds.). 2007. *Managing Weather and Climate Risks in Agriculture.* Berlin: Springer.

Sivakumar, M.V.K., R.P. Motha and H.P. Das (eds.). 2005. *Natural Disasters and Extreme Events in Agriculture: Impacts and Mitigation.* Berlin: Springer.

Stohl A., T. Berg, J.F. Burkhart, A.M. Fjæraa, C. Forster, A. Herber, Ø. Hov, C. Lunder, W.W. McMillan, S. Oltmans, M. Shiobara, D. Simpson, S. Solberg, K. Stebel, J. Ström, K. Tørseth, R. Treffeisen, K. Virkkunen and K.E. Yttri. 2007. Arctic smoke — record high air pollution levels in the European Arctic due to agricultural fires in Eastern Europe in spring 2006. *Atmospheric Chemistry and Physics* 7:511–534.

UNEP. 2000. *Global Environment Outlook 2000: UNEP's Millennium Report on the Environment.* United Kingdom: Earthscan Publishers.

USEPA. 2006. *Global anthropogenic emissions of non-CO_2 greenhouse gases 1990–2020.* United States Environmental Protection Agency Report 430-R-06-003, Washington DC. <http://www.epa.gov/nonco2/econ-inv/international.html>, accessed 1 June 2007.

Van Groenigen K.J., C.W. Osenberg and B.A. Hungate. 2011. Increased soil emissions of potent greenhouse gases under increased atmospheric CO_2. *Nature* 475:214–216.

WMO. 2004. *Management Strategies in Agriculture and Forestry for Mitigation of Greenhouse Gas Emissions and Adaptation to Climate Variability and Climate Change. Technical Note No. 202 (WMO No. 969).* Geneva: World Meteorological Organization.

WMO. 1994. *Climate Variability, Agriculture and Forestry.* Report of the CAgM-IX Working Group on the study of Climate Effects on Agriculture including Forests, and of the Effects of Agriculture and Forest on Climate. Technical Note No. 196 (WMO No. 802). Geneva: World Meteorological Organization.

World Climate Research Programme. 2011. *Evaluation of Satellite-Related Global Climate Datasets. Frascati, Italy, 18–20 April 2011,* WCRP Report No. 33, GCOS Publication No. 153.

Chapter 14

The Agricultural Model Intercomparison and Improvement Project (AgMIP): Integrated Regional Assessment Projects

Cynthia Rosenzweig[*,†], James W. Jones[‡], Jerry L. Hatfield[§], Carolyn Z. Mutter[†], Samuel G.K. Adiku[¶], Ashfaq Ahmad[⁻], Yacob Beletse[**], Babooji Gangwar[††], Dileepkumar Guntuku[‡‡], Job Kihara[§§], Patricia Masikati[¶¶], Ponnusamy Paramasivan[‖], K.P.C. Rao[***], and Lareef Zubair[†††]

[*]*NASA Goddard Institute for Space Studies, New York, NY, USA*
[†]*Columbia University Center for Climate Systems Research, New York, NY, USA*
[‡]*University of Florida, Gainesville, FL, USA*
[§]*National Laboratory for Agriculture and the Environment, USDA-ARS Ames, IA, USA*
[¶]*University of Ghana, Accra, Ghana*
[⁻]*University of Agriculture, Faisalabad, Pakistan*
[**]*South Africa Agricultural Research Council, Roodeplaat, South Africa*
[††]*Indian Council of Agricultural Research, Modipuram, Uttar Pradesh, India*
[‡‡]*International Crop Research Institute for the Semi Arid Tropics, Patancheru, India*
[§§]*International Center for Tropical Agriculture, Nairobi, Kenya*
[¶¶]*International Crop Research Institute for the Semi Arid Tropics, Bulawayo, Zimbabwe*
[‖]*Tamil Nadu Agricultural University, Coimbatore, Tamil Nadu, India*
[***]*International Crop Research Institute for the Semi Arid Tropics, Nairobi, Kenya*
[†††]*Foundation for Environment, Climate and Technology, Central Province, Sri Lanka*
[*]*cynthia.rosenzweig@nasa.gov*

Introduction

The worldwide agricultural sector faces the significant challenge of increasing production to provide food security for a population projected to rise to nine billion by mid-century, while protecting the environment and the sustainable functioning of ecosystems. This challenge is compounded by the need to adapt to climate change by taking advantage of potential benefits, and by minimizing the potentially negative impacts, to agricultural production. The goals of Agricultural Model Intercomparison and Improvement Project (AgMIP) are to improve substantially the characterization of world food security under climate change and to enhance adaptation capacity in both developing and developed countries (www.agmip.org).

To accomplish these goals, AgMIP organizes protocol-based intercomparisons of crop and economic models and ensemble projections to produce enhanced climate change assessments by the crop and economic modeling communities. These assessments focus on research related to climate change agricultural impacts and adaptation (Rosenzweig et al., 2012). This chapter describes the AgMIP regional integrated assessment projects in Sub-Saharan Africa and South Asia.

AgMIP Structure and Scientific Approach

The Agricultural Model Intercomparison and Improvement Project (AgMIP) conducts agricultural model intercomparison and future climate change assessments with participation from multiple crop and economic modeling groups around the world (Fig. 1). AgMIP research activities are organized under four project teams (Climate, Crop Modeling, Economics, and Information Technology), with guidance provided by a Leadership Team as well as a Steering Group and Donor Forum. In addition, there are three cross-cutting themes — Representative Agricultural Pathways, Aggregation and Scaling, and Uncertainty — which span the activities of all teams.

AgMIP Regional Activities

AgMIP Regions are geographical areas in which collaborative efforts are created to implement the protocols and provide outputs for use in the region and for use in global studies. AgMIP regional activities are underway in Sub-Saharan Africa, South Asia, North America, South America, and Europe and in development in Australia and East Asia (see Fig. 1). AgMIP is holding recurring regional workshops to bring together climate scientists, agronomists, and economists from leading regional and international institutions to build capacity and conduct simulations and analyses at field-to-regional scales according to the AgMIP protocols. Participation from scientists in important agricultural regions is crucial to AgMIP goals, as local expertise is vital to establishing grounded simulations for regional agriculture.

*in development

Fig. 1. Location of AgMIP regional activities.

AgMIP Sub-Saharan Africa Regional Projects

In Sub-Saharan Africa, four multi-disciplinary and international teams are undertaking integrated analyses of food production systems with a special focus on how climate contributes to food insecurity in the region, with a fifth team providing coordination across projects (Fig. 2).

Climate change impacts on West African Agriculture (CIWARA) — A regional assessment

Teams of crop, economic, and climate scientists in Western Africa are assessing climate change impacts on agriculture in semi-arid and sub-humid Western Africa (Fig. 3). The project is standardizing an open-access database, comparing

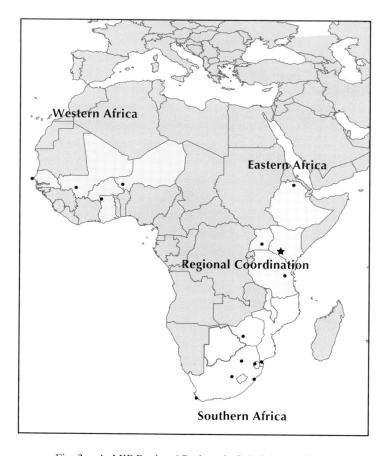

Fig. 2.   AgMIP Regional Projects in Sub-Saharan Africa.

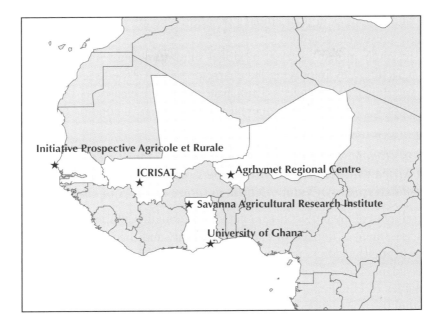

Fig. 3.   AgMIP Western Africa Project region and institutions.

and improving crop model performance, improving spatial coverage of agro-meteorological advisories for smallholder farmers, updating selected policy instruments at the national and regional levels, and building research capacity for integrated climate change impacts assessments.

The project is led and coordinated by Professor Samuel G.K. Adiku of University of Ghana, with assistance from Drs. Ibrahima Hathie of IPAR; Jesse B. Naab of SARI; P. S. Traore of ICRISAT Mali; and Seydou Traore of Agrhymet Regional Centre. Key research staff include Drs. M. Adam of ICRISAT Mali; M. Diancoumba of SARI; D. S. MacCarthy of University of Ghana; and B. Sarr of Agrhymet Regional Centre.

This assessment will increase representation of climate, crop, and economic data in West Africa, enrich agro-biodiversity in crop models, apply representative agricultural pathways to study future regional climate change impacts, and will refine crop calendars and agro-meteorological advisories into location and ecotype specific decision support tools. West African policy-makers will be briefed on possible local impacts of climate variability and change, to inform decisions in adaptation and mitigation.

Capacity building will include graduate student internships, exchange visits, and technical training, as well as outreach activities for the media and policy-makers; data sharing and joint publication. The aim is to create a network of AgMIP alumni from local universities in their country of origin who are continuously involved in the project's research activities.

The research project assesses regional climate change impacts on Western African agriculture by improving the characterization of food security risks due to climate variability and change and enhances the adaptive capacity of West African populations for changing environmental and technological conditions.

Impacts of climate variability and change on agricultural systems in Eastern Africa while enhancing the region's capacity to undertake integrated assessment of vulnerabilities to future changes in climate

The overall goal of the project is to conduct a systematic, comprehensive, and quantitative assessment of impacts of climate variability and change on agricultural systems that explicitly address the farm, local, national, and regional level impacts and identifies adaptation options with due consideration to the interactions between the key variables of climate, crop, and socio-economics in Eastern Africa. Operating from within the framework of the AgMIP global project, it works toward establishing country and regional teams with climate, crop and economic modelers and IT experts, enhancing their skills in the use and application of new science tools and conducting comprehensive assessment of impacts, vulnerabilities and adaptation options, applying cutting-edge science and interdisciplinary knowledge.

Teams of climate, crop, and economic modelers in Ethiopia, Kenya, Tanzania and Uganda are assessing the impact of climate change on important crops utilizing the AgMIP Protocols (Fig. 4). The project utilizes climate, crop and economic data to calibrate and validate models across a wide range of conditions for Eastern Africa. The integrated analysis of model results will enable systematic assessment of variable and changing climate on agricultural systems. Knowledge of likely climate impacts enables decision-makers to consider options for managing negative impacts and for capitalizing on favorable conditions.

The project is coordinated by Dr. K.P.C Rao, Principal Scientist of ICRISAT–Nairobi, with assistance from Dr. M. Tenywa of Makerere University, Uganda. Key research staff include Drs. B. Wafula, of Kenya Agriculture Research Institute; E. Mpeta of Sokoine University of Agriculture, Tanzania; and, A.A. Berhe of Mekelle University, Ethiopia.

In addition to in-country research, regular exchange of information will occur between country teams through training activities aimed to enhance skills needed to undertake the work. This includes hands-on sessions in data assessment and management, use of models and other tools, and write-up of results. It also includes training on methods for regional integrated assessments utilizing information from climate, crop, and economic models.

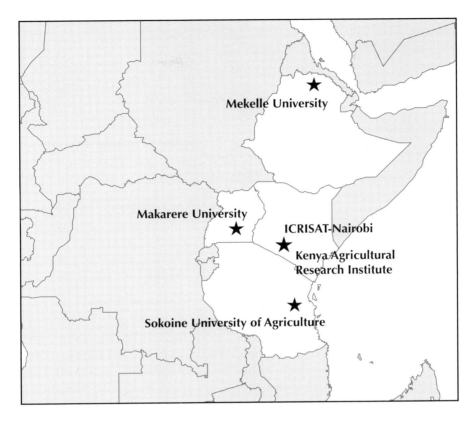

Fig. 4.   AgMIP Eastern Africa Project region and institutions.

Southern Africa Agricultural Model Intercomparison and Improvement Project (SAAMIIP)

A team of climate, crop, economic, and IT research scientists in Southern Africa is evaluating the impact of climate change on the production and prices of important crops (Fig. 5). A simultaneous goal is to build human and institutional capacity to explore and evaluate these impacts and associated field management adaptation strategies on food prices and production.

The project team is testing the accuracy of models for staple crops, using these models to estimate regional-scale food production for the period of 2070–2099 for the IPCC's A2 climate and development scenario, identifying field-level adaptation strategies for maintaining or increasing yields, and evaluating economic impacts of climate change on different farming systems. This knowledge will build capacity across the disciplines of climate, crop, and economic modeling in the region.

The project is coordinated by Dr. Y. Beletse of South Africa Agricultural Research Council–Roodeplaat, with assistance from Drs. O. Crespo of University of Cape

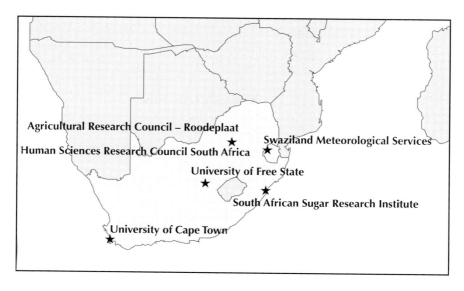

Fig. 5.   AgMIP Southern Africa Project region and institutions.

Town, and S. Walker of University of Free State. Key researchers include Drs. W. Durand of South Africa Agricultural Research Council–Potchefstroom; A. Singels of South African Sugar Research Institute; C. Nhemachena of Human Sciences Research Council South Africa; and M.S. Gamedze of Swaziland Meteorological Services.

The impacts evaluation and capacity-building activities are anticipated to result in validated crop models for maize, sorghum, sugarcane, wheat, and sweet potatoes — the staple and nutritionally important crops in Southern Africa. Inter-comparison of model outputs is likely to lead to improvement of the models. Estimated productivity levels from crop models will be used as inputs to economic models to enable simulation of economic outcomes for different farming systems given a range of climate change scenarios. The integrated analysis of outputs from linked climate, crop and economic modeling enables the assessment of a range of possible future socio-economic pathways. It also builds critical capacity among the team members in methodologies for conducting integrated assessments that will be shared locally through targeted workshops, meetings, training, and stakeholder outreach.

Crop-livestock intensification in the face of climate change: Exploring opportunities to reduce risk and increase resilience in Southern Africa using an integrated multi-modeling approach

Teams of crop, economic, and climate scientists in Southern Africa are exploring opportunities to reduce risk and increase resilience in Southern Africa using

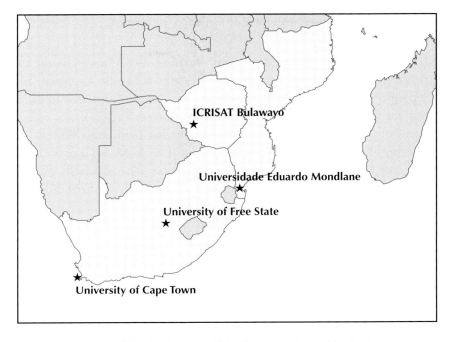

Fig. 6. AgMIP Southeastern Africa Project region and institutions.

an integrated multi-modeling approach (Fig. 6). The project characterizes selected mixed farming systems in Southern Africa in terms of biophysical and socio-economic characteristics, develops and evaluates crop-livestock management and climate change adaptation strategies that increase food production, agro-diversity and economic returns, and explores the interactions and synergies of increased diversity and integration and their contribution to reduce risk and increase system resilience.

The project is led and coordinated by Dr. P. Masikati of ICRISAT Bulawayo, with assistance from Drs. O. Crespo of University of Cape Town, and S. Walker of University of Free State. Key research staff include Drs. S. Homann Kee Tui of ICRISAT Bulawayo; L. Claessens of ICRISAT Nairobi; S. Famba of Universidade Eduardo Mondlane; A. van Rooyen of ICRISAT Bulawayo; and C. Lennard of University of Cape Town.

The project will increase understanding of challenges and opportunities in the current mixed farming systems of Southern Africa for better targeting of interventions to increase systems resilience and reduce climate-induced risk. It will also improve understanding of the interactions and synergies of production system components, such as which combinations bring about profitable production systems and how to use these to facilitate development along sustainable pathways.

Knowledge enhancement for modeling in climate change: Capacity-building in Southern, Western, and Eastern Africa

The goal of the project is to build capacity for trans-disciplinary climate change and agricultural research throughout the Sub-Saharan Africa (SSA) region. The aim is to prepare and publish integrated assessments of climate change impacts and adaptation for Western, Eastern, and Southern Africa (see Fig. 2).

The project is led by Dr. J. Kihara of CIAT. Key collaborating scientists include Drs. J. Huising and S. Koala of CIAT; S. Zingore of IPNI; D.M. Sefakor of University of Ghana; and J. Mangisoni of University of Malawi. The team, in coordination with the AgMIP Leaders (Drs. C. Rosenzweig, J. Jones, J. Hatfield, K. Boote, P. Thorburn, C. Porter, S. Janssen, A. Ruane, J. Antle, R. Valdivia, G. Nelson) will build curricula, develop individual and group learning modules, and teach in climate, crop, economic, information technology, and integrated assessments.

The Regional Coordination Team will liaise with designated members of the AgMIP Sub-Saharan Regional Research Teams, and with AgMIP Leadership, to enable collaboration. Research Teams will be advised on how to organize and prepare development of an inventory of data needed for regional assessments (e.g., weather, site experiment, soil, socio-economic parameters); preparation of presentation materials; guidelines for identifying IT goals of the project (e.g., data management plans, project website, etc.); establishing concise summaries of climate, crop and economic model analyses in planning or underway; and, identification of stakeholders whose engagement is likely to be mutually beneficial.

Workshop reports will summarize presentations and learning activities and will also provide guidance on next steps. This is likely to include consideration of training on specific topics to be advanced in partnership with national and international groups (e.g., NARS, CCAFS and others) that have expressed interest in co-sponsorship of special training sessions of value to multiple climate change and agriculture initiatives in the region.

AgMIP South Asia Regional Projects

In South Asia, four multi-disciplinary and international teams are undertaking integrated analyses of food production systems with a special focus on how climate contributes to food insecurity in the region, with a fifth team providing coordination across projects (Fig. 7).

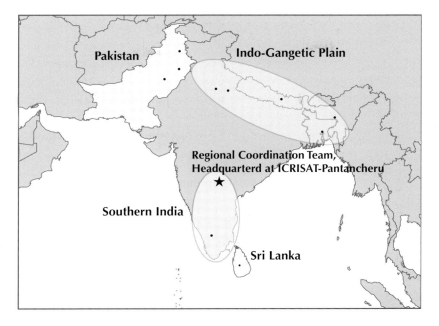

Fig. 7.   AgMIP Regional Projects in South Asia.

Assessing climatic vulnerability and projecting crop productivity using integrated crop and economic modeling techniques

Wheat, rice and cotton are major crops in Pakistan not only in terms of local consumption but also in view of large exports. These crops are grown on approximately 8.81, 2.37 and 2.69 million hectares of land respectively, with a total production of 24.2 million tons of wheat, 4.8 million tons of rice, and 11.5 million bales of cotton. These crops are grown in different agro-ecological zones of Pakistan. Each zone represents diverse soil, social, hydrological and climatic conditions.

The overall goal of the project is the analysis of historic/current climate, as well as crop and economic data to determine the trends of climate change in the region and its likely impact on crop productivity and the economy. This includes calibration and validation of crop models for wheat, rice and cotton, regional economic models, as well as quantification of the spatial and temporal yield variability and yield forecasting under future climate change scenarios.

An expert team of climate, crop, and economic scientists are analyzing the possible impacts of variable and changing climate on wheat, rice and cotton production under the agro-ecological conditions of Pakistan. The AgMIP regional study

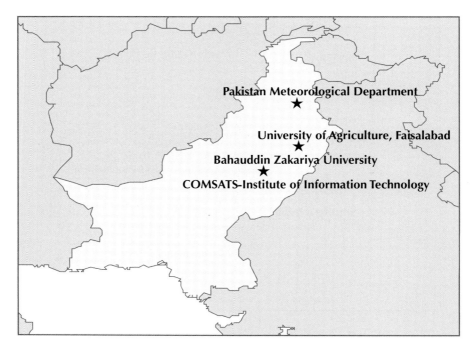

Fig. 8.   AgMIP Pakistan Project region and institutions.

is anticipated to improve confidence in the performance of climate, crop and economic models by comparing them for the same input conditions utilizing the AgMIP
Protocols.

The project is led by Dr. A. Ahmad, Professor at Agro-Climatology lab, University of Agriculture, Faisalabad (UAF). Additional investigators include Drs. S.A.
Wajid, T. Khaliq, M. Ashfaq, and A.R. Sattar, all at UAF; S. Ahmad of Bahaud-
din Zakariya University; G. Rasul of Pakistan Meteorological Department; and W.
Nasim of COMSATS — Institute of Information Technology (Fig. 8).

The project will enable improved confidence in predictions of likely climate
change impacts and will also allow for rigorous analysis and scrutiny of concepts
and assumptions underlying each model. Improved understanding of model behavior
is likely to be helpful when communicating with the farmers and decision-makers
who must plan for the outcomes of changing climate.

In addition to research, regular exchange of information will occur through training activities aimed to enhance skills needed to undertake the work. This includes
hands-on sessions in data assessment and management, use of models and other
tools, and write-up of results. It also includes training on methods for regional integrated assessments utilizing information from climate, crop, and economic models.

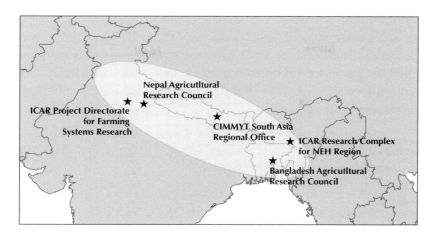

Fig. 9.   AgMIP Indo-Gangetic Plain Project region and institutions.

Strengthening simulation approaches for understanding, projecting, and managing climate risks in stress-prone environments across the central and Eastern Indo-Gangetic Basin

This project focuses on cereal crop production regions in the Central and Eastern Indo-Gangetic Basin (approximated by the oval in Fig. 9). Agricultural simulation and climate modelers are teaming up with socio-economic scientists in Northern India, Nepal and Bangladesh, in partnership with AgMIP. Together, they will establish a multi-model framework of calibrated agricultural impact models to assess climate impacts. Climate, benchmark soil, and crop cultivar data are assembled for a set of crop models in each major production domain, and AgMIP Protocols are followed to simulate contemporary and future climate risks. The intent is to make a lasting contribution to methodologies for adapting agricultural systems to current and projected climate risks in South Asia.

The project is led by Dr. B. Gangwar, Project Director for the Indian Council of Agricultural Research (ICAR) Farming Systems Research Project Directorate (FSRPD). Additional investigators include Drs. N. Subash of ICAR/FSRPD; A. Das of ICAR Research Complex for the NEH Region; S.K. Ghulam Hussain from Bangladesh Agricultural Research Council; R. Darai from Nepal Agricultural Research Council, and A. McDonald of CIMMYT — South Asia Regional Office in Kathmandu Nepal.

This research strengthens simulation approaches for understanding, projecting, and managing climate risks in stress-prone environments across the Central and Eastern Indo-Gangetic Basin. The project team is also evaluating promising adaptation strategies, suggesting areas for model improvement, developing a qualitative

summary of the non-climate drivers of change, and, analyzing different policy scenarios by integrating an economic model with the crop simulation models. The
project includes the creation of a synchronized database on climate, soil and crop
information for the region for other researchers and experts to utilize in their own
integrated analyses. The goal is to strengthen and broaden collaborative research in
agricultural systems for this region.

Integrated assessment of climate change impacts on principal crops and farm household incomes in Southern India

Farmers in Tamil Nadu, Andhra Pradesh, and other Provinces of the Deccan Plateau
are located in the rain-shadow of the Western Ghats, with meager annual rainfall
averaging 500–900 mm (Fig. 10). The region is experiencing steadily decreasing
soil fertility, falling water tables, growing dependence on groundwater for irrigation,
and increasing fallow lands — characteristics that may be impacted by changes in
long-term climate trends. Agriculture sustains over half of the region's population.

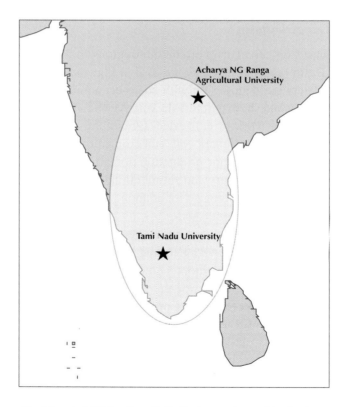

Fig. 10.   AgMIP Southern India Project region and institutions.

Climate, crop, and economic researchers at Tamil Nadu Agricultural Univerity (TNAU) and Ancharya NG Ranga Agricultural University (ANGRAU) and partners are assessing the impact of climate change on agricultural production in Southern India and its implications for farm household income and food security. The team will apply AgMIP Protocols to integrate climate, crop and economic models and assess impacts. The goal of the project is to reduce uncertainty in climate change impacts and adaptation options.

The project is led by Dr. P. Paramasivan, Professor at Agricultural Economics, TNAU; Co-leaders are V. Geethalakshmi, Professor and Head of the Agro Climate Research Centre, TNAU and Dr. M. Gopinath, Professor of Agricultural and Resource Economics at Oregon State University. TNAU team members include A. Lakshmanan, Nano Science and Technology; K. Mahendran, Agricultural and Rural Management; and, R. Krishnan, Remote Sensing and GIS. ANGRAU team members include Dr. Raji Reddy, Head, Agro Climate Research Centre and Dr. D. Murthy, Soil and Crop Modeling, Agro Climate Research Centre.

This research project provides an integrated assessment of climate change impacts on principal crops and farm household incomes in Southern India. Following a characterization of farm household production systems in the region, the project will downscale climate change scenarios, simulate productivity and production impacts of crops under different systems and scenarios using hydrological and crop models, and, integrate climate model outputs with impact models. Operating within the framework of the AgMIP global project, this Southern India project is conducting a comprehensive assessment of impacts, vulnerabilities and adaptation options for the region.

Modeling the impacts of a variable and changing climate on rice and sugarcane agricultural systems in Sri Lanka

A team of climate, crop, economic and IT experts are modeling the impacts of a variable and changing climate on rice and sugarcane agricultural systems in Sri Lanka (Fig. 11). The AgMIP Regional Project will assess the impact of climate change on rice and sugarcane production utilizing the AgMIP Protocols. The project aims to reduce uncertainty in climate change impacts and adaptation options.

The AgMIP Sri Lanka Project is led by Dr. L. Zubair, Principal Scientist at Foundation for Environment, Climate and Technology (FECT), Sri Lanka, with co-leadership by Dr. S.P. Nissanka of University of Peradeniya. Additional investigators include Drs. W.M.W. Weerakoon and B.V.R. Punyawardhene from the Department of Agriculture, Sri Lanka; Dr. A.P. Keerthipala, Ms. B.D. Sandya Ariyawansa, Mr. K. Sanmuganathan, and Mr. A.L.C. DeSilva from the Sugarcane Research Institute; Dr. N. Fernando, Mr. S. Ratnayake, Mr. M. Weerasekera, and Ms. Z. Yahiya

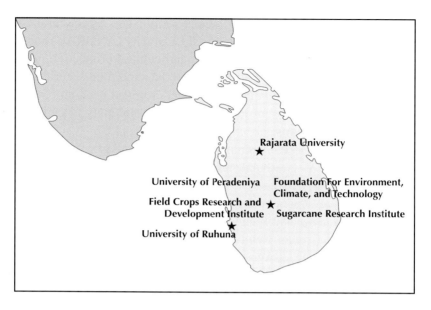

Fig. 11.   AgMIP Sri Lanka Project region and institutions.

of FECT; Professor K.D.N. Weerasinghe from University of Ruhuna; Professor P. Wickramagamage and Dr. W. Athukorale from University of Peradeniya; Dr. W.C.P. Egodawatta and Mr. J. Gunarathna from Rajarata University; and, Ms. R. Bandara of Sabragamuwa University.

Anticipated beneficial outcomes of the project include: (i) advancement of state-of-the-art modeling of the impacts of a variable and changing climate on agriculture and food security at multiple scales to inform policy-making and resource management; (ii) harnessing of high quality data resources and expertise in Sri Lanka to contribute to global efforts to characterize the impacts of a variable and changing climate on agriculture; (iii) development of expertise, infrastructure, data and IT resources for climate, crop and economic modeling in the partner universities, departments and research institutes and; (iv) establishment of a multi-disciplinary network of collaborators in the fields of climate, crops and economics who will foster transdisciplinary research, with special attention given to the fostering of the next generation of scientists with training programs and access to project resources and outputs.

Enhancing capacities of the AgMIP South Asia regional teams through capacity-building workshops and knowledge-sharing platforms

The goal of this project is to strengthen the building of capacity for the AgMIP trans-disciplinary research teams throughout the South Asia region through a partnership between AgMIP and the International Crops Research Institute for the Semi-Arid

Tropics (ICRISAT). The aim is to prepare and publish integrated assessments of climate change impacts and adaptation for Pakistan, the Indo-Gangetic Basin, Southern India, and Sri Lanka (see Fig. 7).

The project is led by Dr. D. Guntuku of IRCISAT Global Leader for Knowledge Sharing and Innovation, with contributions from Drs. C. Bantilan and R. Mula, Mr. S.V. Prasad Rao, and a Research Fellow to be determined (ICRISAT). In addition, the AgMIP Leaders (Drs. C. Rosenzweig, J. Jones, J. Hatfield, K. Boote, P. Thorburn, C. Porter, S. Janssen, A. Ruane, J. Antle, R. Valdivia, G. Nelson) and Drs. P. Singh, S. Nedumaran (ICRISAT) and D. Murthy (ANGRAU) are among those identified as expert contributors who enable curriculum building, development of individual and group learning modules, and teaching in climate, crop, economic, information technology, and integrated assessments.

The Regional Coordination Team will liaise with designated representatives from each AgMIP Regional Research Team, and with AgMIP Leadership, enable collaborative workshop development. Research Teams will be advised on how to organize and prepare for participation in each workshop. This includes development of an inventory of data needed for regional assessments (e.g., weather, site experiment, soil, socio-economic parameters); preparation of materials for presentation at the workshop; guidelines for identifying IT goals of the project (e.g., data management plans, project website, etc.); establishment of concise summaries of climate, crop and economic models analyses in planning or underway; and, identification of stakeholders whose engagement is likely to be mutually beneficial.

Workshop reports will summarize presentations and learning activities and will also provide guidance on next steps. In addition, the team will provide guidance on requested training on specific topics, to be advanced in partnership with national and international entities that have expressed interest in the co-sponsorship of specialized topic training that is needed for other initiatives in the region.

Conclusions

AgMIP has already developed strong international collaborations and research activities are underway in many regions. A major goal of AgMIP is to create capacity-building partnerships around the world, enhancing the ability of researchers in each agricultural region, as well as globally, to evaluate current and future climate impacts and adaptations, and thus to contribute to future food security.

Acknowledgments

The authors gratefully acknowledge the contributions of AgMIP Team Leaders for Climate (Alexander C. Ruane, *NASA Goddard Institute for Space Studies,*

New York, NY, USA), Crop Modeling (Kenneth J. Boote, *University of Florida, Gainesville, FL, USA* and Peter J. Thorburn, *Commonwealth Scientific and Industrial Research Organisation, Brisbane, Australia*), Economics (John M. Antle, *Oregon State University, Corvallis, OR, USA* and Gerald C. Nelson, *International Food Policy Research Institute, Washington, DC, USA*) and Information Technology (Cheryl Porter, *University of Florida, Gainesville, FL, USA* and Sander Janssen, *Alterra, Wageningen University and Research Centre, Wageningen, Netherlands*). The authors express sincere appreciation to Peter Craufurd for contributions in the role of AgMIP Principal Investigator at the *International Crop Research Institute for the Semi Arid Tropics*. Rex Navarro, R. Narsing Rao, M.S. Raju and others at the *International Crop Research Institute for the Semi Arid Tropics*; Tim Johnston, Kelema Jackson, Andrew Thibodeau, and others at *Columbia University*; and, Eileen Herrera and others at the *US Department of Agriculture* are acknowledged for their contributions to project operations and finance. Soyee Chiu and Shari Lifson at *Columbia University* provided technical, editorial, and graphical support. The *Columbia University* coordinated implementation of the AgMIP program for Sub-Saharan Africa and South Asia is possible owing to support from the British *Department for International Development's UK aid*, executed in partnership with the *US Department of Agriculture Agricultural Research Service* through Agreement Number 59-3625-1-745.

Reference

Rosenzweig, C., J.W. Jones, J.L. Hatfield, A.C. Ruane, K.J. Boote, P. Thorburn, J.M. Antle, G.C. Nelson, C. Porter, S. Janssen, S. Asseng, B. Basso, F. Ewert, D. Wallach, G. Baigorria and J.M. Winter. 2012. The Agricultural Model Intercomparison and Improvement Project (AgMIP): Protocols and pilot studies. *Agricultural and Forest Meteorology.*

Conclusion
Agricultural Solutions for Climate Change at Global and Regional Scales

Cynthia Rosenzweig* and Daniel Hillel[†]

*The Earth Institute at Columbia University, New York,
and The NASA Goddard Institute for Space Studies, New York*

cynthia.rosenzweig@nasa.gov
[†] *dh244@columbia.edu*

The chapters in this volume represent a worldwide survey of potential agricultural responses to climate change. Agriculture is here defined in its broadest sense as encompassing all the systems of plant and animal production that humans use to obtain food, feed, natural fiber, and biofuel. Several major conclusions arise from this volume's *tour d'horizon* of agricultural responses to climate change, both in regards to mitigation and adaptation.

First, many authors emphasize that the *linkages between climate and agricultural production are complex* due to the interactions of temporal and spatial scales related to within-season weather, decadal variability, and long-term climate change processes. Climate variations at different time scales (seasonal, decadal, and longer-term) and spatial scales may either spur or retard agricultural production in different regions. Hence regions with large observed components of decadal variability need to factor the uncertainty due to this climate variability into their adaptation plans. Sustained programs of observations, projections, and research are essential to help fulfill these needs.

The second major conclusion is that the *agricultural regions of the world are already responding to current climate change and planning for the future* in regard to both mitigation and adaptation. The regional chapters make clear that climate change is already affecting and will continue to affect different agricultural regions in various ways in the coming decades. The regional chapters indicate that agriculture in developing countries is highly vulnerable to climate change, and that future impacts on developed regions are also important due to their large role in international trade. While mitigation and adaptation strategies can be generalized for agriculture at a global scale, specific measures need to be developed at regional scales, considering the great diversity in agricultural systems and settings. The various chapters in this book demonstrate the myriad of constructive approaches currently available or under development to both mitigate and adapt to climate change at the regional scale.

The third major conclusion is that enhanced *investment in research is essential to further climate change mitigation and adaptation.* In recognition of the importance of climate variability and change to agriculture, a working group of the American Society of Agronomy, Crop Science Society of America, and Soil Science Society of America developed a position statement on climate change (Climate Change Position Statement Working Group, 2011). This document addresses many of the themes explored in this volume and presents a challenge to the research community to fill the knowledge gaps that require immediate attention.

Agriculture — Climate Connections

The connections between climate and agricultural production are multifaceted because of the interactions of temporal and spatial scales of weather and climate processes (Baethgen and Goddard, this volume). In order to better understand these complexities, the appropriate use of climate information takes into account the limitations as well as strengths of both observations and climate models. Important to the provision of information is the quantification of uncertainty, which is difficult given imperfect models and short observational records. The uncertainty measured by differences in projections across climate models or scenarios of greenhouse gas (GHG) increases does not necessarily encompass the full range of uncertainty in future anthropogenic climate changes. Additionally, natural variations in climate will always be part of the experience of climate changes by the agricultural sector.

Observational analyses can provide a context within which global and regional climate models can be interpreted. They can indicate the relative magnitude of past climate variability and can be compared to climate model simulations to determine whether observed climate trends are following model-based expectations of the future. Observations also can help to diagnose whether a recent 20–30 year "trend" has been dominated by low-frequency variability that is possibly part of natural fluctuations rather than caused by human interference.

Observational analyses are often limited by short length of record, coarse spatial resolution, and missing data. Despite limitations of both observed and projected climate information, it is possible to make informed use of the opportunities provided by climate information from both model data and observations to create better plans for the future management of the world's major agricultural regions.

To further that goal, the World Meteorological Organization (WMO) has developed a range of activities, including the Agricultural Meteorology Programme, the Global Atmospheric Watch, the World Climate Research Programme, and the Global Observing System Programme, that aim to contribute to a better understanding of climate change and agriculture (Sivakumar *et al.*, this volume).

In 2009, the WMO launched the Global Framework for Climate Services (GFCS) to strengthen production, availability, delivery, and application of science-based climate prediction and services. As part of this effort, WMO is coordinating a comprehensive observational program called the WMO Integrated Global Observing System (WIGOS). WIGOS builds on and adds value to existing surface and space-based subsystems, while providing a foundation for integration of new and emerging observational technologies.

The goal of WIGOS is to provide compatible, quality-assured, quality-controlled and well-documented long-term observations, as well as improved timely advisories and early warnings on extreme weather and climate events. It will also enable national weather services to improve weather, climate, water resources, and related environmental monitoring and forecast services that can contribute to adaptation and mitigation in the agricultural sector, especially in developing countries.

Regional Responses

Each regional chapter raises important issues regarding the impacts of climate change on local agricultural systems and offers both mitigation and adaptation solutions tailored to the specific circumstances.

North America

In North America, there are very few years in which there is not some effect of a weather anomaly on production amounts of any commodity (Hatfield, this volume). Within the records on crop production, it is relatively easy to identify the factors causing major losses in production; however, it is more difficult to ascertain factors that contribute to the smaller losses (between 5–15%) of the potential yield.

Hatfield presents a framework that links weather, climate, and production inputs through cropping systems to commodity outputs related to both quantity and quality of agricultural production. This framework can be used in North America and elsewhere to help to plan research on climate change mitigation and adaptation programs.

Latin America and the Caribbean

The majority of the existing agricultural production systems in Latin America and the Caribbean, where droughts, floods, storms, and untimely frosts have had — and will have — profound impacts, remain vulnerable to climate change (Baethgen and Goddard, this volume). Year-to-year climate variability, which may be worsened in the presence of trends, will lead to the greatest socio-economic impacts over the near-term decades. Hence, an important focus is the building of resilience to the impacts associated with the range of scales of climate variability and change, taking into account interannual, decadal, and longer-term time-scales.

Europe

The 27 countries of the European Union account for approximately a fifth of both global cereal and meat production (Porter *et al.*, this volume). The variability of crop yields has already increased in Europe as a consequence of extreme climatic events, such as the summer heat of 2003, which led to uninsured economic losses for the agriculture sector in the European Union estimated at 36 billion Euros. While the EU is likely to be able to offset internal reductions in yield under climate change by increasing imports, this could further increase pressure on already food-insecure low-income and low-yielding regions of the world.

Through its land resources, planning and agricultural sciences, Europe is taking a leadership role in adapting to and mitigating climate change. Recognizing that research will be a key to meeting these global challenges in the coming decades, 20 European countries have developed a Joint Programming Initiative on Agriculture, Food Security and Climate Change (FACCE JPI: www.faccejpi.com). The joint programming of EU research is a new process aimed at combining a strategic framework, a bottom-up approach and high-level commitment from member states to work together on important issues such as food security and climate.

Africa

Agriculture in Africa must undergo a significant transformation in order to meet the related challenges of food security and climate change (Naab *et al.*, this volume). Effective climate-smart practices already exist that could be implemented to lift millions of people out of poverty and hunger by increasing the productivity and profitability of small-scale farmers in Africa. There exists a high potential for Africa to contribute to mitigating GHG emissions through soil organic carbon sequestration in cropping and grazing lands, restoration of degraded soils, and land use and cover changes. Although the potential of carbon sequestration in degraded soils and ecosystems of Africa is high, the realization of this potential is challenging. Yet the need to restore degraded soils and ecosystems is urgent and should be given high priority.

Considerable investment is required to fill data and knowledge gaps and to conduct research and development of technologies and methodologies. Strengthened institutional capacity will be required to improve the dissemination of climate information. Adaptation will involve improved risk management through insurance schemes and Information Communication Technologies (ICTs), such as cell phones, which provide farmers with rapid access to information regarding weather forecasts or facilitate their insurance payments. Institutional reforms and financial support will be required to enable small-scale farmers in Africa make the transition to climate-smart agriculture.

Australia and New Zealand

Australia and New Zealand represent a microcosm of the global challenges facing agriculture under climate change (Thorburn *et al.*, this volume). Agriculture is important economically in the two countries and contributes a greater share of emissions to the national GHG inventory than in most developed countries. Intensive livestock production dominates in New Zealand, whereas grains and extensive livestock production dominate in Australia. Impacts will be felt by farmers not only through a drier and warmer climate, but also through a regulatory environment that will soon impose costs upon their GHG-generating activities.

Agriculture in Australia and New Zealand has always been affected by substantial variations in climate, so the agricultural sector has considerable experience in dealing with climate variability. Changes in climate beyond 2°C pose some particular issues for the geographic location of agricultural industries, supply of irrigation water, viability of enterprises, and resilience of rural communities. These issues may cause some industries to cease being viable in their traditional locations, creating the potential for industries to move to new locations as other more resilient agricultural systems emerge in traditional production areas. There are already some examples of such transformation in Australia.

The most promising agricultural opportunities for GHG mitigation in Australia and New Zealand appear to lie in reduction of methane emissions by livestock, although there may be some complex interactions in the fodder-livestock system that limit the achievement of the full mitigation potential. Substantial reductions in emissions of N_2O seem unlikely, with the possible exception of promoting the adoption of optimal N fertilization practices in intensive crops that are currently being over-fertilized. Such a practice change is likely to be consistent with current initiatives to reduce other environmental impacts of cropping in these regions (e.g., nitrate pollution of water), and so may happen independent of, or be accelerated by, climate change mitigation initiatives.

Middle East and North Africa (MENA)

In order to sustain agricultural systems in the MENA region, better water management will be needed now and in the future (Barghouti *et al.*, this volume). With projections showing the potential for increasing water scarcity, it is essential that more efficient agricultural water consumption be implemented. This would include increasing use of marginal waters such as treated wastewater in some countries (e.g., Jordan, Tunisia, Morocco, etc.).

Rainfed agriculture will likely be minimized and more irrigated agriculture will likely be needed to ensure production of many crops currently grown in the MENA region. In addition to enhancing the availability of water, irrigation technologies will require testing so that they can be introduced and adopted throughout the region. It will be equally important that these technologies are simple and cost-effective for the poor farmers of the region. Governments will need to assist in the development and establishment of innovative agricultural technologies as part of their water and food security policies, targeted subsidies, and investment plans.

Biosaline agriculture, defined as crop production on saline soils where, in most cases, desalinated seawater and brackish/saline groundwater are the only sources of irrigation water, will play an important role in many parts of the MENA region, particularly on salt-affected lands.

With changing climate patterns, altering the timing or location of cropping activities will become very important in the region. Furthermore, in order to improve the livelihoods of poor farmers in the Middle East and North Africa, diversifying income through integration of other farming activities such as livestock raising will also likely play a key role.

Israel

In the semi-arid Middle East, including Israel, climate is expected to become warmer while rainfall decreases, thus challenging agriculture with increased potential transpiration and reduced natural water flows (Cohen *et al.*, this volume). Israeli agriculture, which has been developed under the adverse conditions of limited water resources and poor soils, have met these challenges by building a solid research foundation and introducing large-scale desalination of seawater to partly decouple crop water requirements from natural water supplies. These changes have come at the price of increasing energy use, but have also made Israeli agriculture a paradigm for how climate change challenges and risks can be met.

Predicted average temperature increases of 2–3°C will increase irrigation requirements for Israeli farmers. Higher temperatures may allow earlier planting and longer seasons for winter vegetable crops, thus increasing potential profitability. Contemporary temperature increases have already led farmers in Northern Israel to move apple orchards to higher and cooler altitudes to fulfill the chill requirement for proper fruit development. With future warming, this trend will probably continue.

If the predicted decreases in rainfall eventualize, non-irrigated field crops (including winter wheat, hay, legumes, and safflower) now grown in the semiarid south will move north, where the rainfall is greater but the availability of agricultural land is limited. Overall, in that scenario, non-irrigated crop production could be significantly reduced.

In regard to adaptation, experimental work has demonstrated that reduced radiation load by deployment of agricultural screens above crops can prevent extremely high temperature damage. This adaptation is already being deployed in some orchards to protect against sun-scald and thermal damage, hail storms, and strong winds, and to reduce crop water use. In regard to adaptation to reduced water supplies, the national program for wastewater reuse for irrigation shows that this can be practicable, but that response time for development and deployment of such measures is close to a decade.

Increased temperatures may increase the profitability of farming because Israeli farmers may substitute capital for climate by using irrigation and crop covers, especially if international market conditions are favorable and water allocations are guaranteed. Moderate increases in temperature and decreases in rainfall may result in a shift to fruit production with irrigation and net cover. As long as water quotas for irrigation are maintained, changes in precipitation may have only a marginal impact.

China

Over the past several decades, agriculture in China has experienced the effects of some devastating climate extremes (Tang *et al.*, this volume). Many environmental stresses such as high and low temperatures, floods, and droughts have occurred more frequently. A major adaptive response is to introduce and cultivate new stress-resistant cultivars to reduce negative impacts. The two objectives of this crop breeding program are to breed stress-resistant varieties of current crops, as well as additional crops and new varieties that have suitable phenology for the projected climate change conditions.

South Asia

Global climate change is likely to affect the food security and livelihoods of millions of poor farmers in South Asia (Aggarwal *et al.*, this volume). Projected climate change threatens to limit the future capacity of South Asia to remain agriculturally self-sufficient. Urgent steps are needed to increase its adaptive capacity to face current as well as future climatic risks. Efforts to combat hunger and climate change in South Asia need to be complementary and synergized. A win-win solution is to start with such climate mitigation strategies that are needed for sustainable development such as increasing soil organic carbon content.

Research Priorities to Advance Agronomic Solutions to Climate Change

Sustained major investments in research are needed to develop the technologies, decision support tools, information, and effective communication strategies to transform agriculture into a system that is resilient to climate variability and climate change (Wolfe, this volume). With timely and appropriate proactive investment in research, the agriculture sector will have the necessary tools for strategic adaptation to meet the challenges and take advantage of opportunities associated with climate change. Policy-makers will have information to facilitate adaptation and to minimize associated inequities in impacts and costs. Research will also enable farmers to contribute significantly to climate change mitigation by having access to new stratagems, such as new GHG and soil carbon accounting frameworks.

Agricultural decision-makers — regional and national planners as well as farmers — need information that can reduce uncertainty about climate change, its potential impacts, and the effectiveness of adaptation and mitigation options. Future agricultural research needs to balance attention to breeding for high temperature and drought-resistant crops with enhanced focus on agronomy and practical field-based crop management.

At its broadest, the climate change challenge is to evaluate the range of adaptation and mitigation strategies applicable to crop, fruit, fiber, and forage production with an emphasis on the interactions among the factors of water and nutrient management across a range of cropping systems and genetic material. This is often expressed in terms of fully understanding the interactions among genotype, environment, and management (GEM), where management includes technology.

Future research priorities need to include intercomparison and improvement of agricultural models, so that responses to climate change may be predicted and adaptation strategies tested more effectively than they have been in the past (Rosenzweig *et al.*, this volume). These improved models can then be used in rigorous multi-model assessments such as those being coordinated by the Agricultural Model Intercomparison and Improvement Project (AgMIP) (www.agmip.org). AgMIP is also developing a web-based agronomic library that contains crop models and data from crop experiments, as well as climate and soils data. A world-wide effort is needed to populate this library for use by the scientific community to evaluate and compare models across differing environments and management systems under changing climate conditions.

Based on the findings of the chapter authors of this volume, research priorities related to climate change solutions for agriculture include:

- *Climate science*
  Climate science is interdisciplinary and seeks to provide a comprehensive under-standing of land-ocean-atmosphere interactions. To identify the timescales and spatial scales of information needed to assist agricultural decisions and policies, it is useful to characterize the relative magnitudes of observed climate variabil-ity at different temporal scales (seasonal, interannual, decadal, and longer-term) and spatial scales pertaining to different agricultural regions. These can then be used to inform the use of global and regional climate model projections of future climate change and adaptation decisions. It is also important to characterize the broad range of uncertainty in future anthropogenic climate changes since cli-mate models do not fully capture the global climate system and its regional manifestations.

- *Soil and crop responses to high temperature and increasing carbon dioxide*
  Key research needs are to determine how agricultural soils and crops around the world respond to high temperature and increasing carbon dioxide, and to their interactions. For crops, experiments are needed to determine how sensitivity to climate fluctuations varies with stage of development. Maximum temperature thresholds of current varieties need to be defined. Post-harvest treatments will need to be developed and tested at higher temperatures.

  Many more experimental studies are needed to ascertain the responses to ele-vated CO_2 and high temperature for the full range of crop plants (fruits and vegetables as well as grains and legumes), with results reported in a handbook for use by practitioners throughout the world. Among the needed information fac-tors are the upper thresholds of CO_2 levels and temperature for maximum yield response, as well as changes to osmotic adjustment in a CO_2-enriched and warmer world.

Important research questions include (Kirkham, this volume):

1. What are the optimal concentrations of CO_2 in the air and the ambient tempera-tures for the growth of different crops?
2. How do the four principal soil physical factors that affect root growth (water, aeration, temperature, and mechanical impedance) interact with elevated CO_2 to control crop growth? And how do these interact with variable chemical factors (e.g., pH, salinity, and nutrients)?

3. Because elevated CO_2 tends to increase soil water content, how do irrigation recommendations vary for different crops under elevated CO_2 at higher temperatures?
4. Why does elevated CO_2 close stomata?
5. How do elevated CO_2 and higher temperature affect stomatal density?
6. How far away from a canopy is the increase in canopy temperature (due to elevated CO_2) measurable and how does this interact with higher ambient temperatures?
7. What is the change in water use efficiency for different crops under elevated CO_2 in warmer conditions?

- *Genomics*
  Genomics are needed to ensure germplasm is available for productive and efficient crops under changing climate conditions. Examples of specific breeding goals include resistance to high temperatures that cause spikelet sterility and altered phenology so that cropping systems can be redistributed as climate conditions for agro-ecological zones shift.

- *Soil and water management*
  Research is needed to provide detailed understanding of how soil and water processes will respond to increasing atmospheric concentrations of carbon dioxide and changing climate. Key research areas include soil hydraulic conductivity, carbon and nitrogen interactions, and erosion. This is necessary so that the agricultural community can anticipate and act to prevent soil deterioration. Research is also required to improve water-management systems, re-use of wastewater, and irrigation scheduling and application technology. More attention to soil water conservation will be needed in regions that are undergoing drying, and more attention to drainage will be needed in areas that are becoming wetter.

- *Weed and pest control*
  Adaptation strategies for weed, pathogen, and insect pest control are needed, including improvements of regional monitoring and integrated pest management (IPM) of weed, disease, and pest range and migration shifts; accelerated responses to new weeds, pathogens, and insect pests; enhanced real-time weather-based systems for weed, disease, and pest control; development of non-chemical options for new diseases and pests; and preparation of rapid response action plans to control invasive species.

- *Agricultural models*
  All types of agricultural models, including dynamic process and statistical approaches need to be compared and improved to better understand

interrelationships among biotic and abiotic agro-ecosystem components, including genetics, crop physiology, meteorology and climate science, soil science, and economics.

- **Decision support tools for adaptation and mitigation**
  Effective decision support tools are crucial for determining the optimum timing and magnitude of investments for strategic adaptation to climate change. These must consider the need for farmers to maintain and maximize returns over multiple planning horizons. This will require addressing uncertainties in climate model projections regarding precipitation, frequency and severity of extreme events, and temporal and spatial climate variability. It also will require quantifying costs and benefits of adaptation at the farm level for fruit, nut, and vegetable crops and livestock as well as grain production systems.

  To improve mitigation efforts in the agricultural sector, better tools are needed for monitoring, accounting, and managing GHG emissions. This task includes attention to overall energy use in agriculture, as well improved understanding of carbon and nitrogen interactions. Farmers will need support in seeking and applying the most cost-effective management systems for GHG mitigation, considering potential uncertainties in carbon prices, the degree of abatement achieved, and impacts on farm management into the future.

- **Communication strategies**
  Cognitive and cultural factors must be integrated into communication strategies as part of the design of effective adaptation and mitigation programs for agriculture. Advances in decision sciences pertaining to risk perception, temporal discounting, decision-making under uncertainty, participatory processes, decision architecture, equity, and framing should be taken into account.

- **Climate finance and policies**
  Climate finance could provide a major stimulus to improving agriculture and making farmers more climate-resilient. The mechanisms that are being put in place to implement the Kyoto Protocol and follow-on agreements — through carbon emission trading — and prevailing agricultural policies will largely determine whether farmers can engage in activities that enhance carbon sequestration, particularly in Africa and other developing regions. There is also a need to develop policy frameworks for implementing adaptation options so that the farmers are protected against the potentially adverse impacts of climate change.

- **Links to sustainability**
  Agriculture is shifting to a new paradigm that embraces a focus on sustainability and environmental services. This new paradigm is represented by a broad set of practices sometimes described as 'Conservation Agriculture.' The potential for

the new sustainability and environmental services paradigm in agriculture should be more fully explored and researched, and support networks for farmers adopting these practices should be established and enhanced in response to growing needs.

Climate change strategies for agriculture necessitate the simultaneous consideration of globalization, population increase, and income growth, as well as regional, socio-economic, and environmental contexts. Governments must do more to fully integrate climate change concerns into their sustainable development policies. And further steps are needed to encourage all agricultural-sector stakeholders to take part in mitigation and adaptation efforts.

Strategies that maximize synergies between adaptation, mitigation, food production and sustainable development would be most appropriate. For example, increasing the efficiency of fertilizer and water use will lead to higher profits for farmers as well as to greater carbon sequestration, which will in turn improve soil health and contribute to higher production. Such practices would be more readily adopted if farmers could be rewarded for the environmental services that agriculture provides. It is time that society consider such incentives to farmers in the interest of protecting and enhancing the global environment, food security, and rural livelihoods.

Effective capacity-building is an essential component of agricultural responses to climate change. Specialized training activities on climate change and agriculture are needed. Capacity-building is not limited to scientific and technological aspects, but includes human resources development, mobilization, communications, and outreach activities as well. Adoption of both mitigation and adaptation strategies necessitates social empowerment of farmers as well as the development of technical competencies in acquiring knowledge and skills related to the relevant practices.

Worldwide, global cooperation will serve to generate increased attention to climate change mitigation and adaptation research, capacity building, development activities, science-based policy-making, and support of international adaptation and mitigation funds. A unified global perspective is essential to developing effective regional agricultural strategies so that lessons learned can be shared and applied efficiently. This is the way forward to ensure food security for all while sustaining soil, water, and biotic resources in a changing climate.

References

Climate Change Position Statement Working Group. 2011. *Climate Change Position Statement.* American Society of Agronomy, Crop Science Society of America, and Soil Science Society of America. Madison, WI.

Index